Marat V. Markin
Elementary Functional Analysis

Also of Interest

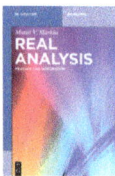

Real Analysis. Measure and Integration
Marat V. Markin, 2019
ISBN 978-3-11-060097-1, e-ISBN (PDF) 978-3-11-060099-5,
e-ISBN (EPUB) 978-3-11-059882-7

Elementary Operator Theory
Marat V. Markin, 2019
ISBN 978-3-11-060096-4, e-ISBN (PDF) 978-3-11-060098-8,
e-ISBN (EPUB) 978-3-11-059888-9

Applied Nonlinear Functional Analysis. An Introduction
Nikolaos S. Papageorgiou, Patrick Winkert, 2018
ISBN 978-3-11-051622-7, e-ISBN (PDF) 978-3-11-053298-2,
e-ISBN (EPUB) 978-3-11-053183-1

Functional Analysis. A Terse Introduction
Gerardo Chacón, Humberto Rafeiro, Juan Camilo Vallejo, 2017
ISBN 978-3-11-044191-8, e-ISBN (PDF) 978-3-11-044192-5,
e-ISBN (EPUB) 978-3-11-043364-7

Complex Analysis. A Functional Analytic Approach
Friedrich Haslinger, 2017
ISBN 978-3-11-041723-4, e-ISBN (PDF) 978-3-11-041724-1,
e-ISBN (EPUB) 978-3-11-042615-1

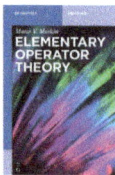

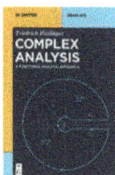

Marat V. Markin

Elementary Functional Analysis

—

DE GRUYTER

Mathematics Subject Classification 2010
46-02, 47-02, 46A22, 46A30, 46A32, 46A35, 46A45, 46B03, 46B04, 46B10, 46C05, 47A30

Author
Prof. Dr. Marat V. Markin
California State University, Fresno
Department of Mathematics
5245 North Backer Avenue, M/S PB 108
Fresno, CA 93740
USA
mmarkin@csufresno.edu

ISBN 978-3-11-061391-9
e-ISBN (PDF) 978-3-11-061403-9
e-ISBN (EPUB) 978-3-11-061409-1

Library of Congress Control Number: 2018950580

Bibliographic information published by the Deutsche Nationalbibliothek
The Deutsche Nationalbibliothek lists this publication in the Deutsche Nationalbibliografie;
detailed bibliographic data are available on the Internet at http://dnb.dnb.de.

© 2018 Walter de Gruyter GmbH, Berlin/Boston
Cover image: Mordolff / Getty Images
Typesetting: VTeX UAB, Lithuania
Printing and binding: CPI books Lecks, GmbH

www.degruyter.com

To my students, past, present, and future.

Preface

> *Functional analysis arose in the early twentieth century and gradually, conquering one stronghold after another, became a nearly universal mathematical doctrine, not merely a new area of mathematics, but a new mathematical world view. Its appearance was the inevitable consequence of the evolution of all of nineteenth-century mathematics, in particular classical analysis and mathematical physics. Its original basis was formed by Cantor's theory of sets and linear algebra. Its existence answered the question of how to state general principles of a broadly interpreted analysis in a way suitable for the most diverse situations.*
> A. M. Vershik

Having at once abandoned futile attempts to say anything better to describe the essence and origins of *functional analysis* than the above epigraph, the author, echoing [32], could not but choose it as a foreword for this book. And yet, a few more words are to be uttered.

Functional analysis

The emergence of functional analysis, a vast and rapidly growing branch of modern mathematics using *"the intuition and language of geometry in the study of functions"* [54], was brought to life by the inherent to mathematics epistemological tendency towards *unification* and *abstraction*. The constantly widening field of applications of functional analysis encompasses ordinary and partial differential equations, numerical analysis, calculus of variations, approximation theory, integral equations, and much more. The profoundly abstract nature and extensive applicability of functional analysis make a course in it to be an indispensable part of a contemporary graduate mathematics curriculum increasing its value not only for graduate students majoring in mathematics, but also for those majoring in physics, science, and engineering.

The purpose of the book and targeted audience

While there is a plethora of excellent, but mostly "tell-it-all" books on the subject (see, e. g., a rather extensive bibliography below), this one is intended to take a unique place in what today seems to be a still wide open niche for an introductory text on the basics of functional analysis to be taught within the existing constraints of the standard, for the United States, *one-semester* graduate curriculum (fifteen weeks with two seventy-five-minute lectures per week). The prerequisites are set intentionally quite low, the students not being assumed to have taken graduate courses in real or complex analysis and general topology, to make the course accessible and attractive to a wider audience of STEM (science, technology, engineer-

https://doi.org/10.1515/9783110614039-201

ing, and mathematics) graduate students or advanced undergraduates with a solid background in calculus and linear algebra. This is why the examples are primarily built around sequence spaces, L_2 spaces being mentioned only tangentially whenever pertinent.

Book's scope and specifics

The book consists of *seven chapters* and an *appendix* taking the reader from the fundamentals of abstract spaces (metric, vector, normed vector, and inner product), through the basics of linear operators and functionals, the *three fundamental principles* (the *Hahn-Banach Theorem* (the extension form), the *Uniform Boundedness Principle*, the *Open Mapping Theorem* and its equivalents: the *Inverse Mapping* and *Closed Graph Theorems*) with their numerous profound implications and certain interesting applications, to the elements of the *duality* and *reflexivity* theory. Chapter 1 outlines some necessary preliminaries, while the Appendix gives a concise discourse on the celebrated *Axiom of Choice*, its equivalents (the *Hausdorff Maximal Principle*, *Zorn's Lemma*, and *Zermello's Well-Ordering Principle*), and *ordered sets*.

Being designed as a text to be used in a classroom, the book constantly calls for the student's actively mastering the knowledge of the subject matter. It contains 112 Problems, which are indispensable for understanding and moving forward. Many important statements, such as *Cantor's Intersection Theorem* (Theorem 2.55), are given as problems, a lot of these are frequently referred to and used in the main body. There are also 376 Exercises throughout the text, including Chapter 1 and the Appendix, which require of the student to prove or verify a statement or an example, fill in necessary details in a proof, or provide an intermediate step or a counterexample. They are also an inherent part of the material. More difficult problems, such as Section 2.19, Problems 6 and 7, are marked with an asterisk, many problem and exercises being supplied with "*existential*" hints.

The book is generous on Examples and contains numerous Remarks accompanying every definition and virtually each statement to discuss certain subtleties, raise questions on whether the converse assertions are true, whenever appropriate, or whether the conditions are essential.

As amply demonstrated by experience, students tend to better remember statements by their names rather than by numbers. Thus, a distinctive feature of this book is that every theorem, proposition, and corollary, unless already possessing a name, is endowed with a descriptive one, making it easier to remember, which, in this author's humble opinion, is quite a bargain when the price for better understanding and retention of the material is a little clumsiness while making a longer reference. Each statement is referred to by its name and not just the number. e. g., the *Representation Theorem for l_p^* ($1 \le p < \infty$)* (Theorem 7.3), as opposed to merely Theorem 7.3.

Acknowledgments

I am eternally grateful, to my mother, Svetlana A. Markina, for her unfailing faith, support, and endless patience, without which this and many other endeavors of mine would be impossible.

My utmost appreciation goes to Mr. Edward Sichel, my pupil and graduate advisee, for his invaluable assistance with proofreading and improving the manuscript. I am also very thankful to Dr. Przemyslaw Kajetanowicz (Department of Mathematics, CSU, Fresno) for his kind assistance with some figures.

My sincere acknowledgments are also due to the following associates of the *Walter de Gruyter GmbH*: Dr. Apostolos Damialis, Acquisitions Editor in Mathematics, for seeing value in my manuscript and making authors his highest priority, Ms. Nadja Schedensack, Project Editor in Mathematics and Physics, for her superb efficiency in managing all project related matters, as well as Ms. Ieva Spudulytė and Ms. Ina Talandienė, VTeX Book Production, for their expert editorial and LaTeX typesetting contributions.

Clovis, California   Marat V. Markin
June 2018

Contents

1 Preliminaries

In this chapter, we outline certain terminology, notations, and preliminary facts essential to our subsequent discourse.

1.1 Set Theoretic Basics

1.1.1 Some Terminology and Notations

- The logic *quantifiers* $\forall$, $\exists$, and $\exists!$ stand for *"for all"*, *"there exist(s)"*, and *"there exists a unique"*, respectively.
- $\mathbb{N} := \{1, 2, 3, \dots\}$ is the set of *natural numbers.*
- $\mathbb{Z} := \{0, \pm 1, \pm 2, \dots\}$ is the set of *integers.*
- $\mathbb{Q}$ is the set of *rational numbers.*
- $\mathbb{R}$ is the set of *real numbers.*
- $\mathbb{C}$ is the set of *complex numbers.*
- $\mathbb{Z}_+$, $\mathbb{Q}_+$, and $\mathbb{R}_+$ are the sets of nonnegative integers, rationals, and reals, respectively.
- $\overline{\mathbb{R}} := [-\infty, \infty]$ is the set of *extended real numbers* (*extended real line*).
- For $n \in \mathbb{N}$, $\mathbb{R}^n$ and $\mathbb{C}^n$ are the *n*-spaces of all *ordered n-tuples* of real and complex numbers, respectively.

Let X be a set. Henceforth, all sets are supposed to be subsets of X.
- $\mathscr{P}(X)$ is the *power set* of X, i. e., the collection of all subsets of X.
- 2^X is the set of all binary functions $f : X \to \{0, 1\}$, provided $X \neq \emptyset$.
- Sets $A, B \subseteq X$ with $A \cap B = \emptyset$ are called *disjoint*.
- Let I be a nonempty indexing set. The sets of a collection $\{A_i\}_{i \in I}$ of subsets of X are said to be *pairwise disjoint* if

$$A_i \cap A_j = \emptyset, \quad i, j \in I, \; i \neq j.$$

- For $A, B \subseteq X$, $A \setminus B := \{x \in X \,|\, x \in A, \text{ but } x \notin B\}$ is the *difference* of A and B, in particular, $A^c := X \setminus A = \{x \in X \,|\, x \notin A\}$ is the *complement* of A and $A \setminus B = A \cap B^c$;
- Let I be a nonempty indexing set and $\{A_i\}_{i \in I}$ be a collection of subsets of X. *De Morgan's laws* state

$$\left(\bigcup_{i \in I} A_i \right)^c = \bigcap_{i \in I} A_i^c \quad \text{and} \quad \left(\bigcap_{i \in I} A_i \right)^c = \bigcup_{i \in I} A_i^c.$$

More generally,

$$B \setminus \bigcup_{i \in I} A_i = \bigcap_{i \in I} B \setminus A_i \quad \text{and} \quad B \setminus \bigcap_{i \in I} A_i = \bigcup_{i \in I} B \setminus A_i.$$

https://doi.org/10.1515/9783110614039-001

- The *Cartesian product* of sets $A_i \subseteq X$, $i = 1,\ldots,n$ ($n \in \mathbb{N}$),

$$A_1 \times \cdots \times A_n := \{(x_1,\ldots,x_n) \mid x_i \in A_i, \ i = 1,\ldots,n\}.$$

1.1.2 Cardinality and Countability

Definition 1.1 (Similarity of Sets). Sets A and B are said to be *similar* if there exists a one-to-one correspondence (bijection) between them.

Notation. $A \sim B$.

Remark 1.1. Similarity is an *equivalence relation* (*reflexive*, *symmetric*, and *transitive*) on the power set $\mathscr{P}(X)$ of a nonempty set X.

Exercise 1.1. Verify.

Thus, in the contualtext, we can use the term *"equivalence"* synonymously to *"similarity"*.

Definition 1.2 (Cardinality). Equivalent sets are said to have the same number of elements or *cardinality*. Cardinality is a characteristic of an equivalence class of similar sets.

Notation. $\mathscr{P}(X) \ni A \mapsto |A|$.

Remark 1.2. Thus, $A \sim B$ *iff* $|A| = |B|$, i. e., two sets are equivalent iff they share the same cardinality.

Examples 1.1.
1. For a nonempty set X, $\mathscr{P}(X) \sim 2^X$.
2. $|\mathbb{N}| = |\mathbb{Z}| = |\mathbb{Q}| := \aleph_0$.
3. $|[0,1]| = |\mathbb{R}| = |\mathbb{C}| := \mathfrak{c}$.

See, e. g., [18, 21].

Definition 1.3 (Domination). If sets A and B are such that A is equivalent to a subset of B, we write

$$A \preceq B$$

and say that B *dominates* A. If, in addition, $A \not\sim B$, we write

$$A \prec B$$

and say that B *strictly dominates* A.

Remark 1.3. The relation $\preceq$ is a *partial order* (*reflexive*, *antisymmetric*, and *transitive*) on the power set $\mathscr{P}(X)$ of a nonempty set X (see Appendix A).

Exercise 1.2. Verify *reflexivity* and *transitivity*.

The *antisymmetry* of $\preceq$ is the subject of the following celebrated theorem.

Theorem 1.1 (Schröder–Bernstein Theorem). *If, for sets A and B, $A \preceq B$ and $B \preceq A$, then $A \sim B$.*[1]

For a proof, see, e. g., [18].

Remark 1.4. The set partial order $\preceq$ defines a partial order $\leq$ on the set of cardinals:

$$|A| \leq |B| \iff A \preceq B.$$

Thus, the *Schröder-Bernstein Theorem* can be equivalently reformulated in terms of the cardinalities as follows:
  If, for sets A and B, $|A| \leq |B|$ and $|B| \leq |A|$, then $|A| = |B|$.

Theorem 1.2 (Cantor's Theorem). *Every set X is strictly dominated by its power set $\mathcal{P}(X)$:*[2]

$$X \prec \mathcal{P}(X).$$

Equivalently,

$$|X| < |\mathcal{P}(X)|.$$

For a proof, see, e. g., [18].
In view of Examples 1.1, we obtain the following

Corollary 1.1. *For a nonempty set X, $X \prec 2^X$, i. e., $|X| < |2^X|$.*

Definition 1.4 (Countable/Uncountable Set). A *countable set* is a set with the same *cardinality* as a subset of the set $\mathbb{N}$ of natural numbers, i. e., *equivalent* to a subset of $\mathbb{N}$.
  A set that is not *countable* is called *uncountable*.

Remarks 1.5.
-   A countable set A is either *finite*, i. e., equivalent to a set of the form $\{1, \ldots, n\} \subset \mathbb{N}$ with some $n \in \mathbb{N}$, in which case, we say that A has n elements, or *countably infinite*, i. e., equivalent to the entire $\mathbb{N}$.
-   For a finite set A of n elements ($n \in \mathbb{N}$),

$$|A| = |\{1, \ldots, n\}| = n,$$

---

1 Ernst Schröder (1841–1902), Felix Bernstein (1878–1956).
2 Georg Cantor (1845–1918).

For a countably infinite set A,

$$|A| = |\mathbb{N}| = \aleph_0$$

(see Examples 1.1).

- In some sources, the term *"countable"* is used in the sense of *"countably infinite"*. To avoid ambiguity, the term *"at most countable"* can be used when finite sets are included in the consideration.

The subsequent statement immediately follows from *Cantor's Theorem* (Theorem 1.2).

Proposition 1.1 (Uncountable Sets). *The sets $\mathscr{P}(\mathbb{N})$ and $2^{\mathbb{N}}$ (the set of all binary sequences) are uncountable.*

Theorem 1.3 (Properties of Countable Sets).
(1) *Every infinite set contains a countably infinite subset (based on the Axiom of Choice (see Appendix A)).*
(2) *Any subset of a countable set is countable.*
(3) *The union of countably many countable sets is countable.*
(4) *The Cartesian product of finitely many countable sets is countable.*

Exercise 1.3. Prove that
(a) the set $\mathbb{Z}$ of all *integers* and the set of all *rational numbers* are countable;
(b) for any $n \in \mathbb{N}$, $\mathbb{Z}^n$ and $\mathbb{Q}^n$ are countable;
(c) the set of all *algebraic numbers* (the roots of polynomials with integer coefficients) is countable.

Subsequently, we also need the following useful result.

Proposition 1.2 (Cardinality of the Collection of Finite Subsets). *The cardinality of the collection of all finite subsets of an infinite set coincides with the cardinality of the set.*

For a proof, see, e. g., [21, 26, 38].

1.2 Terminology Related to Functions

Let X and Y be nonempty sets and $\emptyset \neq D \subseteq X$

$$f : D \to Y.$$

- The set D is called the *domain* (of definition) of f.
- The *value* of f corresponding to an $x \in D$ is designated by $f(x)$.

– The set

$$\{f(x) \mid x \in D\}$$

of all values of f is called the *range* of f (also the *codomain* or *target set*).
– For a set $A \subseteq D$, the set of values of f corresponding to all elements of A

$$f(A) := \{f(x) \mid x \in A\}$$

is called the *image* of A under the function f.
Thus, the *range* of f is the image $f(D)$ of the whole domain D.
– For a set $B \subseteq Y$, the set of all elements of the domain that map to the elements of B

$$f^{-1}(B) := \{x \in D \mid f(x) \in B\}$$

is called the *inverse image* (or *preimage*) of B.

Example 1.2. For $X = Y := \mathbb{R}$, $f(x) := x^2$, and $D := [-1, 2]$,
– $f([-1, 2]) = f([1, 2]) = [1, 4]$.
– $f^{-1}([-2, -1]) = \emptyset$, $f^{-1}([0, 1]) = [-1, 1]$, $f^{-1}([1, 4]) = [-1, 2]$.

Theorem 1.4 (Properties of Inverse Image). *Let X and Y be nonempty sets and*
$\emptyset \neq D \subseteq X$

$$f : D \to Y.$$

Then, for an arbitrary nonempty collection $\{B_i\}_{i \in I}$ of subsets of Y,
(1) $f^{-1}(\bigcup_{i \in I} B_i) = \bigcup_{i \in I} f^{-1}(B_i)$,
(2) $f^{-1}(\bigcap_{i \in I} B_i) = \bigcap_{i \in I} f^{-1}(B_i)$, *and*
(3) *for any $B_1, B_2 \subseteq Y$, $f^{-1}(B_1 \setminus B_2) = f^{-1}(B_1) \setminus f^{-1}(B_2)$,*

i. e., preimage preserves all set operations.

Exercise 1.4.
(a) Prove.
(b) Show that image preserves unions, i. e., for an arbitrary nonempty collection
$\{A_i\}_{i \in I}$ of subsets of D,

$$f\left(\bigcup_{i \in I} A_i\right) = \bigcup_{i \in I} f(A_i),$$

and unions *only*. Give corresponding counterexamples for intersections and differences.

1.3 Upper and Lower Limits

Definition 1.5 (Upper and Lower Limits). Let $(x_n)_{n\in\mathbb{N}}$ (another notation is $\{x_n\}_{n=1}^\infty$) be a sequence of real numbers.

The *upper limit* or *limit superior* of $(x_n)_{n\in\mathbb{N}}$ is defined as follows:

$$\overline{\lim_{n\to\infty}}\, x_n := \lim_{n\to\infty} \sup_{k\ge n} x_k = \inf_{n\in\mathbb{N}} \sup_{k\ge n} x_k \in [-\infty, \infty].$$

The *lower limit* or *limit inferior* of $\{x_n\}_{n=1}^\infty$ is defined as follows:

$$\underline{\lim_{n\to\infty}}\, x_n := \lim_{n\to\infty} \inf_{k\ge n} x_k = \sup_{n\in\mathbb{N}} \inf_{k\ge n} x_k \in [-\infty, \infty].$$

Alternative notations are $\limsup_{n\to\infty} x_n$ and $\liminf_{n\to\infty} x_n$, respectively.

Example 1.3. For

$$x_n := \begin{cases} n, & n \text{ is odd,} \\ -1/n, & n \text{ is even,} \end{cases} \quad n \in \mathbb{N},$$

$$\overline{\lim_{n\to\infty}}\, x_n = \infty \text{ and } \underline{\lim_{n\to\infty}}\, x_n = 0.$$

Exercise 1.5.
(a) Verify.
(b) Explain why the upper and lower limits, unlike the regular limit, are guaranteed to exist for an arbitrary sequence of real numbers.
(c) Show that

$$\underline{\lim_{n\to\infty}}\, x_n \le \overline{\lim_{n\to\infty}}\, x_n.$$

Proposition 1.3 (Characterization of Limit Existence). *For a sequence of real numbers* $\{x_n\}_{n=1}^\infty$,

$$\lim_{n\to\infty} x_n \in [-\infty, \infty]$$

exists iff

$$\underline{\lim_{n\to\infty}}\, x_n = \overline{\lim_{n\to\infty}}\, x_n,$$

in which case

$$\lim_{n\to\infty} x_n = \underline{\lim_{n\to\infty}}\, x_n = \overline{\lim_{n\to\infty}}\, x_n.$$

2 Metric Spaces

In this chapter, we study abstract sets endowed with a notion of *distance*, whose properties mimic those of the regular distance in three-dimensional space. Distance brings to life various topological notions such as *limit, continuity, openness, closedness, compactness, denseness*, and the geometric notion of *boundedness*, as well as the notions of *fundamentality* of sequences and *completeness*. We consider all these and more in depth here.

2.1 Definition and Examples

Definition 2.1 (Metric Space). A *metric space* is a nonempty set X with a *metric* (or *distance function*), i. e., a mapping

$$\rho(\cdot,\cdot) : X \times X \to \mathbb{R}$$

subject to the following *metric axioms*:

1. $\rho(x,y) \geq 0,\ x,y \in X.$ *Nonnegativity*
2. $\rho(x,y) = 0$ iff $x = y.$ *Separation*
3. $\rho(x,y) = \rho(y,x),\ x,y \in X.$ *Symmetry*
4. $\rho(x,z) \leq \rho(x,y) + \rho(y,z),\ x,y,z \in X.$ *Triangle Inequality*

For any fixed $x,y \in X$, the number $\rho(x,y)$ is called the *distance* of x from y, or from y to x, or between x and y.

Notation. (X,ρ).

Remark 2.1. A function $\rho(\cdot,\cdot) : X \times X \to \mathbb{R}$ satisfying the metric axioms of *symmetry*, *triangle inequality*, and the following weaker form of the *separation axiom*:

2w. $\rho(x,y) = 0$ if $x = y$

also necessarily satisfies the axiom of *nonnegativity* and is called a *semimetric* (or *pseudometric*) on X (see the examples to follow).

Exercise 2.1. Verify that 2w, 3, and 4 imply 1.

Examples 2.1.
1. Any nonempty set X is a *metric space* relative to the *discrete metric*

$$X \ni x,y \mapsto \rho_d(x,y) := \begin{cases} 0 & \text{if } x = y, \\ 1 & \text{if } x \neq y. \end{cases}$$

2. The *real line* $\mathbb{R}$ or the *complex plane* $\mathbb{C}$ is a metric space relative to the regular distance function $\rho(x,y) := |x - y|$.

https://doi.org/10.1515/9783110614039-002

3. Let $n \in \mathbb{N}$ and $1 \leq p \leq \infty$. The real/complex n-space $\mathbb{R}^n$ or $\mathbb{C}^n$ is a *metric space* relative to the *p-metric*

$$\rho_p(x,y) = \begin{cases} [\sum_{i=1}^n |x_i - y_i|^p]^{1/p} & \text{if } 1 \leq p < \infty, \\ \max_{1 \leq i \leq n} |x_i - y_i| & \text{if } p = \infty, \end{cases}$$

where $x := (x_1, \ldots, x_n)$ and $y := (y_1, \ldots, y_n)$, designated by $l_p^{(n)}$ (real or complex, respectively).

Remarks 2.2.

- For $n = 1$, all these metrics coincide with $\rho(x,y) = |x - y|$.
- For $n = 2, 3$, and $p = 2$, we have the usual *Euclidean distance*.
- $(\mathbb{C}, \rho) = (\mathbb{R}^2, \rho_2)$.

4. Let $1 \leq p \leq \infty$. The set l_p of all real/complex sequences $\{x_k\}_{k=1}^\infty$ satisfying

$$\sum_{k=1}^\infty |x_k|^p < \infty \quad (1 \leq p < \infty),$$

$$\sup_{k \in \mathbb{N}} |x_k| < \infty \quad (p = \infty)$$

(*p-summable/bounded* sequences, respectively) is a *metric space* relative to the *p-metric*

$$\rho_p(x,y) = \begin{cases} [\sum_{k=1}^\infty |x_k - y_k|^p]^{1/p} & \text{if } 1 \leq p < \infty, \\ \sup_{k \in \mathbb{N}} |x_k - y_k| & \text{if } p = \infty, \end{cases}$$

where $x := \{x_k\}_{k=1}^\infty, y := \{y_k\}_{k=1}^\infty \in l_p$.

Remark 2.3. When it is contextually important to distinguish between the real and complex cases, we use the notations

$$l_p^{(n)}(\mathbb{R}), \, l_p(\mathbb{R}) \quad \text{and} \quad l_p^{(n)}(\mathbb{C}), \, l_p(\mathbb{C}),$$

respectively.

Exercise 2.2. Verify Examples 2.1, 1, 2 and 3, 4 for $p = 1$ and $p = \infty$.

Remark 2.4. While verifying Examples 2.1, 3 and 4 for $p = 1$ and $p = \infty$ is straightforward, proving the *triangle inequality* for $1 < p < \infty$ requires *Minkowski's*[1] *inequality*.

---

[1] Hermann Minkowski (1864–1909).

2.2 Hölder's and Minkowski's Inequalities

Here, we are to prove the celebrated *Hölder's*[2] and *Minkowski's inequalities* for n-tuples ($n \in \mathbb{N}$) and sequences. We use *Hölder's inequality* to prove *Minkowski's inequality* for the case of n-tuples. In turn, to prove *Hölder's inequality* for n-tuples, we rely upon *Young's*[3] *inequality*, and hence, the latter is to be proved first. For the sequential case, *Hölder's* and *Minkowski's inequalities* are proved independently based on their analogues for n-tuples.

2.2.1 Conjugate Indices

Definition 2.2 (Conjugate Indices). We call $1 \le p, q \le \infty$ *conjugate indices* if they are related as follows:

$$\frac{1}{p} + \frac{1}{q} = 1 \quad \text{for } 1 < p, q < \infty,$$
$$q = \infty \quad \text{for } p = 1,$$
$$q = 1 \quad \text{for } p = \infty.$$

Examples 2.2. In particular, $p = 2$ and $q = 2$ are conjugate as well as $p = 3$ and $q = 3/2$.

Remark 2.5. Thus, for $1 < p, q < \infty$,

$$q = \frac{p}{p-1} = 1 + \frac{1}{p-1} \rightarrow \begin{cases} \infty, & p \to 1+, \\ 1, & p \to \infty, \end{cases}$$

$q > 2$ if $1 < p < 2$ and $1 < q < 2$ if $p > 2$, and the following relationships hold:

$$p + q = pq,$$
$$pq - p - q + 1 = 1 \Rightarrow (p-1)(q-1) = 1,$$
$$(p-1)q = p \Rightarrow p - 1 = \frac{p}{q},$$
$$(q-1)p = q \Rightarrow q - 1 = \frac{q}{p}. \tag{2.1}$$

2.2.2 Young's Inequality

Theorem 2.1 (Young's Inequality). *Let $1 < p, q < \infty$ be conjugate indices. Then, for any $a, b \ge 0$,*

$$ab \le \frac{a^p}{p} + \frac{b^q}{q}.$$

---

2 Otto Ludwig Hölder (1859–1937).
3 William Henry Young (1863–1942).

Proof. The inequality is, obviously, true if $a = 0$ or $b = 0$ and for $p = q = 2$.

Exercise 2.3. Verify.

Suppose that $a, b > 0$, $1 < p < 2$ or $p > 2$ and recall that

$$(p - 1)(q - 1) = 1 \quad \text{and} \quad (p - 1)q = p.$$

(see (2.1)).

Comparing the areas in Figure 2.1, which corresponds to the case of $p > 2$, the case of $1 < p < 2$ being symmetric, we conclude that

$$A \le A_1 + A_2,$$

where A is the area of the rectangle $[0, a] \times [0, b]$, the equality being the case *iff* $b = a^{p-1} = a^{p/q}$.

Hence,

$$ab \le \int_0^a x^{p-1}\,dx + \int_0^b y^{q-1}\,dy = \frac{a^p}{p} + \frac{b^q}{q}. \qquad \square$$

Remark 2.6. As Figure 2.1 shows, equality in *Young's Inequality* (Theorem 2.1) holds *iff* $a^p = b^q$.

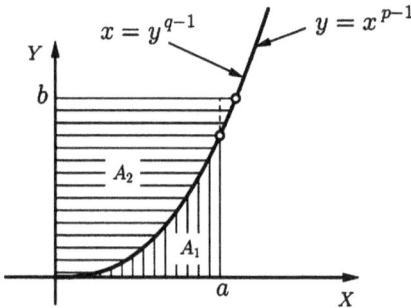

Figure 2.1: The case of $2 < p < \infty$.

2.2.3 The Case of n-Tuples

Definition 2.3 (*p*-Norm of an *n*-Tuple). Let $n \in \mathbb{N}$. For an *n*-tuple $x := (x_1, \dots, x_n) \in \mathbb{C}^n$ ($1 \le p \le \infty$), the *p-norm* of x is the distance of x from the *zero n-tuple* $0 := (0, \dots, 0)$ in $l_p^{(n)}$:

$$\|x\|_p := \rho_p(x, 0) = \begin{cases} [\sum_{i=1}^n |x_i|^p]^{1/p} & \text{if } 1 \le p < \infty, \\ \max_{1 \le i \le n} |x_i| & \text{if } p = \infty. \end{cases}$$

Remarks 2.7.

- For an $x := (x_1, \ldots, x_n) \in \mathbb{C}^n$, $\|x\|_p = 0 \Leftrightarrow x = 0$.
- Observe that, for any $x := (x_1, \ldots, x_n)$, $y := (y_1, \ldots, y_n) \in \mathbb{C}^n$,

$$\rho_p(x, y) = \|x - y\|_p,$$

where

$$x - y = (x_1, \ldots, x_n) - (y_1, \ldots, y_n) := (x_1 - y_1, \ldots, x_n - y_n).$$

Exercise 2.4. Verify.

Theorem 2.2 (Hölder's Inequality for n-Tuples). *Let $n \in \mathbb{N}$ and $1 \le p, q \le \infty$ be conjugate indices. Then, for any $x := (x_1, \ldots, x_n)$, $y := (y_1, \ldots, y_n) \in \mathbb{C}^n$,*

$$\sum_{i=1}^{n} |x_i y_i| \le \|x\|_p \|y\|_q.$$

Proof. The symmetric cases of $p = 1$, $q = \infty$ and $p = \infty$, $q = 1$ are trivial.

Exercise 2.5. Verify.

Suppose that $1 < p, q < \infty$ and let $x := (x_1, \ldots, x_n)$, $y := (y_1, \ldots, y_n) \in \mathbb{C}^n$ be arbitrary.

If

$$\|x\|_p = 0 \quad \text{or} \quad \|y\|_q = 0,$$

which (see Remarks 2.7) is equivalent to $x = 0$ or $y = 0$, respectively, *Hölder's inequality* is, obviously, true.

If

$$\|x\|_p \neq 0 \quad \text{and} \quad \|y\|_q \neq 0,$$

applying *Young's Inequality* (Theorem 2.1) to the nonnegative numbers

$$a_j = \frac{|x_j|}{\|x\|_p} \quad \text{and} \quad b_j = \frac{|y_j|}{\|y\|_q}$$

for each $j = 1, \ldots, n$, we have:

$$\frac{|x_j y_j|}{\|x\|_p \|y\|_q} \le \frac{1}{p} \frac{|x_j|^p}{\sum_{i=1}^{n} |x_i|^p} + \frac{1}{q} \frac{|y_j|^q}{\sum_{i=1}^{n} |y_i|^q}, \, j = 1, \ldots, n.$$

We obtain *Hölder's inequality* by adding the above n inequalities:

$$\frac{\sum_{j=1}^{n} |x_j y_j|}{\|x\|_p \|y\|_q} \le \frac{1}{p} \frac{\sum_{j=1}^{n} |x_j|^p}{\sum_{i=1}^{n} |x_i|^p} + \frac{1}{q} \frac{\sum_{j=1}^{n} |y_j|^q}{\sum_{i=1}^{n} |y_i|^q} = \frac{1}{p} + \frac{1}{q} = 1$$

and multiplying through by $\|x\|_p \|y\|_q$. $\quad\square$

Theorem 2.3 (Minkowski's Inequality for n-Tuples). *Let* $1 \leq p \leq \infty$. *Then, for any* $(x_1, \ldots, x_n), (y_1, \ldots, y_n) \in \mathbb{C}^n$,

$$\|x + y\|_p \leq \|x\|_p + \|y\|_p,$$

where

$$x + y = (x_1, \ldots, x_n) + (y_1, \ldots, y_n) := (x_1 + y_1, \ldots, x_n + y_n).$$

Proof. The cases of $p = 1$ or $p = \infty$ are trivial.

Exercise 2.6. Verify.

For an arbitrary $1 < p < \infty$, we have:

$$\sum_{i=1}^{n} |x_i + y_i|^p = \sum_{i=1}^{n} |x_i + y_i|^{p-1} |x_i + y_i|$$

$$\text{since} \quad |x_i + y_i| \leq |x_i| + |y_i|, \ i = 1, \ldots, n;$$

$$\leq \sum_{i=1}^{n} |x_i + y_i|^{p-1} |x_i| + \sum_{i=1}^{n} |x_i + y_i|^{p-1} |y_i|$$

$$\text{by } \textit{Hölder's inequality} \text{ (Theorem 2.2)};$$

$$\leq \left[\sum_{i=1}^{n} |x_i + y_i|^{(p-1)q} \right]^{1/q} \left[\sum_{i=1}^{n} |x_i|^p \right]^{1/p} + \left[\sum_{i=1}^{n} |x_i + y_i|^{(p-1)q} \right]^{1/q} \left[\sum_{i=1}^{n} |y_i|^p \right]^{1/p}$$

$$\text{since} \quad (p-1)q = p;$$

$$= \left[\sum_{i=1}^{n} |x_i + y_i|^p \right]^{1/q} \left[\sum_{i=1}^{n} |x_i|^p \right]^{1/p} + \left[\sum_{i=1}^{n} |x_i + y_i|^p \right]^{1/q} \left[\sum_{i=1}^{n} |y_i|^p \right]^{1/p}$$

$$= \left[\sum_{i=1}^{n} |x_i + y_i|^p \right]^{1/q} \left(\left[\sum_{i=1}^{n} |x_i|^p \right]^{1/p} + \left[\sum_{i=1}^{n} |y_i|^p \right]^{1/p} \right).$$

Considering that, for $\sum_{i=1}^{n} |x_i + y_i|^p = 0$, *Minkowski's inequality* trivially holds, suppose that $\sum_{i=1}^{n} |x_i + y_i|^p > 0$. Then, dividing through by $[\sum_{i=1}^{n} |x_i + y_i|^p]^{1/q}$, in view of $1 - \frac{1}{q} = \frac{1}{p}$, we see that, in this case, *Minkowski's inequality* holds as well, which completes the proof. □

2.2.4 Sequential Case

Definition 2.4 (p-Norm of a Sequence). For $x := \{x_k\}_{k=1}^{\infty} \in l_p$ ($1 \leq p \leq \infty$), the *norm* of x is the distance of x from the *zero sequence* $0 := \{0, 0, 0, \ldots\}$ in l_p:

$$\|x\|_p := \rho_p(x, 0) = \begin{cases} [\sum_{k=1}^{\infty} |x_k|^p]^{1/p} & \text{if } 1 \leq p < \infty, \\ \sup_{k \geq 1} |x_k| & \text{if } p = \infty. \end{cases}$$

Remarks 2.8.
- For an $x := \{x_k\}_{k=1}^{\infty} \in l_p$, $\|x\|_p = 0 \Leftrightarrow x = 0$.
- Observe that, for any $x := \{x_k\}_{k=1}^{\infty}, y := \{y_k\}_{k=1}^{\infty} \in l_p$,

$$\rho_p(x,y) = \|x - y\|_p,$$

where

$$x - y = \{x_k\}_{k=1}^{\infty} - \{y_k\}_{k=1}^{\infty} := \{x_k - y_k\}_{k=1}^{\infty}.$$

Exercise 2.7. Verify.

Theorem 2.4 (Minkowski's Inequality for Sequences). *Let* $1 \le p \le \infty$. *Then, for any* $x := \{x_k\}_{k=1}^{\infty}, y := \{y_k\}_{k=1}^{\infty} \in l_p$, $x + y := \{x_k + y_k\}_{k=1}^{\infty} \in l_p$ *and*

$$\|x + y\|_p \le \|x\|_p + \|y\|_p.$$

Proof. The cases of $p = 1$ or $p = \infty$ are trivial.

Exercise 2.8. Verify.

For an arbitrary $1 < p < \infty$ and each $n \in \mathbb{N}$, by *Minkowski's inequality* for n-tuples,

$$\left[\sum_{k=1}^{n} |x_k + y_k|^p\right]^{1/p} \le \left[\sum_{k=1}^{n} |x_k|^p\right]^{1/p} + \left[\sum_{k=1}^{n} |y_k|^p\right]^{1/p} \le \left[\sum_{k=1}^{\infty} |x_k|^p\right]^{1/p} + \left[\sum_{k=1}^{\infty} |y_k|^p\right]^{1/p}.$$

Passing to the limit as $n \to \infty$, we infer both the convergence for the series $\sum_{k=1}^{\infty} |x_k + y_k|^p$, i.e., the fact that $x + y \in l_p$, and the desired inequality. $\square$

Exercise 2.9. Applying *Minkowski's inequality* (Theorems 2.3 and 2.4), verify Examples 2.1, 3 and 4, for $1 < p < \infty$.

Theorem 2.5 (Hölder's Inequality for Sequences). *Let* $1 \le p, q \le \infty$ *be conjugate indices. Then, for any* $x = \{x_k\}_{k=1}^{\infty} \in l_p$ *and any* $y = \{y_k\}_{k=1}^{\infty} \in l_q$, *the product sequence* $\{x_k y_k\}_{k=1}^{\infty} \in l_1$ *and*

$$\sum_{k=1}^{\infty} |x_k y_k| \le \|x\|_p \|y\|_q.$$

Exercise 2.10. Prove based on *Minkowski's Inequality for n-Tuples* (Theorem 2.2) similarly to proving *Minkowski's Inequality for Sequences* (Theorem 2.4).

Remark 2.9. The important special cases of *Hölder's inequality* with $p = q = 2$:

$$\sum_{k=1}^{n} |x_k y_k| \le \left[\sum_{k=1}^{n} |x_k|^2\right]^{1/2}\left[\sum_{k=1}^{n} |y_k|^2\right]^{1/2} \quad (n \in \mathbb{N}),$$

$$\sum_{k=1}^{\infty} |x_k y_k| \le \left[\sum_{k=1}^{\infty} |x_k|^2\right]^{1/2}\left[\sum_{k=1}^{\infty} |y_k|^2\right]^{1/2} \tag{2.2}$$

are known as the *Cauchy*[4]*–Schwarz*[5] *inequalities* (2.2) for n-tuples and sequences, respectively.

2.3 Subspaces of a Metric Space

Definition 2.5 (Subspace of a Metric Space). If (X, ρ) is a *metric space* and $Y \subseteq X$, then the restriction of the metric $\rho(\cdot, \cdot)$ to $Y \times Y$ is a metric on Y and the metric space (Y, ρ) is called a *subspace* of (X, ρ).

Examples 2.3.
1. Any nonempty subset of $\mathbb{R}^n$ or $\mathbb{C}^n$ ($n \in \mathbb{N}$) is a metric space relative to the p-metric $(1 \le p \le \infty)$.
2. The sets of real/complex sequences

$$c_{00} := \left\{ x := \{x_k\}_{k=1}^{\infty} \mid \exists N \in \mathbb{N} : x_k = 0, \ k \ge N \right\} \quad (\textit{eventually zero sequences}),$$

$$c_0 := \left\{ x := \{x_k\}_{k=1}^{\infty} \mid \lim_{k \to \infty} x_k = 0 \right\} \quad (\textit{vanishing sequences}),$$

$$c := \left\{ x := \{x_k\}_{k=1}^{\infty} \mid \lim_{k \to \infty} x_k \in \mathbb{R} \ (\text{or } \mathbb{C}) \right\} \quad (\textit{convergent sequences})$$

endowed with the *supremum metric*

$$x := \{x_k\}_{k=1}^{\infty}, y := \{y_k\}_{k=1}^{\infty} \mapsto \rho_{\infty}(x, y) := \sup_{k \in \mathbb{N}} |x_k - y_k|$$

are *subspaces* of the space of *bounded sequences* l_{∞} due to the set-theoretic inclusions

$$c_{00} \subset l_p \subset l_q \subset c_0 \subset c \subset l_{\infty},$$

where $1 \le p < q < \infty$.

Exercise 2.11. Verify the inclusions and show that they are *proper*.

2.4 Function Spaces

The following gives more examples of metric spaces.

Examples 2.4.
1. Let T be a *nonempty set*. The set $M(T)$ of all real/complex-valued functions *bounded* on T, i. e., all functions $f : T \to \mathbb{R}$ (or $\mathbb{C}$) such that

$$\sup_{t \in T} |f(t)| < \infty,$$

---

4 Augustin-Louis Cauchy (1789–1857).
5 Karl Hermann Amandus Schwarz (1843–1921).

is a *metric space* relative to the *supremum metric* (or *uniform metric*)

$$M(T) \ni f,g \mapsto \rho_\infty(f,g) := \sup_{t\in T} |f(t) - g(t)|.$$

Remark 2.10. The spaces $l_\infty^{(n)}$ ($n \in \mathbb{N}$) and l_∞ are particular cases of $M(T)$ with $T = \{1,\ldots,n\}$ and $T = \mathbb{N}$, respectively.

2. The set $C[a,b]$ of all real/complex-valued functions *continuous* on an interval $[a,b]$ $(-\infty < a < b < \infty)$ is a *metric space* relative to the *maximum metric* (or *uniform metric*)

$$C[a,b] \ni f,g \mapsto \rho_\infty(f,g) = \max_{a\le t\le b} |f(t) - g(t)|$$

as a *subspace* of $M[a,b]$.

Exercise 2.12. Answer the following questions:
(a) Why can max can be used instead of sup for $C[a,b]$?
(b) Can sup be replaced with max for the *uniform metric* on $M[a,b]$?

3. For any $-\infty < a < b < \infty$, the set P of all *polynomials* with real/complex coefficients and the set P_n of all *polynomials* of *degree at most* n ($n = 0,1,2,\ldots$) are metric spaces as *subspaces* of $(C[a,b],\rho_\infty)$ due to the following (proper) inclusions:

$$P_m \subset P_n \subset P \subset C[a,b] \subset M[a,b],$$

where $0 \le m < n$.

4. The set $C[a,b]$ $(-\infty < a < b < \infty)$ is a *metric space* relative to the *integral metric*

$$C[a,b] \ni f,g \mapsto \rho_1(f,g) = \int_a^b |f(t) - g(t)|\, dt.$$

However, the extension of the latter to the wider set $R[a,b]$ of real-/complex-valued functions *Riemann integrable* on $[a,b]$ is only a *semimetric*.

Exercise 2.13. Verify both statements.

5. The set $BV[a,b]$ $(-\infty < a < b < \infty)$ of real/complex-valued functions of *bounded variation* on $[a,b]$ $(-\infty < a < b < \infty)$, i. e., all functions $f : [a,b] \to \mathbb{R}$ (or $\mathbb{C}$) such that the *total variation* of f over $[a,b]$

$$V_a^b(f) := \sup_P \sum_{k=1}^n |f(t_k) - f(t_{k-1})| < \infty,$$

where the *supremum* is taken over all partitions P: $a = t_0 < t_1 < \cdots < t_n = b$ of $[a,b]$, is a *metric space* relative to the metric

$$BV[a,b] \ni f,g \mapsto \rho(f,g) := |f(a) - g(a)| + V_a^b(f - g).$$

Exercise 2.14.

(a) Verify.

(b) For which functions in $BV[a, b]$ is $V_a^b(f) = 0$?

(c) Show that $d(f, g) := V_a^b(f - g), f, g \in BV[a, b]$, is only a *semimetric* on $BV[a, b]$.

Remark 2.11. When contextually important to distinguish between the real and complex cases, we can use the notations

$$M(T, \mathbb{R}), \ C([a, b], \mathbb{R}), \ BV([a, b], \mathbb{R}) \text{ and } M(T, \mathbb{C}), \ C([a, b], \mathbb{C}), \ BV([a, b], \mathbb{C}),$$

respectively.

2.5 Further Properties of Metric

Theorem 2.6 (Generalized Triangle Inequality). *In a metric space (X, ρ), for any finite collection of points $\{x_1, \ldots, x_n\} \subseteq X$ $(n \in \mathbb{N}, n \geq 3)$,*

$$\rho(x_1, x_n) \leq \rho(x_1, x_2) + \rho(x_2, x_3) + \cdots + \rho(x_{n-1}, x_n).$$

Exercise 2.15. Prove by induction.

Theorem 2.7 (Inverse Triangle Inequality). *In a metric space (X, ρ), for arbitrary $x, y, z \in X$,*

$$|\rho(x, y) - \rho(y, z)| \leq \rho(x, z).$$

Proof. Let $x, y, z \in X$ be arbitrary.

On one hand, by the *triangle inequality* and *symmetry*,

$$\rho(x, y) \leq \rho(x, z) + \rho(z, y) = \rho(x, z) + \rho(y, z),$$

which implies

$$\rho(x, y) - \rho(y, z) \leq \rho(x, z). \tag{2.3}$$

On the other hand, in the same manner,

$$\rho(y, z) \leq \rho(y, x) + \rho(x, z) = \rho(x, y) + \rho(x, z),$$

which implies

$$\rho(y, z) - \rho(x, y) \leq \rho(x, z). \tag{2.4}$$

Jointly, inequalities (2.3) and (2.4) are equivalent to the desired one. $\qquad\square$

Theorem 2.8 (Quadrilateral Inequality). *In a metric space* (X,ρ), *for arbitrary* $x, y, u, v \in X$,

$$|\rho(x,y) - \rho(u,v)| \le \rho(x,u) + \rho(y,v).$$

Proof. For any $x, y, u, v \in X$,

$$|\rho(x,y) - \rho(u,v)| = |\rho(x,y) - \rho(y,u) + \rho(y,u) - \rho(u,v)|$$
$$\le |\rho(x,y) - \rho(y,u)| + |\rho(y,u) - \rho(u,v)|$$

by the *Inverse Triangle Inequality* (Theorem 2.7);

$$\le \rho(x,u) + \rho(y,v). \qquad \square$$

Remark 2.12. With only the *symmetry* axiom and the *triangle inequality* used in the proofs of this section's statements, they are, obviously, true for a *semimetric*.

2.6 Convergence and Continuity

The notion of metric brings to life the important concepts of *limit* and *continuity*.

2.6.1 Convergence of a Sequence

Definition 2.6 (Limit and Convergence of a Sequence). A sequence of points $\{x_n\}_{n=1}^{\infty}$ in a *metric space* (X,ρ) is said to *converge* (to be *convergent*) to a point $x \in X$ if

$$\forall \varepsilon > 0 \; \exists N \in \mathbb{N} \; \forall n \ge N : \rho(x_n, x) < \varepsilon,$$

i. e.,

$$\lim_{n\to\infty} \rho(x_n, x) = 0 \; (\rho(x_n, x) \to 0, \; n \to \infty).$$

We write in this case

$$\lim_{n\to\infty} x_n = x \text{ or } x_n \to x, \; n \to \infty$$

and say that x is the *limit* of $\{x_n\}_{n=1}^{\infty}$.

A sequence $\{x_n\}_{n=1}^{\infty}$ in a *metric space* (X,ρ) is called *convergent* if it converges to some $x \in X$ and *divergent* otherwise.

Theorem 2.9 (Uniqueness of a Limit). *The limit of a convergent sequence* $\{x_n\}_{n=1}^{\infty}$ *in a metric space* (X,ρ) *is unique.*

Exercise 2.16. Prove.

Examples 2.5 (Convergence in Concrete Metric Spaces).
1. A sequence is convergent in a discrete space (X, ρ_d) *iff* it is *eventually constant*.
2. Convergence of a sequence in the space $l_p^{(n)}$ ($n \in \mathbb{N}$ and $1 \le p \le \infty$) is equivalent to *componentwise convergence*, i. e.,

$$(x_1^{(k)}, \ldots, x_n^{(k)}) \to (x_1, \ldots, x_n), \ k \to \infty \ \Leftrightarrow \ \forall i = 1, \ldots, n : \ x_i^{(k)} \to x_i, \ k \to \infty.$$

3. Convergence of a sequence in the space l_p ($1 \le p \le \infty$)

$$\left\{ x_k^{(n)} \right\}_{k=1}^{\infty} =: x^{(n)} \to x := \{ x_k \}_{k=1}^{\infty}, \ n \to \infty,$$

implies *termwise convergence*, i. e.,

$$\forall k \in \mathbb{N} : \ x_k^{(n)} \to x_k, \ n \to \infty.$$

4. Convergence in $(M(T), \rho_\infty)$ (see Examples 2.4) is the *uniform convergence* on T, i. e.,

$$f_n \to f, \ n \to \infty, \ \text{in} \ (M(T), \rho_\infty)$$

iff

$$\forall \varepsilon > 0 \ \exists N \in \mathbb{N} \ \forall n \ge N \ \forall t \in T : \ |f_n(t) - f(t)| < \varepsilon.$$

The same is true for l_∞ as a particular case of $M(T)$ with $T = \mathbb{N}$ and for $(C[a, b], \rho_\infty)$ as a subspace of $(M[a, b], \rho_\infty)$.

Exercise 2.17.
(a) Verify each statement and give corresponding examples.
(b) Give an example showing that the converse to 3 is not true, i. e., termwise convergent is necessary, but not sufficient for convergence in l_p ($1 \le p \le \infty$) (cf. Section 2.19, Problems 6 and 7).

2.6.2 Continuity, Uniform Continuity, and Lipschitz Continuity

Here, we introduce three notions of *continuity* in the order of increasing strength: regular, *uniform*, and *Lipschitz*.

Definition 2.7 (Continuity of a Function). Let (X, ρ) and (Y, σ) be metric spaces. A function $f : X \to Y$ is called *continuous* at a point $x_0 \in X$ if

$$\forall \varepsilon > 0 \ \exists \delta > 0 \ \forall x \in X \ \text{with} \ \rho(x, x_0) < \delta : \ \sigma(f(x), f(x_0)) < \varepsilon.$$

A function $f : X \to Y$ is called *continuous* on X if it is continuous at every point of X. The set of all such functions is designated as $C(X, Y)$ and we write $f \in C(X, Y)$.

Remarks 2.13.
- When X and Y are subsets of $\mathbb{R}$ with the regular distance, we obtain the familiar calculus (ε,δ)-definitions.
- When $Y = \mathbb{R}$ or $Y = \mathbb{C}$, the shorter notation $C(X)$ is often used.

It is convenient to characterize continuity in terms of sequences.

Theorem 2.10 (Sequential Characterization of Local Continuity). *Let (X,ρ) and (Y,σ) be metric spaces. A function $f : X \to Y$ is continuous at a point $x_0 \in X$ iff, for each sequence $\{x_n\}_{n=1}^{\infty}$ in X such that*

$$\lim_{n\to\infty} x_n = x_0 \text{ in } (X,\rho),$$

we have:

$$\lim_{n\to\infty} f(x_n) = f(x_0) \text{ in } (Y,\sigma).$$

Exercise 2.18. Prove.

Hint. The *necessity* is proved directly. Prove the *sufficiency* by *contrapositive*.

Using the *Sequential Characterization of Local Continuity* (Theorem 2.10), one can easily prove the following two theorems.

Theorem 2.11 (Properties of Numeric Continuous Functions). *Let (X,ρ) be a metric space and $Y = \mathbb{R}$ or $Y = \mathbb{C}$ with the regular distance.*
If f and g are continuous at a point $x_0 \in X$, then
(1) $\forall c \in \mathbb{R}$ (or $c \in \mathbb{C}$), cf is continuous at x_0,
(2) $f + g$ is continuous at x_0,
(3) $f \cdot g$ is continuous at x_0,
(4) *Provided $g(x_0) \neq 0$, $\frac{f}{g}$ is continuous at x_0.*

Theorem 2.12 (Continuity of Composition). *Let (X,ρ), (Y,σ), and (Z,τ), $f : X \to Y$ and $g : Y \to Z$.*
If for some $x_0 \in X$ f is continuous at x_0 and g is continuous at $y_0 = f(x_0)$, then the composition $g(f(x))$ is continuous at x_0.

Exercise 2.19. Prove Theorems 2.11 and 2.12 using the *sequential approach*.

Remark 2.14. The statements of Theorems 2.11 and 2.12 are naturally carried over to functions continuous on the whole space (X,ρ).

Definition 2.8 (Uniform Continuity). *Let (X,ρ) and (Y,σ) be metric spaces. A function $f : X \to Y$ is said to be uniformly continuous on X if*

$$\forall \varepsilon > 0 \ \exists \delta > 0 \ \forall x', x'' \in X \text{ with } \rho(x',x'') < \delta : \ \sigma(f(x'),f(x'')) < \varepsilon.$$

Remark 2.15. As the following example shows, a uniformly continuous function is continuous, but not vice versa.

Example 2.6. For $X = Y = (0, \infty)$, $f(x) = x$ is uniformly continuous and $f(x) = \frac{1}{x}$ is continuous, but not uniformly.

Exercise 2.20. Verify.

Definition 2.9 (Lipschitz Continuity). Let (X, ρ) and (Y, σ) be metric spaces. A function $f : X \to Y$ is said to be *Lipschitz continuous* on X with *Lipschitz constant* L if[6]

$$\exists L \geq 0 \, \forall x', x'' \in X : \sigma(f(x'), f(x'')) \leq L\rho(x', x'').$$

Remarks 2.16.
- The smallest Lipschitz constant is called the *best Lipschitz constant*.
- A constant function is Lipschitz continuous with the best Lipschitz constant $L = 0$.
- By the *Mean Value Theorem*, a real-valued differentiable function f on an interval $I \subseteq \mathbb{R}$ is Lipschitz continuous on I iff its derivative f' is bounded on I.
  In particular, all functions in $C^1[a, b]$ $(-\infty < a < b < \infty)$ are Lipschitz continuous on $[a, b]$.
- As the following example shows, a Lipschitz continuous function is uniformly continuous, but not vice versa.

Example 2.7. For $X = Y = [0, 1]$ with the regular distance, the function $f(x) := \sqrt{x}$ is uniformly continuous on $[0, 1]$, as follows from the *Heine–Cantor Uniform Continuity Theorem* (Theorem 2.44), but is *not* Lipschitz continuous on $[0, 1]$.

Exercise 2.21. Verify.

2.7 Balls, Separation, and Boundedness

The geometric concepts of *balls* and *spheres*, generalizing their familiar counterparts, are rather handy as well as is a generalized notion of *boundedness*.

Definition 2.10 (Balls and Spheres). Let (X, ρ) be a *metric space* and $r \geq 0$.
- The *open ball* of radius r centered at a point $x_0 \in X$ is the set

$$B(x_0, r) := \{x \in X \mid \rho(x, x_0) < r\}.$$

- The *closed ball* of radius r centered at a point $x_0 \in X$ is the set

$$\overline{B}(x_0, r) := \{x \in X \mid \rho(x, x_0) \leq r\}.$$

---

6 Sigismund Lipschitz (1832–1903).

- The *sphere* of radius r centered at a point $x_0 \in X$ is the set

$$S(x_0, r) := \{x \in X \mid \rho(x, x_0) = r\} = \overline{B}(x_0, r) \setminus B(x_0, r).$$

Remarks 2.17.
- When contextually important to indicate which space the balls/spheres are considered in, the letter designating the space in question is added as a subscript. E. g., for (X, ρ), we use the notations

$$B_X(x_0, r), \quad \overline{B}_X(x_0, r), \quad \text{and} \quad S_X(x_0, r), \quad x_0 \in X, r \geq 0.$$

- As is easily seen, for an arbitrary $x_0 \in X$,

$$B(x_0, 0) = \emptyset \quad \text{and} \quad \overline{B}(x_0, 0) = S(x_0, 0) = \{x_0\} \quad (\textit{trivial cases}).$$

Exercise 2.22.
(a) Explain the latter.
(b) Describe balls and spheres in $\mathbb{R}$ and $\mathbb{C}$ with the regular distance, give some examples.
(c) Sketch the *unit sphere* $S(0, 1)$ in $(\mathbb{R}^2, \rho_1)$, $(\mathbb{R}^2, \rho_2)$, and $(\mathbb{R}^2, \rho_\infty)$.
(d) Describe balls and spheres in $(C[a, b], \rho_\infty)$.
(e) Let (X, ρ_d) be a *discrete* metric space and $x_0 \in X$ be arbitrary. Describe $B(x_0, r)$, $\overline{B}(x_0, r)$, and $S(x_0, r)$ for different values of $r \geq 0$.

Proposition 2.1 (Separation Property). *Let (X, ρ) be metric space. Then*

$$\forall x, y \in X, \ x \neq y \ \exists r > 0 : \ B(x, r) \cap B(y, r) = \emptyset.$$

i. e., distinct points in a metric space can be separated by disjoint balls.

Exercise 2.23. Prove.

The definitions of convergence and continuity can be naturally reformulated in terms of balls.

Definition 2.11 (Equivalent Definitions of Convergence and Continuity). A sequence of points $\{x_n\}_{n=1}^\infty$ in a *metric space* (X, ρ) is said to *converge* (to be *convergent*) to a point $x \in X$ if

$$\forall \varepsilon > 0 \ \exists N \in \mathbb{N} \ \forall n \geq N : \ x_n \in B(x_0, \varepsilon),$$

in which case we say that the sequence $\{x_n\}_{n=1}^\infty$ is *eventually* in the ε-ball $B(x, \varepsilon)$.

Let (X, ρ) and (Y, σ) be metric spaces. A function $f : X \to Y$ is called *continuous* at a point $x_0 \in X$ if

$$\forall \varepsilon > 0 \ \exists \delta > 0 : \ f(B_X(x_0, \delta)) \subseteq B_Y(f(x_0), \varepsilon).$$

Definition 2.12 (Bounded Set). Let (X, ρ) be a *metric space*. A nonempty set $A \subseteq X$ is called *bounded* if

$$\mathrm{diam}(A) := \sup_{x,y \in A} \rho(x, y) < \infty.$$

The number $\mathrm{diam}(A)$ is called the *diameter* of A.

Remark 2.18. The *empty set* $\emptyset$ is regarded to be bounded with $\mathrm{diam}(\emptyset) := 0$.

Examples 2.8.
1. In a metric space (X, ρ), an open/closed ball of radius $r > 0$ is a bounded set of diameter *at most 2r*.
2. In $(\mathbb{R}, \rho)$, the sets $(0, 1]$, $\{1/n\}_{n \in \mathbb{N}}$ are bounded and the sets $(-\infty, 1)$, $\{n^2\}_{n \in \mathbb{N}}$ are not.
3. In l_∞, the set $\{(x_n)_{n \in \mathbb{N}} \mid |x_n| \leq 1,\ n \in \mathbb{N}\}$ is bounded and, in l_p ($1 < p < \infty$), it is not.
4. In $(C[0, 1], \rho_\infty)$, the set $\{t^n\}_{n \in \mathbb{Z}_+}$ is bounded and, in $(C[0, 2], \rho_\infty)$, it is not.

Exercise 2.24.
(a) Verify.
(b) Show that a set A is *bounded* iff it is contained in some (closed) ball, i. e.,

$$\exists x \in X\ \exists r \geq 0 :\ A \subseteq \bar{B}(x, r).$$

(c) Describe all bounded sets in a *discrete* metric space (X, ρ_d).
(d) Give an example of a metric space (X, ρ), in which, for a ball $\bar{B}(x, r)$ with some $x \in X$ and $r > 0$,

$$\mathrm{diam}(\bar{B}(x, r)) < 2r.$$

Theorem 2.13 (Properties of Bounded Sets). *The bounded sets in a metric space (X, ρ) have the following properties:*
(1) *a subset of a bounded set is bounded;*
(2) *an arbitrary intersection of bounded sets is bounded;*
(3) *a finite union of bounded sets is bounded.*

Exercise 2.25.
(a) Prove.
(b) Give an example showing that an *infinite union* of bounded sets need not be bounded.

Definition 2.13 (Bounded Function). Let T be a nonempty set and (X, ρ) be a *metric space*. A function $f : T \to X$ is called *bounded* if the *set of its values* $f(T)$ is bounded in (X, ρ).

Remark 2.19. As a particular with for $T = \mathbb{N}$, we obtain the definition of a *bounded sequence*.

2.8 Interior Points, Open Sets

Now, we are ready to define *openness* and *closedness* for sets.

Definition 2.14 (Interior Point). Let (X, ρ) be a metric space. A point $x \in X$ is called an *interior point* of a nonempty set $A \subseteq X$ if A contains a nontrivial open ball centered at x, i. e.,

$$\exists r > 0 : \ B(x, r) \subseteq A.$$

Examples 2.9.
1. In an arbitrary metric space (X, ρ), any point $x \in X$ is, obviously, an interior point of an open ball $B(x, r)$ or a closed ball $\bar{B}(x, r)$ with an arbitrary $r > 0$.
2. For the set $[0, 1)$ in $\mathbb{R}$ with the regular distance, the points $0 < x < 1$ are interior and the point $x = 0$ is not.
3. A singleton $\{x\}$ in $\mathbb{R}$ with the regular distance, has no interior points.

Exercise 2.26. Verify.

Definition 2.15 (Interior of a Set). The *interior* of a nonempty set A in a metric space (X, ρ) is the set of all interior points of A.

Notation. $\text{int}(A)$.

Remark 2.20. Thus, we have always the inclusion

$$\text{int}(A) \subseteq A,$$

the prior examples showing that the inclusion may be *proper* and that $\text{int}(A)$ may be *empty*.

Exercise 2.27. Give some examples.

Definition 2.16 (Open Set). A *nonempty set A* in a metric space (X, ρ) is called *open* if each point of A is its *interior point*, i. e., $A = \text{int}(A)$.

Remark 2.21. The *empty set* $\emptyset$ is regarded to be open and the whole space X is trivially open as well.

Exercise 2.28.
(a) Verify that, in $\mathbb{R}$ with the regular distance, the intervals of the form (a, ∞), $(-\infty, b)$, and (a, b) $(-\infty < a < b < \infty)$ are open sets.
(b) Prove that, in a *metric space* (X, ρ), an *open ball* $B(x_0, r)$ $(x_0 \in X, r \geq 0)$ is an *open set*.
(c) Describe all open sets in a *discrete* metric space (X, ρ_d).

The concept of openness in a metric space can be characterized sequentially.

Theorem 2.14 (Sequential Characterizations of Open Sets). *Let* (X,ρ) *be a metric space. A set* $A \subseteq X$ *is open in* (X,ρ) *iff, for any sequence* $\{x_n\}_{n=1}^{\infty} \subseteq X$ *convergent to a point* $x_0 \in A$, $\{x_n\}_{n=1}^{\infty}$ *is eventually in* A.

Exercise 2.29. Prove.

Theorem 2.15 (Properties of Open Sets). *The open sets in a metric space* (X,ρ) *have the following properties:*

(1) $\emptyset$ *and* X *are open sets;*
(2) *an arbitrary union of open sets is open;*
(3) *an arbitrary finite intersection of open sets is open.*

Exercise 2.30.
(a) Prove.
(b) Give an example showing that an *infinite intersection* of open sets need not be open.

Definition 2.17 (Metric Topology). The collection $\mathscr{G}$ of all open sets in a metric space (X,ρ) is called the *metric topology* generated by the metric ρ.

2.9 Limit Points, Closed Sets

Definition 2.18 (Limit Point/Derived Set). Let (X,ρ) be a metric space. A point $x \in X$ is called a *limit point* (also an *accumulation point* or a *cluster point*) of a set A in X if every open ball centered at x contains a point of A *distinct* from x, i. e.,

$$\forall r > 0 : B(x,r) \cap (A \setminus \{x\}) \neq \emptyset.$$

The set A' of all limit points of A is called the *derived set* of A.

Remarks 2.22.

– A *limit point* x of a set A need not belong to A. It may even happen that none of them does, i. e.,

$$A' \subseteq A^c.$$

– Each open ball centered at a limit point x of a set A in a metric space (X,ρ) contains *infinitely many* points of A distinct from x_0.
– To have a *limit point*, a set A in a metric space (X,ρ) must necessarily be *nonempty* and even *infinite*. However, an infinite set need not have limit points.

Exercise 2.31.
(a) Verify and give corresponding examples.
(b) Describe the situation in a *discrete* metric space (X,ρ_d).

(c) Give examples showing that an *interior point* of a set need not be its *limit point* and vise versa.

Limit points can be characterized sequentially as follows.

Theorem 2.16 (Sequential Characterization of Limit Points). *Let (X,ρ) be a metric space. A point $x \in X$ is a limit point of a set $A \subseteq X$ iff A contains a sequence of points $\{x_n\}_{n=1}^{\infty}$ distinct from x convergent to x, i. e.,*

$$x \in A' \iff \exists \{x_n\}_{n=1}^{\infty} \subseteq A \setminus \{x\} : x_n \to x, n \to \infty.$$

Exercise 2.32. Prove.

Definition 2.19 (Isolated Point). Let (X,ρ) be a *metric space*. A point x of a set A in X, which is not its limit point, is called an *isolated point* of A, i. e., there is an open ball centered at x containing no other points of A, but x.

Exercise 2.33. Show that a function f from a metric space (X,ρ) to a metric space (Y,σ) is *continuous* at isolated points of X, if any.

Definition 2.20 (Closure of a Set). The *closure $\overline{A}$* of a set A in a metric space (X,ρ) is the set consisting of all points, which are either points of A or limit points of A, i. e.,

$$\overline{A} := A \cup A'.$$

Remarks 2.23.
- Obviously $\emptyset' = \emptyset$, and hence, $\overline{\emptyset} = \emptyset$.
- We always have the inclusion

$$A \subseteq \overline{A},$$

which may be *proper*.
- A point $x \in \overline{A}$ iff every nontrivial open ball centered at x contains a point of A (not necessarily *distinct* from x).

Exercise 2.34. Verify and give a corresponding example.

By the definition and the *Sequential Characterization of Limit Points* (Theorem 2.16), we obtain the following

Theorem 2.17 (Sequential Characterization of a Closure). *Let (X,ρ) be a metric space. For a set $A \subseteq X$,*

$$x \in \overline{A} \iff \exists \{x_n\}_{n=1}^{\infty} \subseteq A : x_n \to x, n \to \infty.$$

Exercise 2.35.

(a) Prove.

(b) Is it true that, in any metric space (X,ρ), for each $x \in X$ and $r > 0$,

$$\overline{B(x,r)} = \overline{B}(x,r)?$$

(c) Let A be a set in a metric space (X,ρ). Prove that, for each *open set* O in (X,ρ),

$$O \cap \overline{A} \neq \emptyset \iff O \cap A \neq \emptyset.$$

Definition 2.21 (Closed Set)**.** Let (X,ρ) be a metric space. A set A in X is called *closed* if it contains all its limit points, i. e., $A' \subseteq A$, and hence, $A = \overline{A}$.

Remarks 2.24.

– The whole space X is trivially closed.

– Also closed are the sets with no limit points, in particular, *finite sets*, including the empty set $\emptyset$.

– A set in a metric space (X,ρ), which is simultaneously closed and open is called *clopen*. There are always at least two (*trivial*) clopen sets: $\emptyset$ and X. However, there can exist nontrivial ones.

Exercise 2.36.

(a) Verify that in $\mathbb{R}$ with the regular distance the intervals of the form $[a, \infty)$, $(-\infty, b]$, and $[a, b]$ $(-\infty < a < b < \infty)$ are closed sets.

(b) Verify that the sets $(0, 1)$ and $\{2\}$ are clopen in the metric space $(0, 1) \cup \{2\}$ with the regular distance.

(c) Describe all closed sets in a *discrete* metric space (X, ρ_d).

Theorem 2.18 (Characterizations of Closed Sets)**.** *Let (X,ρ) be a metric space and $A \subseteq X$. Then the following statements are equivalent:*

1. *A is closed in (X,ρ).*

2. *The complement A^c is open.*

3. *(Sequential Characterization) For any sequence $\{x_n\}_{n=1}^{\infty} \subseteq A$ convergent in (X,ρ), $\lim_{n \to \infty} x_n \in A$, i. e., A contains the limits of all its convergent sequences.*

Exercise 2.37.

(a) Prove.

(b) Show in *two different ways* that, in a *metric space* (X,ρ), a *closed ball* $\overline{B}(x, r)$ $(x \in X,$ $r \geq 0)$ is a *closed set*.

The properties of the closed sets follow immediately from the properties of the open sets via *de Morgan's laws* (see Section 1.1.1) considering the fact that the closed and open sets are complementary.

Theorem 2.19 (Properties of Closed Sets). *The closed sets in a metric space (X, ρ) have the following properties:*

(1) *$\emptyset$ and X are closed sets;*

(2) *an arbitrary intersection of closed sets is closed;*

(3) *a finite union of closed sets is closed.*

Exercise 2.38.

(a) Prove.

(b) Give an example showing that an *infinite union* of closed sets need not be closed.

2.10 Dense Sets and Separable Spaces

Here, we consider the notions of the *denseness* of a set in a metric space, i. e., of a set's points being able to be found arbitrarily close to all points of the space, and of the *separability* of a metric space, i. e., of a space's containing a countable such a set.

Definition 2.22 (Dense Set). A set A in a metric space (X, ρ) is called *dense* if $\overline{A} = X$.

Remark 2.25. Thus, a set A is *dense* in a metric space (X, ρ) *iff* an arbitrary nontrivial open ball contains a point of A (see Remarks 2.23).

Example 2.10. The set $\mathbb{Q}$ of the rational numbers is dense in $\mathbb{R}$.

Exercise 2.39. Verify.

From the *Sequential Characterization of Closure* (Theorem 2.17), immediately follows

Theorem 2.20 (Sequential Characterization of Dense Sets). *A set A is dense in a metric space (X, ρ) iff*

$$\forall x \in X \; \exists \; \{x_n\}_{n=1}^{\infty} \subseteq A : x_n \to x, \; n \to \infty.$$

Definition 2.23 (Separable Metric Space). A metric space (X, ρ) containing a *countable dense subset* is a called *separable*.

Remark 2.26. Any countable metric space is, obviously, separable. However, as the following examples show, a metric space need not be countable to be separable.

Examples 2.11.

1. The spaces $l_p^{(n)}$ are *separable* for $(n \in \mathbb{N}, 1 \le p \le \infty)$, which includes the cases of $\mathbb{R}$ and $\mathbb{C}$ with the regular distances.

   Indeed, as a countable dense set here, one can consider that of all ordered n-tuples with (real/complex) rational components.

2. The spaces l_p are *separable* for $1 \le p < \infty$.

Indeed, as a countable dense set here, one can consider that of all eventually zero sequences with (real/complex) rational terms.

3. The space $(C[a, b], \rho_\infty)$ is separable, which follows from *Weierstrass Approximation Theorem* (Theorem 2.49) when we consider as a countable dense set that of all polynomials with rational coefficients.

4. More examples of separable spaces can be built based on the fact that any subspace of a separable metric space is separable (Proposition 2.19) (see Section 2.19, Problem 17).

5. The space l_∞ is *not separable*.

Exercise 2.40.

(a) Verify 1–3 using the *Properties of Countable Sets* (Theorem 1.3).

(b) Prove 5.

> **Hint.** In l_∞, consider the *uncountable* set B of all *binary sequences*, i. e., the sequences whose only entries are 0 or 1 (see the *Uncountable Sets Proposition* (Proposition 1.1)). The balls of the *uncountable* collection
>
> $$\{B(x, 1/2) \mid x \in B\}$$
>
> are *pairwise disjoint*. Explain the latter and, assuming that there exists a countable dense subset in l_∞, arrive at a *contradiction*.

(c) Prove that the spaces $(C[a, b], \rho_p)$ $(1 \le p < \infty)$ (see Section 2.19, Problem 4) are separable.

(d) In which case is a *discrete* metric space (X, ρ_d) separable?

More facts on separability are stated as problems in Section 2.19.

2.11 Exterior and Boundary

Relative to a set A in a metric space (X, ρ), each point of the space falls into one of the tree pairwise disjoint classes: *interior*, *exterior*, or *boundary*. The interior points having been defined above (see Definition 2.14), it remains to define the exterior and boundary ones.

Definition 2.24 (Exterior and Boundary Points). Let A be a set in a metric space (X, ρ).

– We say that $x \in X$ is an *exterior point* of A if it is an interior point of the complement $A^c := X \setminus A$, i. e., there is an open ball centered at x_0 contained in A^c.
  All exterior points of a set A form its *exterior* ext(A).

– We say that $x \in X$ is a *boundary point* of A if it is neither interior nor exterior point of A, i. e., every open ball centered at x contains both a point of A and a point of A^c.
  All boundary points of a set A form its *boundary* ∂A.

Remark 2.27. By definition, for each set A in a metric space (X,ρ), $\text{int}(A)$, $\text{ext}(A)$, and ∂A form a *partition* of X, i. e., are *pairwise disjoint* and

$$\text{int}(A) \cup \partial A \cup \text{ext}(A) = X.$$

Exercise 2.41.
(a) In an arbitrary metric space (X,ρ), determine $\text{int}(\emptyset)$, $\text{ext}(\emptyset)$, $\partial \emptyset$ and $\text{int}(X)$, $\text{ext}(X)$, ∂X.
(b) Determine $\text{int}(A)$, $\text{ext}(A)$, and ∂A of a nonempty proper subset A of a *discrete* metric space (X,ρ_d).
(c) Determine $\text{int}(\mathbb{Q})$, $\text{ext}(\mathbb{Q})$, and $\partial \mathbb{Q}$ in $\mathbb{R}$.

2.12 Equivalent Metrics, Homeomorphisms and Isometries

2.12.1 Equivalent Metrics

Definition 2.25 (Equivalent Metrics). Two metrics ρ_1 and ρ_2 on a nonempty set X are called *equivalent* if they generate the same *metric topology*.

Exercise 2.42. Show that two metrics ρ_1 and ρ_2 on a nonempty set X are equivalent *iff*
(a) for an arbitrary $x \in X$, each open ball centered at x relative to ρ_1 contains an open ball centered at x relative to ρ_2 and *vice versa*;
(b) any sequence $\{x_n\}_{n=1}^{\infty}$ convergent relative to ρ_1 converges to the same limit relative to ρ_2 and *vice versa*.

Exercise 2.43.
(a) Show that the equivalence of metrics on a nonempty set X is an *equivalence relation* (*reflexive*, *symmetric*, and *transitive*) on the set of all metrics on X.
(b) Show that, for any metric space (X,ρ), ρ is equivalent to the *standard bounded metric* $d(x,y) := \min(\rho(x,y),1)$, $x,y \in X$ (see Section 2.19, Problem 3).
(c) Show that, if, for two metrics ρ_1 and ρ_2 on a nonempty set X,

$$\exists\, c, C > 0: \quad c\rho_1(x,y) \le \rho_2(x,y) \le C\rho_1(x,y), \ x,y \in X, \tag{2.5}$$

ρ_1 and ρ_2 are equivalent.
Use the prior example to show that the *converse* is not true.
(d) Show that, on the n-space $\mathbb{R}^n$ or $\mathbb{C}^n$ ($n \in \mathbb{N}$), all *p-metrics* ($1 \le p \le \infty$) (see Examples 2.1) are *equivalent* in the sense of (2.5).

Hint. Show the equivalence of any *p-metric* ρ_p ($1 \le p < \infty$) in the sense of (2.5) to ρ_∞.

2.12.2 Homeomorphisms and Isometries

Definition 2.26 (Homeomorphism of Metric Spaces). A *homeomorphism* of a metric space (X,ρ) to a metric space (Y,σ) is a mapping

$$T : X \to Y,$$

which is *bijective* and *bicontinuous*, i. e., $T \in C(X, Y)$ and the inverse $T^{-1} \in C(Y,X)$. The space (X,ρ) is said to be *homeomorphic* to (Y,σ).

Remarks 2.28.
- The relation of *being homeomorphic to* is an *equivalence relation* on the set of all metric spaces, and thus, we can say that homeomorphism is *between* the spaces.
- Homeomorphic metric spaces are *topologically indistinguishable*, i. e., have the same *topological properties* such as *separability* or the existence of nontrivial *clopen* sets (*disconnectedness* (see, e. g., [38, 41])).
- Two metrics ρ_1 and ρ_2 on a nonempty set X are equivalent *iff* the *identity mapping*

$$Ix := x, \ x \in X,$$

is a *homeomorphism* between the spaces (X,ρ_1) and (X,ρ_2).

Exercise 2.44. Verify.

To show that two metric spaces are homeomorphic, one needs to specify a homeomorphism between them; whereas to show that they are not homeomorphic, one only needs to specify a topological property not shared by them.

Examples 2.12.
1. Any open bounded interval (a, b) $(-\infty < a < b < \infty)$ is homeomorphic to $(0, 1)$ via

$$f(x) := \frac{x - a}{b - a}.$$

2. An open interval of the form $(-\infty, b)$ in $\mathbb{R}$ is homeomorphic to $(-b, \infty)$ via

$$f(x) := -x.$$

3. An open interval of the form (a, ∞) in $\mathbb{R}$ is homeomorphic to $(1, \infty)$ via

$$f(x) := x - a + 1.$$

4. The interval $(1, \infty)$ is homeomorphic to $(0, 1)$ via

$$f(x) := \frac{1}{x}.$$

5. The interval $(0, 1)$ is homeomorphic to $\mathbb{R}$ via

$$f(x) := \cot(\pi x).$$

6. The interval $(0, 1)$ is *not* homeomorphic to the set $(0, 1) \cup \{2\}$ with the regular distance, the latter having nontrivial clopen sets (see Examples 2.36).
7. The spaces l_1 and l_∞ are *not* homeomorphic, the former being separable and the latter being nonseparable.

Remark 2.29. Thus, all open intervals in $\mathbb{R}$ (bounded or not) are homeomorphic.

The following is a very important case of a homeomorphism.

Definition 2.27 (Isometry of Metric Spaces). Let (X, ρ) and (Y, σ) be metric spaces.
An *isometry of X to Y* is a *one-to-one* (i. e., *injective*) mapping $T : X \to Y$, which is *distance preserving*, i. e.,

$$\sigma(Tx, Ty) = \rho(x, y), \ x, y \in X.$$

It is said to *isometrically embed X in Y*.
If an isometry $T : X \to Y$ is *onto* (i. e., *surjective*), it is called an *isometry between X and Y* and the spaces are called *isometric*.

Remarks 2.30.
–   The relation of *being isometric to* is an *equivalence relation* on the set of all metric spaces, and thus, we can say that isometry is *between* the spaces.
–   An isometry between metric spaces (X, ρ) and (Y, σ) is, obviously, a homeomorphism between them but not vice versa (see Examples 2.12).
–   Isometric metric spaces are *metrically indistinguishable*.

Exercise 2.45. Identify isometries in Examples 2.12.

2.13 Completeness and Completion

Let us now deal with the concept of *completeness*, which is a fundamental property of metric spaces underlying many important facts.

2.13.1 Cauchy/Fundamental Sequences

Definition 2.28 (Cauchy/Fundamental Sequence). A sequence $\{x_n\}_{n=1}^{\infty}$ in a metric space (X, ρ) is called a *Cauchy sequence*, or a *fundamental sequence*, if

$$\forall \, \varepsilon > 0 \, \exists N \in \mathbb{N} \, \forall m, n \geq N : \rho(x_m, x_n) < \varepsilon.$$

Remark 2.31. The latter is equivalent to

$$\rho(x_m, x_n) \to 0, \ m, n \to \infty$$

or to

$$\sup_{k \in \mathbb{N}} \rho(x_{n+k}, x_n) \to 0, \ n \to \infty.$$

Examples 2.13.
1. A sequence is fundamental in a discrete space (X, ρ_d) iff it is *eventually constant*.
2. The sequence $\{1/n\}_{n=1}^{\infty}$ is *fundamental* in $\mathbb{R}$ and the sequence $\{n\}_{n=1}^{\infty}$ is not.
3. The sequence $\{x_n := \{1, 1/2, \dots, 1/n, 0, 0, \dots\}\}_{n=1}^{\infty}$ is *fundamental* in (c_{00}, ρ_{∞}) and in l_p $(1 < p < \infty)$, but not in l_1.
4. The sequence $\{e_n := \{\delta_{nk}\}_{k=1}^{\infty}\}_{n=1}^{\infty}$, where δ_{nk} is the *Kronecker delta*, is fundamental neither in (c_{00}, ρ_{∞}) nor in l_p $(1 \le p \le \infty)$.

Exercise 2.46. Verify.

Theorem 2.21 (Properties of Fundamental Sequences). *In a metric space (X, ρ),*
(1) *every fundamental sequence $\{x_n\}_{n=1}^{\infty}$ is bounded,*
(2) *every convergent sequence $\{x_n\}_{n=1}^{\infty}$ is fundamental,*
(3) *if a sequence $\{x_n\}_{n=1}^{\infty}$ is fundamental, then any sequence $\{y_n\}_{n=1}^{\infty}$ asymptotically equivalent to $\{x_n\}_{n=1}^{\infty}$ in the sense that*

$$\rho(x_n, y_n) \to 0, \ n \to \infty,$$

is also fundamental.

Exercise 2.47. Prove.

Remark 2.32. A fundamental sequence need not converge. Thus, the sequence $\{x_n = \{1, 1/2, \dots, 1/n, 0, 0, \dots\}\}_{n=1}^{\infty}$ is fundamental, but divergent in (c_{00}, ρ_{∞}). It does converge to $x = \{1/n\}_{n=1}^{\infty}$, however, in the wider space (c_0, ρ_{∞}).

Exercise 2.48. Verify.

Proposition 2.2 (Fundamentality and Uniform Continuity). *Let (X, ρ) and (Y, σ) be metric spaces and a function $f : X \to Y$ be uniformly continuous on X. If $\{x_n\}_{n=1}^{\infty}$ is a fundamental sequence in (X, ρ), then $\{f(x_n)\}_{n=1}^{\infty}$ is a fundamental sequence in (Y, σ), i.e., a uniformly continuous function maps a fundamental sequence to a fundamental sequence.*

Exercise 2.49.
(a) Prove.
(b) Give an example showing that a continuous function need not preserve fundamentality.

2.13.2 Complete Metric Spaces

Definition 2.29 (Complete Metric Space). A metric space (X, ρ), in which every Cauchy/fundamental sequence converges, is called *complete* and *incomplete* otherwise.

Examples 2.14.
1. The spaces $\mathbb{R}$ and $\mathbb{C}$ are *complete* relative to the regular distance as is known from analysis courses.
2. The spaces $\mathbb{R} \setminus \{0\}$, $(0, 1)$, and $\mathbb{Q}$ are *incomplete* as subspaces of $\mathbb{R}$.
3. A *discrete* metric space (X, ρ_d) is *complete*.

   Exercise 2.50. Verify 2 and 3.

4.

   Theorem 2.22 (Completeness of the n-Space). *The (real or complex) space $l_p^{(n)}$ $(n \in \mathbb{N}, 1 \le p \le \infty)$ is complete.*

   Exercise 2.51. Prove.

   Hint. Considering the equivalence of all p-metrics on the n-space in the sense of (2.5) (see Exercise 2.43) and Exercise 2.54, it suffices to show the completeness of the n-space relative to ρ_∞.

5.

   Theorem 2.23 (Completeness of l_p $(1 \le p < \infty)$). *The (real or complex) space l_p $(1 \le p < \infty)$ is complete.*

   Proof. Let $1 \le p < \infty$ and

   $$x^{(n)} := \{x_k^{(n)}\}_{k=1}^\infty, \ n \in \mathbb{N}.$$

   be an arbitrary fundamental sequence in l_p.
   Since

   $$\forall \varepsilon > 0 \ \exists N \in \mathbb{N}: \ \rho_p(x_m, x_n) < \varepsilon, \ m, n \ge N,$$

   for each $k \in \mathbb{N}$,

   $$\left| x_k^{(m)} - x_k^{(n)} \right| \le \left[\sum_{i=1}^\infty \left| x_i^{(m)} - x_i^{(n)} \right|^p \right]^{1/p} = \rho_p(x_m, x_n) < \varepsilon, \ m, n \ge N,$$

   which implies that, for every $k \in \mathbb{N}$, the numeric sequence $\{x_k^{(n)}\}_{n=1}^\infty$ of the kth terms is *fundamental*, and hence, *converges*, i. e.,

   $$\forall k \in \mathbb{N} \ \exists x_k \in \mathbb{R} \ (\text{or } \mathbb{C}): \ x_k^{(n)} \to x_k, \ n \to \infty.$$

Let us show that $x := \{x_k\}_{k=1}^{\infty} \in l_p$.

For any $K \in \mathbb{N}$ and arbitrary $m, n \geq N$,

$$\sum_{k=1}^{K} \left|x_k^{(m)} - x_k^{(n)}\right|^p \leq \sum_{k=1}^{\infty} \left|x_k^{(m)} - x_k^{(n)}\right|^p = \rho_p(x_m, x_n)^p < \varepsilon^p.$$

Whence, fixing arbitrary $K \in \mathbb{N}$ and $n \geq N$ and passing to the limit as $m \to \infty$, we obtain

$$\sum_{k=1}^{K} \left|x_k - x_k^{(n)}\right|^p \leq \varepsilon^p, \quad K \in \mathbb{N}, \; n \geq N.$$

Now, for an arbitrary $n \geq N$, passing to the limit as $K \to \infty$, we arrive at

$$\sum_{k=1}^{\infty} \left|x_k - x_k^{(n)}\right|^p \leq \varepsilon^p, \quad n \geq N, \tag{2.6}$$

which, in particular, implies that

$$x - x_N \in l_p,$$

and hence, by *Minkowski's Inequality for Sequences* (Theorem 2.4),

$$x = (x - x_N) + x_N \in l_p.$$

Whence, in view (2.6), we infer that

$$\rho_p(x, x_n) \leq \varepsilon, \; n \geq N,$$

which implies that

$$x_n \to x, \; n \to \infty,$$

in l_p and completes the proof. □

6.

Theorem 2.24 (Completeness of $(M(T), \rho_\infty)$). *The (real or complex) space $(M(T), \rho_\infty)$ is complete.*

Proof. Let $\{f_n\}_{n=1}^{\infty}$ be an arbitrary fundamental sequence in $(M(T), \rho_\infty)$. Since

$$\forall \varepsilon > 0 \; \exists N \in \mathbb{N} : \rho_\infty(f_m, f_n) < \varepsilon, \; m, n \geq N,$$

for each $t \in T$,

$$|f_m(t) - f_n(t)| \leq \sup_{s \in T} |f_m(s) - f_n(s)| = \rho_\infty(f_m, f_n) < \varepsilon, \; m, n \geq N,$$

which implies that, for every $t \in T$, the numeric sequence $\{f_n(t)\}_{n=1}^{\infty}$ of the values at t is *fundamental*, and hence, *converges*, i. e.,

$$\forall t \in T \; \exists f(t) \in \mathbb{R} \text{ (or } \mathbb{C}) : f_n(t) \to f(t), \; n \to \infty.$$

Let us show that function $T \ni t \mapsto f(t)$ belongs to $M(T)$.
Indeed, for any $t \in T$,

$$|f_m(t) - f_n(t)| < \varepsilon, \ m, n \geq N.$$

Whence, fixing arbitrary $t \in T$ and $n \geq N$ and passing to the limit as $m \to \infty$, we have:

$$|f(t) - f_n(t)| \leq \varepsilon, \ t \in T, \ n \geq N,$$

i. e.,

$$\sup_{t \in T} |f(t) - f_n(t)| \leq \varepsilon, \ n \geq N. \tag{2.7}$$

Therefore,

$$\sup_{t \in T} |f(t)| \leq \sup_{t \in T} |f(t) - f_N(t)| + \sup_{t \in T} |f_N(t)| \leq \sup_{t \in T} |f_N(t)| + \varepsilon < \infty,$$

which implies that $f \in M(T)$.
Whence, in view (2.7), we infer that

$$\rho_\infty(f, f_n) \leq \varepsilon, \ n \geq N,$$

which implies that

$$f_n \to f, \ n \to \infty,$$

in $(M(T), \rho_\infty)$ and completes the proof. □

As a particular case with $T = \mathbb{N}$, we obtain the following

Corollary 2.1 (Completeness of l_∞). *The (real or complex) space l_∞ is complete.*

7.

Theorem 2.25 (Completeness of $(C[a,b], \rho_\infty)$ $(-\infty < a < b < \infty)$). *The (real or complex) space $(C[a,b], \rho_\infty)$ $(-\infty < a < b < \infty)$ is complete.*

Exercise 2.52. Prove (cf. [16, Section 1.7, Exercise (i)])

8. The space P of all *polynomials* with real/complex coefficients is *incomplete* as a subspace of $(C[a,b], \rho_\infty)$.

Exercise 2.53. Give a corresponding *counterexample*.

9.

Proposition 2.3 (Incompleteness of $(C[a,b], \rho_p)$ $(1 \leq p < \infty)$). *The space $(C[a,b], \rho_p)$ $(1 \leq p < \infty)$ is incomplete.*

Proof. let $1 \leq p < \infty$. For a fixed $c \in (a, b)$ (say, $c := (a + b)/2$), consider the sequence $\{f_n\}$ in $(C[a, b], \rho_p)$ defined for all $n \in \mathbb{N}$ sufficiently large so that $a < c - 1/n$ as follows:

$$
f_n(t) := \begin{cases} 0 & \text{for } a \leq t \leq c - 1/n, \\ [n(t - c + 1/n)]^{1/p} & \text{for } c - 1/n < t < c, \\ 1 & \text{for } c \leq t \leq b. \end{cases}
$$

The sequence $\{f_n\}$ is *fundamental* in $(C[a, b], \rho_p)$. Indeed, for all sufficiently large $m, n \in \mathbb{N}$ with $m \leq n$,

$$
\rho_p(f_m, f_n) = \left[\int_{c-1/m}^{c} |f_m(t) - f_n(t)|^p \, dt \right]^{1/p} \qquad \text{by *Minkowski's inequality*;}
$$

$$
\leq \left[\int_{c-1/m}^{c} |f_m(t)|^p \, dt \right]^{1/p} + \left[\int_{c-1/m}^{c} |f_n(t)|^p \, dt \right]^{1/p}
$$

$$
= \left[\int_{c-1/m}^{c} m(t - c + 1/m) \, dt \right]^{1/p} + \left[\int_{c-1/n}^{c} n(t - c + 1/n) \, dt \right]^{1/p} = \frac{1}{m^p} + \frac{1}{n^p}.
$$

Since, for any $f \in C[a, b]$ and all sufficiently large $n \in \mathbb{N}$,

$$
\rho_p(f_n, f) = \left[\int_{a}^{c-1/n} |f(t)|^p \, dt + \int_{c-1/n}^{c} |f_n(t) - f(t)|^p \, dt + \int_{c}^{b} |1 - f(t)|^p \, dt \right]^{1/p},
$$

this jointly with the assumption that

$$
f_n \to f, \ n \to \infty,
$$

in $(C[a, b], \rho_p)$ would imply that

$$
\int_{a}^{c} |f(t)|^p \, dt + \int_{c}^{b} |1 - f(t)|^p \, dt = 0.
$$

Whence, by the *continuity* of f, we infer that

$$
f(t) = 0, \ t \in [a, c) \quad \text{and} \quad f(t) = 1, \ t \in (c, b],
$$

which *contradicts* the continuity of f at $x = c$.

The obtained contradiction implies that sequence $\{f_n\}$ cannot converge in $(C[a, b], \rho_p)$. Hence, the metric space $(C[a, b], \rho_p)$ is *incomplete*. $\qquad\square$

Remark 2.33. The property of completeness is *isometrically invariant* (see Section 2.19, Problem 27), but is *not homeomorphically invariant*.

For instance, the complete space $\mathbb{R}$ is homeomorphic to the incomplete space $(0,1)$ (see Examples 2.12 and 2.14).

A more sophisticated example is as follows: the set $X := \{1/n\}_{n\in\mathbb{N}}$ has the same *discrete* metric topology relative to the regular metric ρ as relative to the discrete metric ρ_d, which implies that the two metrics are *equivalent* on X (see Section 2.12.1), i. e., the spaces (X,ρ) and (X,ρ_d) and *homeomorphic* relative the identity mapping $Ix := x$ (see Remarks 2.28). However, the former space is incomplete whereas the latter is complete (see Examples 2.14).

Exercise 2.54.
(a) Explain.
(b) Prove that, if metrics ρ_1 and ρ_2 on a set X are *equivalent* in the sense of (2.5), (X,ρ_1) is complete *iff* (X,ρ_2) is complete.

2.13.3 Subspaces of Complete Metric Spaces

Proposition 2.4 (Characterization of Completeness). *Let (X,ρ) be complete metric space and $Y \subseteq X$. The subspace (Y,ρ) is complete iff the set Y is closed in (X,ρ).*

Exercise 2.55. Prove.

Exercise 2.56. Apply Proposition 2.4 to show that
(a) the spaces $\mathbb{R} \setminus \{0\}$, $(0,1)$, and $\mathbb{Q}$ are *incomplete* as subspaces of $\mathbb{R}$,
(b) the space (c_{00},ρ_∞) is *incomplete* as a subspace of (c_0,ρ_∞),
(c) the spaces (c_0,ρ_∞) and (c,ρ_∞) are *complete* as subspaces of l_∞,
(d) the space $(C[a,b],\rho_\infty)$ is *complete* as a subspace of $(M[a,b],\rho_\infty)$, and
(e) the space P of all *polynomials* with real/complex coefficients is *incomplete* as a subspace of $(C[a,b],\rho_\infty)$.

2.13.4 Nested Balls Theorem

The celebrated *Nested Intervals Theorem* (a form of the *completeness* of the *real numbers*) allows the following generalization.

Theorem 2.26 (Nested Balls Theorem). *A metric space (X,ρ) is complete iff, for every sequence of closed balls*

$$\{B_n := \overline{B}(x_n,r_n)\}_{n=1}^\infty$$

such that

(1) $B_{n+1} \subseteq B_n$, $n \in \mathbb{N}$, and
(2) $r_n \to 0$, $n \to \infty$,

the intersection $\bigcap_{n=1}^{\infty} B_n$ is a singleton.

Proof. "If" (*Sufficiency*) part. Under the *sufficiency* conditions, let $\{x_n\}_{n=1}^{\infty}$ be an arbitrary *fundamental sequence* in (X, ρ). Then there exists a *subsequence* $\{x_{n(k)}\}_{k=1}^{\infty}$ ($n(k) \in \mathbb{N}$ and $n(k) < n(k+1)$ for $k \in \mathbb{N}$) such that

$$\forall k \in \mathbb{N} : \rho(x_m, x_{n(k)}) \le \frac{1}{2^{k+1}}, \quad m \ge n(k). \tag{2.8}$$

Exercise 2.57. Explain.

Consider the sequence of closed balls $\{B_k := \bar{B}(x_{n(k)}, 1/2^k)\}_{k=1}^{\infty}$.
It is *nested*, i. e.,

$$B_{k+1} \subseteq B_k, \quad k \in \mathbb{N},$$

since, for any $y \in B_{k+1}$, by the *triangle inequality* in view of (2.8),

$$\rho(y, x_{n(k)}) \le \rho(y, x_{n(k+1)}) + \rho(x_{n(k+1)}, x_{n(k)}) \le \frac{1}{2^{k+1}} + \frac{1}{2^{k+1}} = \frac{1}{2^k}, \quad k \in \mathbb{N},$$

i. e., $y \in B_k$.
As for the *radii*, we have:

$$1/2^k \to 0, \quad k \to \infty.$$

Whence, by the premise, we infer that

$$\exists x \in X : \bigcap_{k=1}^{\infty} B_k = \{x\},$$

and hence,

$$0 \le \rho(x_{n(k)}, x) \le 1/2^k, \quad k \in \mathbb{N},$$

which, by the *Squeeze Theorem*, implies that

$$x_{n(k)} \to x, \quad k \to \infty, \text{ in } (X, \rho).$$

Since the *fundamental sequence* $\{x_n\}_{n=1}^{\infty}$ contains a *subsequence* $\{x_{n(k)}\}_{n=1}^{\infty}$ convergent to x, by the *Fundamental Sequence with Convergent Subsequence Proposition* (Proposition 2.22) (see Section 2.19, Problem 24), it also converges to x:

$$x_n \to x, \quad n \to \infty, \text{ in } (X, \rho),$$

which proves the *completeness* of the space (X, ρ).

"Only if" (*Necessity*) part. Suppose that the metric space (X, ρ) is *complete* and let $\{B_n := \bar{B}(x_n, r_n)\}_{n=1}^{\infty}$ be a sequence of closed balls satisfying the above conditions.

Then the sequence of the centers $\{x_n\}_{n=1}^{\infty}$ is *fundamental* in (X, ρ). Indeed, for all $n, k \in \mathbb{N}$, since $B_{n+k} \subseteq B_n$,

$$0 \le \rho(x_{n+k}, x_n) \le r_n,$$

which, in view of $r_n \to 0$, $n \to \infty$, by the *Squeeze Theorem*, implies that

$$\sup_{k \in \mathbb{N}} \rho(x_{n+k}, x_n) \to 0, \ n \to \infty.$$

By the *completeness* of (X, ρ),

$$\exists x \in X : \ x_n \to x, \ n \to \infty, \ \text{in } (X, \rho).$$

Let us show that

$$\bigcap_{n=1}^{\infty} B_n = \{x\}.$$

Indeed, since

$$\forall n \in \mathbb{N} : \ x_m \in B_n, \ m \ge n,$$

in view of the *closedness* of B_n, by the *Sequential Characterization of Closed Sets* (Theorem 2.18),

$$\forall n \in \mathbb{N} : \ x = \lim_{m \to \infty} x_m \in B_n,$$

and hence, we have the inclusion

$$\{x\} \subseteq \bigcap_{n=1}^{\infty} B_n.$$

On the other hand, for any $y \in \bigcap_{n=1}^{\infty} B_n$, since $x, y \in B_n$ for each $n \in \mathbb{N}$, by the *triangle inequality*,

$$0 \le \rho(x, y) \le \rho(x, x_n) + \rho(x_n, y) \le 2r_n,$$

which, in view of $r_n \to 0$, $n \to \infty$, by the *Squeeze Theorem*, implies that

$$\rho(x, y) = 0,$$

and hence, by the *separation axiom*, $x = y$.

Thus, the inverse inclusion

$$\{x\} \supseteq \bigcap_{n=1}^{\infty} B_n.$$

holds as well, and we conclude that

$$\bigcap_{n=1}^{\infty} B_n = \{x\}.$$ □

Remark 2.34. Each of the *four necessity conditions* in the *Nested Balls Theorem* (the completeness of the space, the closedness of the balls, and conditions (1) and (2)) is essential and cannot be relaxed or dropped.

Exercise 2.58. Verify by providing corresponding *counterexamples*.

A more general version of the *Nested Balls Theorem*, which can be proved by mimicking the proof of the latter, is the following

Theorem 2.27 (Generalized Nested Balls Theorem). *A metric space (X,ρ) is complete iff for every sequence of nonempty closed sets $\{B_n\}_{n=1}^{\infty}$ such that*
(1) $B_{n+1} \subseteq B_n$, $n \in \mathbb{N}$, *and*
(2) $\operatorname{diam}(B_n) := \sup_{x,y \in B_n} \rho(x,y) \to 0$, $n \to \infty$,

the intersection $\bigcap_{n=1}^{\infty} B_n$ is a singleton.

2.13.5 Completion

The following beautiful construct of obtaining a complete metric space from an arbitrary one is crucial in functional analysis.

Theorem 2.28 (Completion Theorem for Metric Spaces). *An arbitrary metric space (X,ρ) can be isometrically embedded as a dense subspace in a complete metric space $(\tilde{X},\tilde{\rho})$ called a completion of (X,ρ).*
    Any two completions of (X,ρ) are isometric.

Proof. On the set $\mathcal{X}$ of all Cauchy sequences in (X,ρ), let us define the *asymptotic equivalence relation* as follows:

$$\{x_n\}_{n=1}^{\infty} \sim \{y_n\}_{n=1}^{\infty} \iff \rho(x_n, y_n) \to 0, \ n \to \infty.$$

Exercise 2.59. Verify the equivalence axioms.

The collection of *all equivalence classes* $\tilde{X}$ is a metric space relative to the mapping

$$\tilde{X} \ni [x], [y] \mapsto \tilde{\rho}([x], [y]) := \lim_{n \to \infty} \rho(x_n, y_n), \tag{2.9}$$

where $x := \{x_n\}_{n=1}^{\infty}$ and $y := \{y_n\}_{n=1}^{\infty}$ are *arbitrary* representatives of the classes $[x]$ and $[y]$, respectively.

Exercise 2.60.

(a) Use the *Quadrilateral Inequality* (Theorem 2.8) to prove that the mapping $\tilde{\rho}$ is *well defined*, i. e., the limit *exists* and is *independent* of the choice of the representatives of the equivalence classes.

(b) Prove that $\tilde{\rho}$ is a *metric* on $\tilde{X}$.

The space $(\tilde{X}, \tilde{\rho})$ is *complete*. Indeed, let $\{[x_n]\}_{n=1}^{\infty}$ be a *Cauchy sequence* in $(\tilde{X}, \tilde{\rho})$, with $x_n := \{x_k^{(n)}\}_{k=1}^{\infty}$ being a representative of the class $[x_n]$, $n \in \mathbb{N}$.

Since for each $n \in \mathbb{N}$, the sequence $\{x_k^{(n)}\}_{k=1}^{\infty}$ is *fundamental* in (X, ρ),

$$\forall n \in \mathbb{N} \; \exists k(n) \in \mathbb{N} \; \forall p \geq k(n) : \rho\left(x_p^{(n)}, x_{k(n)}^{(n)}\right) < \frac{1}{n}. \tag{2.10}$$

Let show that the sequence $\{x_{k(n)}^{(n)}\}_{n=1}^{\infty}$ is *fundamental* in (X, ρ).

For any $m, n, p \in \mathbb{N}$, by the *triangle inequality*,

$$\rho\left(x_{k(m)}^{(m)}, x_{k(n)}^{(n)}\right) \leq \rho\left(x_{k(m)}^{(m)}, x_p^{(m)}\right) + \rho\left(x_p^{(m)}, x_p^{(n)}\right) + \rho\left(x_p^{(n)}, x_{k(n)}^{(n)}\right). \tag{2.11}$$

Since the sequence $\{[x_n]\}_{n=1}^{\infty}$ is *fundamental* in $(\tilde{X}, \tilde{\rho})$,

$$\forall \varepsilon > 0 \; \exists N \in \mathbb{N} \; \forall m, n \geq N : \tilde{\rho}([x_m], [x_n]) = \lim_{p \to \infty} \rho\left(x_p^{(m)}, x_p^{(n)}\right) < \frac{\varepsilon}{2},$$

and hence,

$$\forall \varepsilon > 0 \; \exists N \in \mathbb{N} \; \forall m, n \geq N \; \exists K(m, n) \in \mathbb{N} \; \forall p \geq K(m, n) :$$

$$\rho\left(x_p^{(m)}, x_p^{(n)}\right) < \frac{\varepsilon}{2}. \tag{2.12}$$

Without loss of generality, we can regard $N \in \mathbb{N}$ to be large enough so that

$$\frac{1}{n} < \frac{\varepsilon}{4}, \; n \geq N.$$

Let us fix arbitrary $m, n \geq N$. Then for all

$$p \geq \max\left[k(m), k(n), K(m, n)\right],$$

in view of the choice of $k(m)$ and $k(n)$ (see (2.10)), we have:

$$\rho\left(x_{k(m)}^{(m)}, x_p^{(m)}\right) < \frac{1}{m} < \frac{\varepsilon}{4} \quad \text{and} \quad \rho\left(x_p^{(n)}, x_{k(n)}^{(n)}\right) < \frac{1}{n} < \frac{\varepsilon}{4}. \tag{2.13}$$

From (2.11)–(2.13), we infer that

$$\rho\left(x_{k(m)}^{(m)}, x_{k(n)}^{(n)}\right) < \varepsilon, \; m, n \geq N,$$

and hence, the sequence $\{x_{k(n)}^{(n)}\}_{n=1}^{\infty}$ is *fundamental* in (X, ρ).

Let us show now that the sequence $\{[x_n]\}_{n=1}^{\infty}$ converges in $(\tilde{X}, \tilde{\rho})$ to the *equivalence class* $[x]$ represented by the sequence $x := \{x_{k(n)}^{(n)}\}_{n=1}^{\infty}$. Indeed, for any fixed $n \in \mathbb{N}$,

$$\tilde{\rho}([x_n],[x]) = \lim_{p\to\infty} \rho\left(x_p^{(n)}, x_{k(p)}^{(p)}\right) \qquad\qquad \text{by the } \textit{triangle inequality};$$

$$\leq \lim_{p\to\infty} \rho\left(x_p^{(n)}, x_{k(n)}^{(n)}\right) + \lim_{p\to\infty} \rho\left(x_{k(n)}^{(n)}, x_{k(p)}^{(p)}\right) \qquad\qquad \text{by the choice of } k(n) \text{ (see (2.10))};$$

$$\leq \frac{1}{n} + \lim_{p\to\infty} \rho\left(x_{k(n)}^{(n)}, x_{k(p)}^{(p)}\right). \tag{2.14}$$

Since the sequence $\{x_{k(n)}^{(n)}\}_{n=1}^{\infty}$ is *fundamental* in (X, ρ),

$$\forall \varepsilon > 0 \; \exists N \in \mathbb{N} \; \forall n, p \geq N : \; \rho\left(x_{k(n)}^{(n)}, x_{k(p)}^{(p)}\right) < \frac{\varepsilon}{2}.$$

For each fixed $n \geq N$, passing to the limit as $p \to \infty$, we arrive at

$$\lim_{p\to\infty} \rho\left(x_{k(n)}^{(n)}, x_{k(p)}^{(p)}\right) \leq \frac{\varepsilon}{2}, \; n \geq N. \tag{2.15}$$

Without loss of generality, we can regard $N \in \mathbb{N}$ to be large enough so that

$$\frac{1}{n} < \frac{\varepsilon}{2}, \; n \geq N.$$

Considering this, we infer from inequalities (2.14) and (2.15) that

$$\tilde{\rho}([x_n],[x]) < \varepsilon, \; n \geq N,$$

and hence,

$$\lim_{n\to\infty}[x_n] = [x] \text{ in } (\tilde{X}, \tilde{\rho}),$$

which proves the *completeness* of $(\tilde{X}, \tilde{\rho})$.

The *isometric embedding* $T : X \to \tilde{X}$ is obtained by associating with each $x \in X$ the equivalence class $Tx \in \tilde{X}$ represented by the constant sequence $\{x, x, \ldots, x, \ldots\}$, and hence, consisting of all sequences *convergent* to x in (X, ρ).

Exercise 2.61. Verify.

Thus, $T(X)$ is the set of all equivalence classes of sequences convergent in (X, ρ).

To show that $T(X)$ is *dense* in $\tilde{X}$, consider an arbitrary equivalence class $[x] \in \tilde{X}$ represented by a Cauchy sequence $\{x_n\}_{n=1}^{\infty}$.

Since

$$\forall \varepsilon > 0 \; \exists N \in \mathbb{N} \; \forall m, n \geq N : \; \rho(x_m, x_n) < \varepsilon,$$

we have:

$$\forall n \geq N : \; \tilde{\rho}([x], Tx_n) = \lim_{m\to\infty} \rho(x_m, x_n) \leq \varepsilon,$$

i. e.,

$$\lim_{n \to \infty} Tx_n = [x] \text{ in } (\tilde{X}, \tilde{\rho}),$$

which, by the *Sequential Characterization of Dense Sets* (Theorem 2.20), implies that

$$\overline{T(X)} = \tilde{X}.$$

Thus, $(\tilde{X}, \tilde{\rho})$ is a *completion* of (X, ρ).

It remains now to prove the *uniqueness* of completion *up to an isometry*.

Observe that a completion $(\tilde{X}, \tilde{\rho})$ of (X, ρ) encompasses a complete metric space $(\tilde{X}, \tilde{\rho})$ along with an *isometric dense embedding* T of (X, ρ) in $(\tilde{X}, \tilde{\rho})$, i. e., forms a triple $(\tilde{X}, \tilde{\rho}, T)$.

Let $(\tilde{X}, \tilde{\rho}, T)$ and $(\tilde{Y}, \tilde{\sigma}, S)$ be two arbitrary completions of (X, ρ).

We obtain an *isometry* of $\tilde{X}$ onto $\tilde{Y}$ by continuously extending the isometry ST^{-1} of $T(X)$, which is dense in $\tilde{X}$, *onto* $S(X)$, which is dense in $\tilde{Y}$.

Exercise 2.62. Describe the extension process. □

Remark 2.35. It follows immediately that, if X_0 is a *dense subset* of a *complete* metric space (X, ρ), the latter is a *completion* of the space (X_0, ρ).

Examples 2.15.
1. Since $\overline{\mathbb{Q}} = \mathbb{R}$, $\mathbb{R}$ is a *completion* of $\mathbb{Q}$.
2. More generally, since $\mathbb{Q}^n$ is a *dense subspace* of the real $l_p^{(n)}$ ($n \in \mathbb{N}$, $1 \le p \le \infty$), the latter is a *completion* of $(\mathbb{Q}^n, \rho_p)$.

   Remark 2.36. The complex counterpart is obvious.

3. Since the set c_{00} is *dense* in the complete space (c_0, ρ_∞) (see Exercise 2.56), the latter is a *completion* of (c_{00}, ρ_∞). The same is true for the pair l_p ($1 \le p < \infty$) and (c_{00}, ρ_p).

   Exercise 2.63. Verify.

4. Since the set P of all *polynomials* with real/complex coefficients is *dense subspace* of $(C[a, b], \rho_\infty)$, the latter is a *completion* of (P, ρ_∞).

Exercise 2.64. Construct *completions* of $\mathbb{R} \setminus \{0\}$ and $(0, 1)$ relative to the usual metric.

2.14 Category and Baire Category Theorem

2.14.1 Nowhere Denseness

First, we are to introduce and study *nowhere dense sets*, i. e., sets in a metric space, which are, in a certain sense, "scarce".

Definition 2.30 (Nowhere Dense Set). A set A in a metric space (X,ρ) is called *nowhere dense* if the interior of its closure $\overline{A}$ is empty:

$$\text{int}(\overline{A}) = \emptyset,$$

i. e., $\overline{A}$ contains no nontrivial open balls.

Remarks 2.37.

– A set A is nowhere dense in a metric space (X,ρ) *iff* its closure $\overline{A}$ is nowhere dense in (X,ρ) (see Section 2.19, Problem 15).
– For a closed set A in a metric space (X,ρ), since $A = \overline{A}$, its nowhere denseness is simply the emptiness of its interior:

$$\text{int}(A) = \emptyset.$$

Examples 2.16.

1. The empty set $\emptyset$ is nowhere dense in any metric space (X,ρ) and only the empty set is nowhere dense in a *discrete* metric space (X,ρ_d).
2. Finite sets, the sets $\mathbb{Z}$ of all integers and $\{1/n\}_{n\in\mathbb{N}}$ are nowhere dense in $\mathbb{R}$.
3. The celebrated *Cantor set* (see, e. g., [21, 23, 24]) is a closed nowhere dense set in $\mathbb{R}$ as well as its two-dimensional analogue, the *Sierpinski[7] carpet*, in $\mathbb{R}^2$ with the Euclidean distance (i. e., in $l_2^{(2)}(\mathbb{R})$) (see, e. g., [38]).
4. The set $\mathbb{R}$ of the reals is nowhere dense in $l_2^{(2)}(\mathbb{R})$.
5. An arbitrary *dense* set in a metric space (X,ρ) is *not* nowhere dense. In particular, the sets $\mathbb{Q}$ of the rationals and the set $\mathbb{Q}^c$ of the irrationals are *not* nowhere dense in $\mathbb{R}$.
6. However, a set need not be dense in a metric space not to be nowhere dense. Thus, an nontrivial proper interval I in $\mathbb{R}$ is neither dense nor nowhere dense.

Exercise 2.65. Verify.

Remark 2.38. Formulated entirely in terms of closure and interior, *nowhere denseness* is a *topological property*, i. e., it is preserved by a *homeomorphism*.

Proposition 2.5 (Characterization of Nowhere Denseness). *A set A is nowhere dense in a metric space (X,ρ) iff its exterior* $\text{ext}(A)$ *is dense in* (X,ρ).

Proof. In view of the fact that, for any $B \subseteq X$,

$$\text{ext}(B) = X \setminus \overline{B}$$

---

7 Waclaw Sierpinski (1882–1969).

(see Section 2.19, Problem 21), the statement immediately follows from the following disjoint union representation (first, setting $B := X \setminus \overline{A}$, then setting $B := A$):

$$X = \text{ext}(X \setminus \overline{A}) \cup \overline{X \setminus \overline{A}} = \text{int}(\overline{A}) \cup \overline{\text{ext}(A)}.$$

Exercise 2.66. Explain. □

Since for a closed set A in a metric space (X,ρ),

$$\text{ext}(A) = A^c$$

(see Section 2.19, Problem 21), we immediately obtain the following

Corollary 2.2 (Characterization of Nowhere Denseness of Closed Sets). *A closed set A in a metric space (X,ρ) is nowhere dense iff its open complement A^c is dense in (X,ρ).*

By the complementary nature of closed and open sets (see Theorem 2.18), we can equivalently reformulate the prior statement as follows:

Corollary 2.3 (Characterization of Denseness of Open Sets). *An open set A in a metric space (X,ρ) is dense iff its closed complement A^c is nowhere dense in (X,ρ).*

Example 2.17. The complement of the *Cantor set* is an open dense set in $\mathbb{R}$.

Theorem 2.29 (Properties of Nowhere Dense Sets). *The nowhere dense sets in a metric space (X,ρ) have the following properties:*
(1) *a subset of a nowhere dense set is nowhere dense;*
(2) *an arbitrary intersection of nowhere dense sets is nowhere dense;*
(3) *a finite union of nowhere dense sets is nowhere dense.*

Exercise 2.67. Prove (cf. *Properties of Bounded Sets* (Theorem 2.13)).

Hint. To prove (3), first show that, for any sets A and B in (X,ρ),

$$\overline{A \cup B} = \overline{A} \cup \overline{B},$$

which is equivalent to

$$\text{ext}(A \cup B) = \text{ext}(A) \cap \text{ext}(B)$$

(see Section 2.19, Problem 21), and then exploit the *denseness* jointly with *openness* (cf. *Finite Intersections of Open Dense Sets Proposition* (Proposition 2.24) (Section 2.19, Problem 29)).

Remark 2.39. An *infinite* union of nowhere dense sets in a metric space (X,ρ) need not be nowhere dense.

For instance, any *singleton* is nowhere dense in $\mathbb{R}$. However, $\mathbb{Q}$, being a countably infinite union of singletons, is *dense* in $\mathbb{R}$.

2.14.2 Category

Here, based on the concept of *nowhere denseness*, we quite naturally divide the sets of a metric space into two categories, *category* giving a sense of a set's "fullness" and being closely related to the notion of *completeness*.

Definition 2.31 (First/Second-Category Set). A set A in a metric space (X,ρ) is said to be of the *first category* if it can be represented as a *countable union* of *nowhere dense* sets.

Otherwise, A is said to be of the *second category* in (X,ρ).

Examples 2.18.
1. A *nowhere dense set* in a metric space (X,ρ) is a *first-category set*.
   In particular, the set $\mathbb{N}$ of all naturals is of the first category in $\mathbb{R}$.
2. A *dense set* in a metric space (X,ρ) may be of the first-category as well. For instance, the set $\mathbb{Q}$ of all rationals in $\mathbb{R}$.
3. The space (c_{00},ρ_∞) is of the *first category* in itself (also in (c_0,ρ_∞), (c,ρ_∞), and l_∞).
   Indeed,

$$c_{00} = \bigcup_{n=1}^{\infty} U_n,$$

where

$$U_n := \{x := \{x_1,\ldots,x_n,0,0,\ldots\}\}, \ n \in \mathbb{N}.$$

Each U_n is *closed*, being an *isometric embedding* of the *complete* space $(l_\infty^{(n)},\rho_\infty)$ into (c_{00},ρ_∞), and *nowhere dense* in (c_{00},ρ_∞) (also in (c_0,ρ_∞), (c,ρ_∞), and l_∞) since

$$\forall n \in \mathbb{N} \ \forall x := \{x_1,\ldots,x_n,0,0,\ldots\} \in U_n \ \forall \varepsilon > 0$$
$$\exists y := \{x_1,\ldots,x_n,\varepsilon/2,0,0\ldots\} \in c_{00} \setminus U_n : \rho_\infty(x,y) = \varepsilon/2 < \varepsilon,$$

which implies that $\text{int}(U_n) = \emptyset$, $n \in \mathbb{N}$.
4. As follows from the *Baire*[8] *Category Theorem* (Theorem 2.31), every complete metric space (X,ρ) is of the second category in itself.
   In particular, the complete spaces $\mathbb{R}$ and $\mathbb{C}$ are of the second category in themselves, the former being of the first category (*nowhere dense*, to be precise) in the latter (cf. Examples 2.16).
5. Every nonempty set in a *discrete* metric space (X,ρ_d) is a *second-category set*.

---

[8] René-Louis Baire (1874–1932).

Exercise 2.68. Explain 5.

Remark 2.40. Formulated entirely in terms of nowhere denseness and union, *first category* is a *topological property*, i. e., it is preserved by a *homeomorphism*, and hence, so is *second category*.

Example 2.19. Thus, every open interval I in $\mathbb{R}$, being homeomorphic to $\mathbb{R}$ (see Remark 2.29), is of the second category in itself as a subspace of $\mathbb{R}$ and, since the interior of a set in I coincides with its interior in $\mathbb{R}$, is also of the second category in $\mathbb{R}$.

Theorem 2.30 (Properties of First-Category Sets). *The first-category sets in a metric space (X,ρ) have the following properties:*
(1) *a subset of a first-category set is a first-category set;*
(2) *an arbitrary intersection of first-category sets is a first-category set;*
(3) *an arbitrary countable union of first-category sets is a first-category set.*

Exercise 2.69. Prove.

We immediately obtain the following

Corollary 2.4 (Set With a Second-Category Subset). *A set A in a metric space (X,ρ) containing a second-category subset B is of the second category.*

Examples 2.20.
1. Every nontrivial (open or closed) interval I in $\mathbb{R}$ is of the *second category* in $\mathbb{R}$ and in itself as a subspace of $\mathbb{R}$ since it contains an open interval, which is a second-category set in $\mathbb{R}$ (see Example 2.19).
2. More generally, for the same reason, any set $A \subseteq \mathbb{R}$ with $\operatorname{int}(A) \neq \emptyset$ is of the *second category* in $\mathbb{R}$ and in itself as a subspace of $\mathbb{R}$.

2.14.3 Baire Category Theorem

The *Baire Category Theorem*, already referred to in the prior section, is one of the most important facts about complete metric spaces critical for proving a number of fundamental statements such as the *Uniform Boundedness Principle* (Theorem 6.7) and the *Open Mapping Theorem* (Theorem 6.13). We are to prove the celebrated statement now.

Theorem 2.31 (Baire Category Theorem). *A complete metric space (X,ρ) is of the second category in itself.*

Proof. Let us prove the statement *by contradiction* assuming that there is a complete metric space (X,ρ) of the first category in itself, i. e., X can be represented as a countable union of nowhere dense sets:

$$X = \bigcup_{n=1}^{\infty} U_n.$$

Without loss of generality, we can regard all U_n, $n \in \mathbb{N}$, to be *closed*.

Exercise 2.70. Explain.

Being nowhere dense, the closed set U_1 is a *proper subset* of X, and hence, by the *Characterization of Nowhere Denseness of Closed Sets* (Corollary 2.2), its *open* complement U_1^c is *dense* in (X, ρ), and the more so, *nonempty*. Therefore,

$$\exists x_1 \in U_1^c \ni 0 < \varepsilon_1 < 1 : \overline{B}(x_1, \varepsilon_1) \subseteq U_1^c,$$

i. e.,

$$\overline{B}(x_1, \varepsilon_1) \cap U_1 = \emptyset,$$

Since the open ball $B(x_1, \varepsilon_1/2)$ is not contained in the closed nowhere dense set U_2,

$$\exists x_2 \in B(x_1, \varepsilon_1/2) \ni 0 < \varepsilon_2 < 1/2 : \overline{B}(x_2, \varepsilon_2) \cap U_2 = \emptyset$$

and

$$\overline{B}(x_2, \varepsilon_2) \subseteq B(x_1, \varepsilon_1/2).$$

Continuing inductively, we obtain a sequence of closed balls $\{\overline{B}(x_n, \varepsilon_n)\}_{n=1}^{\infty}$ such that

(1) $\overline{B}(x_{n+1}, \varepsilon_{n+1}) \subseteq \overline{B}(x_n, \varepsilon_n)$, $n \in \mathbb{N}$.
(2) $0 < \varepsilon_n < 1/2^{n-1}$, $n \in \mathbb{N}$.
(3) $\overline{B}(x_n, \varepsilon_n) \cap U_n = \emptyset$, $n \in \mathbb{N}$.

From (1) and (2), by the *Nested Balls Theorem* (Theorem 2.26), we infer that

$$\bigcap_{n=1}^{\infty} \overline{B}(x_n, \varepsilon_n) = \{x\}$$

with some $x \in X$.

Since $x \in \overline{B}(x_n, \varepsilon_n)$ for each $n \in \mathbb{N}$, by (3), we conclude that

$$x \notin \bigcup_{n=1}^{\infty} U_n = X,$$

which is a *contradiction* proving the statement. ☐

Examples 2.21.
1. By the *Baire Category Theorem*, the complete spaces $\mathbb{R}$ and $\mathbb{C}$ are of the second category in themselves (see Examples 2.18).
2. Any set $A \subseteq \mathbb{R}$ with $\text{int}(A) \neq \emptyset$ is of the *second category* in $\mathbb{R}$ and in itself as a subspace of $\mathbb{R}$ (cf. Examples 2.20).

3. However, a set need not have a nonempty interior to be of the second category. Indeed, as follows from the *Baire Category Theorem*, the set $\mathbb{Q}^c$ of all irrationals with $\mathrm{int}(\mathbb{Q}^c) = \emptyset$ is of the *second category* in $\mathbb{R}$, as well as the complement A^c of any first-category set A in a complete metric space (X,ρ) (see Section 2.19, Problem 31).
4. By the *Baire Category Theorem* and the *Characterization of Completeness* (Proposition 2.4), the set of all integers $\mathbb{Z}$ and the *Cantor set* are of the *second category* in themselves as closed subspaces of the complete space $\mathbb{R}$, but are of the *first category* (*nowhere dense*, to be precises) in $\mathbb{R}$ (cf. Examples 2.16).

Remarks 2.41.
- The converse to the *Baire Category Theorem* is not true, i. e., there exist incomplete metric spaces of the second category in themselves.
  For instance, an open interval I in $\mathbb{R}$ is of the *second category* in itself (see Example 2.19), but is *incomplete* as a subspace of $\mathbb{R}$.
- The proof of the *Baire Category Theorem* requires a weaker form of the *Axiom of Choice* (see Appendix A), the *Axiom of Dependent Choices* (see, e. g., [22, 36]).

Corollary 2.5 (Second-Category Properties of Complete Metric Spaces). *In a complete metric space* (X,ρ),
(1) *any representation of X as a countable union of its subsets*

$$X = \bigcup_{n=1}^{\infty} U_n$$

contains at least one subset U_N, which is not nowhere dense, i. e.,

$$\exists N \in \mathbb{N}: \ \mathrm{int}(\overline{U_N}) \neq \emptyset;$$

(2) *any countable intersection $\bigcap_{n=1}^{\infty} U_n$ of open dense sets is nonempty, i. e.,*

$$\bigcap_{n=1}^{\infty} U_n \neq \emptyset.$$

Exercise 2.71. Prove.

Hint. To prove (2) use the *Characterization of Nowhere Denseness of Closed Sets* (Corollary 2.2) (cf. the *Finite Intersections of Open Dense Sets Proposition* (Proposition 2.24)).

2.15 Compactness

The notion of *compactness*, naturally emerging from our spatial intuition, is of utmost importance both in theory and applications. We are to study it and related concepts here.

2.15.1 Total Boundedness

2.15.1.1 Definitions and Examples

Total boundedness is a notion inherent to metric spaces that is stronger than *boundedness*, but weaker than *precompactness*.

Definition 2.32 (ε-Net). Let (X,ρ) be a metric space and $\varepsilon > 0$. A set $N_\varepsilon \subseteq X$ is called an *ε-net for a set* $A \subseteq X$ if A can be *covered* by the collection $\{B(x,\varepsilon) \mid x \in N_\varepsilon\}$ of all open ε-balls centered at the points of N_ε, i. e.,

$$A \subseteq \bigcup_{x \in N_\varepsilon} B(x,\varepsilon).$$

Examples 2.22.
1. For an arbitrary nonempty set A in a metric space (X,ρ) and any $\varepsilon > 0$, A is an ε-net for itself.
2. A *dense set* A in a *metric space* (X,ρ) is an ε-net for the entire X with any $\varepsilon > 0$. In particular, for any $\varepsilon > 0$, $\mathbb{Q}$ and $\mathbb{Q}^c$ are ε-nets for $\mathbb{R}$.
3. For any $\varepsilon > 0$, $\mathbb{Q}$ is an ε-net for $\mathbb{Q}^c$.
4. For any $n \in \mathbb{N}$ and $\varepsilon > 0$, the set

$$N_\varepsilon := \left\{ \left(\frac{\varepsilon k_1}{\sqrt{n}}, \ldots, \frac{\varepsilon k_n}{\sqrt{n}} \right) \,\Big|\, k_i \in \mathbb{Z}, \ i = 1, \ldots, n \right\}$$

is an ε-net for $\mathbb{R}^n$ with the *Euclidean metric*, i. e., for $l_2^{(n)}(\mathbb{R})$.

Exercise 2.72. Verify and make a drawing for $n = 2$.

Remark 2.42. Example 4 can be modified to furnish an ε-net for $\mathbb{R}^n$ endowed with any p-metric $(1 \le p \le \infty)$, i. e., for $l_p^{(n)}(\mathbb{R})$ and can be naturally stretched to the *complex* n-space $\mathbb{C}^n$ relative to any p-metrics $(1 \le p \le \infty)$, i. e., to any space $l_p^{(n)}(\mathbb{C})$.

Exercise 2.73. Verify for $n = 2$, $p = 1$ and for $n = 2$, $p = \infty$.

Remarks 2.43. As the prior examples demonstrate, an ε-net N_ε for a set A in a metric space (X,ρ) need not consist of points of A. It may even happen that A and N_ε are *disjoint*.

Definition 2.33 (Total Boundedness). A set A in a metric space (X,ρ) is called *totally bounded* if, for any $\varepsilon > 0$, there is a *finite* ε-net for A:

$$\forall \varepsilon > 0 \, \exists N \in \mathbb{N}, \ \exists \{x_1, \ldots, x_N\} \subseteq X : \ A \subseteq \bigcup_{n=1}^{N} B(x_n, \varepsilon),$$

i. e., A can be covered by a finite number of ε-balls, however small their radii.

A metric space (X,ρ) is called *totally bounded* if the set X is totally bounded in (X,ρ).

Remarks 2.44.
- A set A is totally bounded in a metric space (X,ρ) *iff* totally bounded is its closure $\bar{A}$.
- For a totally bounded set A in a metric space (X,ρ) and any $\varepsilon > 0$, a finite ε-net for A can be chosen to consist entirely of points of A.
- The total boundedness of a nonempty set A in a metric space (X,ρ) is equivalent to the total boundedness of (A,ρ) as a subspace of (X,ρ).

Exercise 2.74. Verify.

Hint. Consider a finite $\varepsilon/2$-net $\{x_1,\dots,x_N\} \subseteq X$ ($N \in \mathbb{N}$) for A and construct an ε-net $\{y_1,\dots,y_N\} \subseteq A$.

Examples 2.23.
1. *Finite sets*, including the empty set $\emptyset$, are *totally bounded* in an arbitrary metric space (X,ρ) and only finite sets are totally bounded in a *discrete* metric space (X,ρ_d).
2. A *bounded* set A in the n-space $\mathbb{R}^n$ ($n \in \mathbb{N}$) with the *Euclidean metric*, i. e., in $l_2^{(n)}(\mathbb{R})$, is *totally bounded*.

   Indeed, being *bounded*, the set A is contained in a *hypercube*

   $$J_m = [-m,m]^n$$

   with some $m \in \mathbb{N}$. Then, for any $\varepsilon > 0$, $N'_\varepsilon := N_\varepsilon \cap J_m$, where N_ε is the ε-net for $l_2^{(n)}(\mathbb{R})$ from Examples 2.22, is a *finite ε-net* for A.

   Remark 2.45. The same is true for any (real or complex) $l_p^{(n)}$ ($1 \le p \le \infty$) (see Remark 2.42).

   In particular, $(0,1]$ and $\{1/n\}_{n\in\mathbb{N}}$ are *totally bounded* sets in $\mathbb{R}$.
3. The set $E := \{e_n := \{\delta_{nk}\}_{k=1}^\infty\}_{n\in\mathbb{N}}$, where δ_{nk} is the *Kronecker delta*, is *bounded*, but *not totally bounded* in (c_{00},ρ_∞) (also in (c_0,ρ_∞), (c,ρ_∞), and l_∞) since there is no finite $1/2$-net for A.

   The same example works in l_p ($1 \le p < \infty$).

Exercise 2.75. Verify.

2.15.1.2 Properties
Theorem 2.32 (Properties of Totally Bounded Sets). *The totally bounded sets in a metric space (X,ρ) have the following properties:*
(1) *a totally bounded set is necessarily bounded, but not vice versa;*
(2) *a subset of a totally bounded set is totally bounded;*
(3) *an arbitrary intersection of totally bounded sets is totally bounded;*
(4) *a finite union of totally bounded sets is totally bounded.*

Exercise 2.76.
(a) Prove.
(b) Give an example showing that an infinite union of totally bounded sets need not be totally bounded.

Exercise 2.77. Using the set E from Examples 2.23, show that a nontrivial sphere/ball in (c_{00}, ρ_∞) (also in (c_0, ρ_∞), (c, ρ_∞), and l_p $(1 \le p \le \infty)$) is *not totally bounded*.

From the prior proposition and Examples 2.23, we obtain

Corollary 2.6 (Characterization of Total Boundedness in the n-Space). *A set A is totally bounded in the (real or complex) space $l_p^{(n)}$ $(n \in \mathbb{N}, 1 \le p \le \infty)$ iff it is bounded.*

It is remarkable that *total boundedness* in a metric space can be characterized in terms of *fundamentality* as follows.

Theorem 2.33 (Characterization of Total Boundedness). *A nonempty set A is totally bounded in a metric space (X, ρ) iff every sequence $\{x_n\}_{n=1}^\infty$ in A contains a fundamental subsequence $\{x_{n(k)}\}_{k=1}^\infty$.*

Proof. "*Only if*" part. Suppose a set A is totally bounded in a metric space (X, ρ) and let $\{x_n\}_{n=1}^\infty$ be an arbitrary sequence in A.

If the set A is *finite*, i. e., $A = \{y_1, \ldots, y_N\}$ with some $N \in \mathbb{N}$, then $\{x_n\}_{n=1}^\infty$ necessarily assumes the same value y_i with some $i = 1, \ldots, N$ for infinitely many indices $n \in \mathbb{N}$ (i. e., "*frequently*").

Exercise 2.78. Explain.

Hence, $\{x_n\}_{n=1}^\infty$ contains a *constant* subsequence, which is *fundamental*.

Now, assume that the set A is *infinite* and let $\{x_n\}_{n=1}^\infty$ be an arbitrary sequence in A. If $\{x_n\}_{n=1}^\infty$ assumes only a finite number of distinct values, we arrive at the prior case.

Suppose that $\{x_n\}_{n=1}^\infty$ assumes infinite many distinct values. Then, without loss of generality, we can regard that

$$x_m \ne x_n, \quad m, n \in \mathbb{N}.$$

By the total boundedness of A, since it is coverable by a finite number of 1-balls, there must exist a 1-ball $B(y_1, 1)$ with some $y_1 \in X$, which contains *infinitely many* terms of $\{x_n\}_{n=1}^\infty$, and hence, a subsequence $\{x_{1,n}\}_{n=1}^\infty$ of $\{x_n\}_{n=1}^\infty$.

Similarly, there must exist a 1/2-ball $B(y_2, 1/2)$ with some $y_2 \in X$, which contains *infinitely many* terms of $\{x_{1,n}\}_{n=1}^\infty$, and hence, a subsequence $\{x_{2,n}\}_{n=1}^\infty$ of $\{x_{1,n}\}_{n=1}^\infty$.

Continuing inductively, we obtain a countable collection of sequences

$$\{\{x_{m,n}\}_{n=1}^\infty \,|\, m \in \mathbb{Z}_+\},$$

such that, for each $m \in \mathbb{N}$, $\{x_{m,n}\}_{n=1}^\infty$ is a subsequence of $\{x_{(m-1),n}\}_{n=1}^\infty$, with

$$\{x_{0,n}\}_{n=1}^\infty := \{x_n\}_{n=1}^\infty,$$

and

$$\{x_{m,n}\}_{n=1}^{\infty} \subseteq B(y_m, 1/m)$$

with some $y_m \in X$.

Then the *"diagonal subsequence"* $\{x_{n,n}\}_{n=1}^{\infty}$ is a *fundamental subsequence* of $\{x_n\}_{n=1}^{\infty}$ since, by the *triangle inequality*,

$$\forall n \in \mathbb{N}, \forall m \geq n : \rho(x_{m,m}, x_{n,n}) \leq \rho(x_{m,m}, y_n) + \rho(y_n, x_{n,n}) < 1/n + 1/n = 2/n.$$

"If" part. Let us prove this part *by contrapositive* assuming that a set A is *not* totally bounded in a metric space (X, ρ). Then there is an $\varepsilon > 0$ such that there does not exist a finite ε-net for A.

In particular, for an arbitrary $x_1 \in A$ the ε-ball $B(x_1, \varepsilon)$ does not cover A, and hence,

$$\exists x_2 \in A : x_2 \notin B(x_1, \varepsilon).$$

Similarly, the ε-balls $B(x_1, \varepsilon)$ and $B(x_2, \varepsilon)$ do not cover A, and hence,

$$\exists x_3 \in A : x_3 \notin B(x_1, \varepsilon) \cup B(x_2, \varepsilon).$$

Continuing inductively, we obtain a sequence $\{x_n\}_{n=1}^{\infty}$ of points of A such that

$$x_n \notin \bigcup_{k=1}^{n-1} B(x_k, \varepsilon), \ n \geq 2,$$

i. e., all of which are at least at distance ε from each other.

Thus, we have found the sequence $\{x_n\}_{n=1}^{\infty}$ in A with no fundamental subsequence, which completes the proof by contrapositive. □

Based on the prior characterization and *Fundamentality and Uniform Continuity Proposition* (Proposition 2.2), one can easily prove the following statement.

Proposition 2.6 (Total Boundedness and Uniform Continuity). *Let* (X, ρ) *and* (Y, σ) *be metric spaces and a function* $f : X \to Y$ *be uniformly continuous on* X. *If* A *is a totally bounded set in* (X, ρ), *then* $f(A)$ *is a totally bounded set in* (Y, σ), *i. e., a uniformly continuous function maps totally bounded sets to totally bounded sets.*

Exercise 2.79.

(a) Prove.

    Hint. Use the *Fundamentality and Uniform Continuity Proposition* (Proposition 2.2).

(b) Give an example showing that a continuous function need not preserve total boundedness.

Proposition 2.7 (Total Boundedness and Separability). *A totally bounded metric space is separable.*

Proof. Let (X, ρ) be a totally bounded metric space. Then, for any $n \in \mathbb{N}$, there is a finite $1/n$-net N_n for X.

The union $\bigcup_{n=1}^{\infty} N_n$ is a *countable dense set* in (X, ρ).

Exercise 2.80. Explain. □

Remarks 2.46.

– Thus, a nonseparable metric space, e. g., l_∞, is not totally bounded.
– As the example of the space $\mathbb{R}$ with the usual metric shows, the converse statement is not true, i. e., a separable metric space need not be totally bounded.
– A totally bounded metric space need not be complete. For instance, the space $X :=$ $\{1/n\}_{n \in \mathbb{N}}$ with the usual metric ρ is totally bounded, but incomplete. The same is true for the space

$$X := \left\{ \sum_{k=0}^{n} \frac{x^k}{k!} \,\middle|\, n \in \mathbb{N}, x \in [a, b] \right\}$$

of the partial sums of the Maclauren series of e^x as a subspace of $(C[a, b], \rho_\infty)$ $(-\infty < a < b < \infty)$.
– Similarly to boundedness and completeness, total boundedness is *not a topological property* of the space, i. e., is not preserved by a homeomorphism.
  Indeed, as is discussed in Remark 2.33, for $X := \{1/n\}_{n \in \mathbb{N}}$, the spaces (X, ρ) and (X, ρ_d), where ρ is the regular metric and ρ_d is the discrete metric, are *homeomorphic* relative the identity mapping $Ix := x$. However, the former space is totally bounded whereas the latter is not.

Exercise 2.81. Explain.

2.15.2 Compactness, Precompactness

2.15.2.1 Definitions and Examples
Definition 2.34 (Cover, Subcover, Open Cover). A collection $\mathscr{C} = \{C_i\}_{i \in I}$ of subsets of a nonempty set X is said to be a *cover* of a set $A \subseteq X$, or to *cover A*, if

$$A \subseteq \bigcup_{i \in I} C_i. \tag{2.16}$$

A subcollection $\mathscr{C}'$ of a cover $\mathscr{C}$ of A, which is also a cover of A, is called a *subcover* of $\mathscr{C}$.

If (X, ρ) is a metric space, a cover of a set $A \subseteq X$ consisting of open sets is called an *open cover* of A.

Remark 2.47. In particular, when $A = X$, (2.16) acquires the form

$$X = \bigcup_{i \in I} C_i.$$

Examples 2.24.
1. The collection $\{[n, n+1)\}_{n \in \mathbb{Z}}$ is a *cover* for $\mathbb{R}$.
2. The collection $\{(n, n+1)\}_{n \in \mathbb{Z}}$ is *not* a *cover* for $\mathbb{Z}$.
3. The collection of all concentric open balls in a metric space (X, ρ) centered at a fixed point $x \in X$

$$\{B(x, r) \mid r > 0\}$$

is an *open cover* of X, the subcollection

$$\{B(x, n) \mid n \in \mathbb{N}\}$$

being its countable *subcover*.
4. Let A be a *dense set* in a metric space (X, ρ). For any $\varepsilon > 0$, the collection of ε-balls

$$\{B(x, \varepsilon) \mid x \in A\},$$

is an open cover of X, i. e., A is an ε-net for X (see Examples 2.22).
5. Let $\{r_n\}_{n \in \mathbb{N}}$ be a countably infinite subset of $\mathbb{R}$. The collection of intervals

$$\left\{ \left[r_n - 1/2^{n+1}, r_n + 1/2^{n+1} \right] \,\middle|\, n \in \mathbb{N} \right\}$$

does not cover $\mathbb{R}$. This is true even when the set is *dense* in $\mathbb{R}$, as is the case, e. g., for $\mathbb{Q}$.

Exercise 2.82. Verify.

Definition 2.35 (Compactness). A set A is said to be *compact* in a metric space (X, ρ) if each open cover $\mathcal{O}$ of A contains a finite subcover $\mathcal{O}'$.
   A metric space (X, ρ) is called *compact* if the set X is compact in (X, ρ).

Remarks 2.48.
– Compactness in the sense of the prior definition is also called compactness in the *Heine*[9]*–Borel*[10] *sense*.
– Formulated entirely in terms of open sets, *compactness* is a *topological property*, i. e., is preserved by a *homeomorphism*.

---

9 Heinrich Heine (1821–1881).
10 Émile Borel (1871–1956).

– The compactness of a nonempty set A in a metric space (X,ρ) is equivalent to the compactness of (A,ρ) as a subspace of (X,ρ).

Examples 2.25.
1. *Finite sets*, including the empty set $\emptyset$, are *compact* in an arbitrary metric space (X,ρ) and only finite sets are compact in a *discrete* metric space (X,ρ_d).
2. The sets $[0,\infty)$, $(0,1]$, and $\{1/n\}_{n\in\mathbb{N}}$ are *not compact* in $\mathbb{R}$ and the set $\{0\}\cup\{1/n\}_{n\in\mathbb{N}}$ is.
3. The set $E := \{e_n := \{\delta_{nk}\}_{k=1}^{\infty}\}_{n\in\mathbb{N}}$, where δ_{nk} is the *Kronecker delta*, is *closed* and *bounded*, but *not compact* in (c_{00},ρ_∞) (also in (c_0,ρ_∞), (c,ρ_∞), and l_∞) since its open cover by 1/2-balls

$$\{B(e_n, 1/2)\}_{n\in\mathbb{N}},$$

has no finite subcover.
Observe that E is also *not totally bounded* (see Examples 2.23).
The same example works in l_p $(1 \le p < \infty)$.

Exercise 2.83. Verify.

Definition 2.36 (Precompactness). A set A is said to be *precompact* (also *relatively compact*) in a metric space (X,ρ) if its closure $\overline{A}$ is compact.

Remark 2.49. For a closed set in a metric space (X,ρ), in particular for X, precompactness is equivalent to compactness.

Examples 2.26.
1. The set $(0,\infty)$ is *not precompact* in $\mathbb{R}$.
2. The set $\{1/n\}_{n\in\mathbb{N}}$ is *precompact*, but not compact in $\mathbb{R}$ and the same is true for the set $\{\sum_{k=0}^{n} \frac{x^k}{k!} \mid n \in \mathbb{N}, x \in [a,b]\}$ of the partial sums of the Maclaurin series of e^x in $(C[a,b],\rho_\infty)$ $(-\infty < a < b < \infty)$.

Exercise 2.84. Verify.

2.15.2.2 Properties
Theorem 2.34 (Properties of Compact Sets). *The compact sets in a metric space (X,ρ) have the following properties:*
(1) *a compact set is necessarily totally bounded, but not vice versa;*
(2) *a compact set is necessarily closed, but not vice versa;*
(3) *a closed subset of a compact set is compact, in particular, a closed set in a compact metric space (X,ρ) is compact;*
(4) *an arbitrary intersection of compact sets is compact;*
(5) *a finite union of compact sets is compact.*

Proof. Here, we are to prove properties (2) and (3) only, proving properties (1), (4), and (5) left to the reader as an exercise.

(2) Let A be an arbitrary compact set in a metric space (X, ρ).

If $A = \emptyset$ or $A = X$, then A is, obviously, closed.

Suppose that A is a nontrivial proper subset of X and let $x \in A^c$ be arbitrary. Then, for any $y \in A$, by the *Separation Property* (Proposition 2.1), there are disjoint open balls $B(x, r(y))$ and $B(y, r(y))$ (e. g., $r(y) = \rho(x, y)/2$).

By the *compactness* of A, its open cover

$$\{B(y, r(y))\}_{y \in A}$$

contains a finite subcover $\{B(y_1, r(y_1)), \ldots, B(y_N, r(y_N))\}$ with some $N \in \mathbb{N}$.

Then the open set

$$V := \bigcup_{k=1}^{N} B(y_k, r(y_k))$$

contains A and is *disjoint* from the open ball centered at x

$$B(x, r) = \bigcap_{k=1}^{N} B(x, r(y_k)), \quad \text{where } r = \min_{1 \le k \le n} r(y_k).$$

Hence, $B(x, r) \cap A = \emptyset$, i. e., x is an *exterior* point of A, which proves that the complement A^c is *open* in (X, ρ), i. e., by the *Characterizations of Closed Sets* (Theorem 2.18), the set A is *closed*.

A closed set in a metric space need not be compact. For instance, the set $[0, \infty)$ is closed, but not compact in $\mathbb{R}$ (see Examples 2.25).

(3) Let B be a closed subset of a compact set A in (X, ρ) and $\mathcal{O} = \{O_i\}_{i \in I}$ be an arbitrary *open cover* of B.

Then $\mathcal{O} \cup \{B^c\}$ is an open cover of A. By the compactness of A, there is a *finite subcover* of A. Hence, there is a *finite subcollection* $\mathcal{O}'$ of $\mathcal{O}$ such that $\mathcal{O}' \cup \{B^c\}$ is a cover of A, which implies that $\mathcal{O}'$ is a finite subcover of B, proving its compactness. $\qquad\square$

Exercise 2.85.

(a) Prove (1), (4), and (5).

(b) Give an example showing that an infinite union of compact sets need not be compact.

Remark 2.50. From property (1) and Exercise 2.77, we infer that a nontrivial sphere/ball in (c_{00}, ρ_∞) (also in (c_0, ρ_∞), (c, ρ_∞), and l_p $(1 \le p \le \infty)$) is *not compact*.

From property (1) in view of Remarks 2.44, we infer

Proposition 2.8 (Precompactness and Total Boundedness). *A precompact set A in metric space (X,ρ) is totally bounded, but not vice versa.*

Exercise 2.86. Give an example showing that a totally bounded set need not be precompact.

From property (2), we also arrive at the following

Proposition 2.9 (Characterization of Compactness). *A set A in a metric space (X,ρ) is compact iff it is closed and precompact.*

From property (1) and the *Total Boundedness and Separability Proposition* (Proposition 2.7), we immediately obtain the following statement.

Proposition 2.10 (Compactness and Separability). *A compact metric space is separable.*

Furthermore,

Proposition 2.11 (Compactness and Completeness). *A compact metric space is complete.*

Proof. Let us prove the statement *by contradiction* and suppose that a there is a *compact* metric space (X,ρ), which is *incomplete*. Then there exists a *fundamental* sequence $\{x_n\}_{n=1}^{\infty}$ with no limit in (X,ρ), i. e.,

$$\forall x \in X \, \exists \varepsilon = \varepsilon(x) > 0 \, \forall N \in \mathbb{N} \, \exists n \geq N : \rho(x,x_n) \geq \varepsilon. \tag{2.17}$$

On the other hand, by the fundamentality of $\{x_n\}_{n=1}^{\infty}$,

$$\exists M = M(\varepsilon(x)) \in \mathbb{N} \, \forall m,n \geq M : \rho(x_m,x_n) < \varepsilon/2. \tag{2.18}$$

Fixing an $n \geq \max(M,N)$, for which (2.17) holds, by (2.17), (2.18), and *triangle inequality*, we have:

$$\forall m \geq M : \varepsilon \leq \rho(x,x_n) \leq \rho(x,x_m) + \rho(x_m,x_n) < \rho(x,x_m) + \varepsilon/2,$$

and hence,

$$\forall m \geq M : \rho(x,x_m) \geq \varepsilon/2.$$

Thus, for each $x \in X$, there is a ball $B(x,\varepsilon(x)/2)$ such that the sequence $\{x_n\}_{n=1}^{\infty}$ is *eventually* not in it.

By the *compactness* of (X,ρ), the open cover

$$\{B(x,\varepsilon(x)/2)\}_{x \in X}$$

of X contains a finite subcover, which implies that

$$\exists K \in \mathbb{N} \, \forall k \geq K : x_k \notin X.$$

Exercise 2.87. Explain.

The latter is a *contradiction* proving the statement. $\square$

Remarks 2.51.
- Thus, a metric space, which is *nonseparable*, as l_∞, or is *incomplete*, as (c_{00}, ρ_∞), is *not compact*.
- As the example of the space $\mathbb{R}$ with the usual metric, which is both separable and complete, shows, the converses to Proposition 2.10 and Proposition 2.11 are not true, i. e., a separable and complete metric space need not be compact.

Observe that, when proving property (2), we have, in fact, proved the following *separation property*:

Proposition 2.12 (Separation Property for a Compact Set and a Point). *A compact set and a point disjoint from it in a metric space (X, ρ) can be separated by disjoint open sets.*

The latter allows the following generalization:

Proposition 2.13 (Separation Property for Compact Sets). *Disjoint compact sets in a metric space (X, ρ) can be separated by disjoint open sets.*

Proof. Let A and B be arbitrary disjoint compact sets in a metric space (X, ρ).
The statement is trivially true if at least one of the sets is *empty*.

Exercise 2.88. Explain.

Suppose that $A, B \neq \emptyset$. Then, by the *Separation Property for a Compact Set and a Point* (Proposition 2.12), for any $x \in A$, there are disjoint open sets U_x and V_x such that $x \in U_x$ and $B \subseteq V_x$.
By the compactness of A, its open cover $\{U_x\}_{x \in A}$ contains a finite subcover $\{U_{x_1}, \ldots, U_{x_N}\}$ with some $N \in \mathbb{N}$.
Then the open set

$$U := \bigcup_{k=1}^{N} U_{x_k}$$

containing A is *disjoint* from the open set

$$V := \bigcap_{k=1}^{N} V_{x_k}.$$

containing B. $\square$

2.15.3 Hausdorff Criterion

The following essential statement establishes the equivalence of total boundedness and precompactness in a complete metric space and allows us to describe compactness in certain complete metric spaces. The *Heine–Borel Theorem* (Theorem 2.36), which characterizes (pre)compactness in the n-space, is one of its immediate implications.

Theorem 2.35 (Hausdorff Criterion). *In a complete metric space (X,ρ),*[11]
(1) *a set A is precompact iff it is totally bounded;*
(2) *a set A is compact iff it is closed and totally bounded.*

Proof.
(1) *"Only If"* part. This part immediately follows from the *Precompactness and Total Boundedness Corollary* (Proposition 2.8), which is a more general statement not requiring the completeness of the space.
    "If" part. Let us prove this part *by contradiction* and suppose that there is a *totally bounded* set A in a *complete* metric space (X,ρ), which is *not precompact*, i. e., the closure $\overline{A}$ is *not compact*. Then there is an *open cover* $\mathscr{O} = \{O_i\}_{i \in I}$ of the closure $\overline{A}$ with no finite subcover.
    Due to the *total boundedness* of A, and hence, of $\overline{A}$ (see Remarks 2.44), for each $n \in \mathbb{N}$, there is a *finite* $1/2^{n-1}$-net N_n for $\overline{A}$.
    At least one of the sets

$$\{B(x,1) \cap \overline{A}\}_{x \in N_1}$$

cannot be covered by a finite number of sets from $\mathscr{O}$.

Exercise 2.89. Explain.

Suppose this is a set

$$A_1 := B(x_1,1) \cap \overline{A}$$

with some $x_1 \in N_1$.
Similarly, at least one of the sets

$$\{B(x,1/2) \cap A_1\}_{x \in N_2}$$

cannot be covered by a finite number of sets from $\mathscr{O}$.

Exercise 2.90. Explain.

---

[11] Felix Hausdorff (1868–1942).

Suppose this is a set

$$A_2 := B(x_2, 1/2) \cap A_1$$

with some $x_2 \in N_2$.

Continuing inductively, we obtain a sequence of sets $\{A_n\}_{n=1}^{\infty}$ such that

(a) $A_{n+1} \subseteq A_n \subseteq \overline{A}, n \in \mathbb{N}$;

(b) $A_n \subseteq B(x_n, 1/2^{n-1}), n \in \mathbb{N}$, with some $x_n \in N_n$;

(c) for each $n \in \mathbb{N}$, the set A_n cannot be covered by a finite number of sets from $\mathcal{O}$, which, in particular, implies that A_n is *infinite*.

Consider a sequence $\{y_n\}_{n=1}^{\infty}$ of elements of $\overline{A}$ chosen as follows:

$$y_1 \in A_1, \quad y_n \in A_n \setminus \{y_1, \dots, y_{n-1}\}, \; n = 2, 3, \dots.$$

The sequence $\{y_n\}_{n=1}^{\infty}$ is *fundamental* in (X, ρ) since for any $m, n \in \mathbb{N}$ with $n \geq m$,

$$\rho(y_m, y_n) \qquad\qquad\qquad\qquad \text{by the } \textit{triangle inequality};$$

$$\leq \rho(y_m, x_m) + \rho(x_m, y_n) \qquad \text{by the inclusions} \quad A_n \subseteq A_m \subseteq B(x_m, 1/2^{m-1});$$

$$< \frac{1}{2^{m-1}} + \frac{1}{2^{m-1}} = \frac{1}{2^{m-2}}.$$

Hence, by the *completeness* of (X, ρ), there is an element $y \in X$ such that

$$y_n \to y, \; n \to \infty, \; \text{in } (X, \rho).$$

In view of the *closedness* of $\overline{A}$, by the *Sequential Characterization of Closed Sets* (Theorem 2.18), we infer that

$$y \in \overline{A}.$$

Since $\mathcal{O} = \{O_i\}_{i \in I}$ is an open cover of $\overline{A}$,

$$\exists j \in I : y \in O_j,$$

and hence, by the *openness* of O_j,

$$\exists \delta > 0 : B(y, \delta) \subseteq O_j.$$

Choosing an $n \in \mathbb{N}$ sufficiently large so that

$$\frac{1}{2^{n-1}} < \frac{\delta}{3} \quad \text{and} \quad \rho(y_n, y) < \frac{\delta}{3},$$

for any $x \in B(x_n, 1/2^{n-1})$, we have:

$$\rho(x, y) \qquad\qquad\qquad\qquad \text{by the } \textit{triangle inequality};$$

$$\leq \rho(x, x_n) + \rho(x_n, y_n) + \rho(y_n, y) \qquad \text{since } y_n \in A_n \subseteq B(x_n, 1/2^{n-1});$$

$$< \frac{1}{2^{n-1}} + \frac{1}{2^{n-1}} + \rho(y_n, y) < \frac{\delta}{3} + \frac{\delta}{3} + \frac{\delta}{3} = \delta,$$

and hence, we also have the inclusions

$$A_n \subseteq B(x_n, 1/2^{n-1}) \subseteq B(y, \delta) \subseteq O_j,$$

which imply that the set A_n is covered by *one* set $O_j \in \mathcal{O}$.

The obtained *contradiction* proves that $\overline{A}$ is *compact*, and hence, A is *precompact*.

(2) Part (2) immediately follows from part (1) and the *Characterization of Compactness* (Proposition 2.9). □

Remarks 2.52.

- Thus, in a complete metric space, *"totally bounded"* is synonymous to *"precompact"*.

- As the example of the totally bounded, but not precompact set $\{1/n\}_{n \in \mathbb{N}}$ in the incomplete metric space $(\mathbb{R} \setminus \{0\}, \rho)$, where ρ is the regular metric, shows, the requirement of the completeness on the space is essential and cannot be dropped.

The following is an immediate corollary of the *Properties of Compact Sets* (Theorem 2.34), the *Compactness and Completeness Proposition* (Proposition 2.11), and the *Hausdorff Criterion* (Theorem 2.35)

Corollary 2.7 (Characterization of Compactness of a Metric Space). *A metric space is compact iff it is complete and totally bounded.*

2.15.4 Compactness in Certain Complete Metric Spaces

The following descriptions of (pre)compactness in certain complete metric spaces are direct implications of the *Hausdorff Criterion* (Theorem 2.35).

Theorem 2.36 (Heine–Borel Theorem (Compactness in the n-Space)). *Let $n \in \mathbb{N}$ and $1 \leq p \leq \infty$.*

(1) *A set A is precompact in the (real or complex) space $l_p^{(n)}$ iff it is bounded.*

(2) *A set A is compact in the (real or complex) space $l_p^{(n)}$ iff it is closed and bounded.*

Proof. The statement immediately follows, by the *Hausdorff Criterion*, considering the *completeness* of $l_p^{(n)}$ (see Examples 2.14) and the equivalence of boundedness and total boundedness in it (Corollary 2.6). □

Theorem 2.37 (Compactness in l_p $(1 \leq p < \infty)$). *A set A is precompact (a closed set A is compact) in l_p $(1 \leq p < \infty)$ if*

1. *A is bounded;*

2. *$\forall \varepsilon > 0 \, \exists N \in \mathbb{N} \, \forall x := \{x_n\}_{n=1}^{\infty} \in A : \sum_{n=N+1}^{\infty} |x_n|^p < \varepsilon.$*

Proof. In view of the *completeness* of l_p ($1 \le p < \infty$) (Theorem 2.23), by the *Hausdorff Criterion* (Theorem 2.35), we are to show that the two conditions imply the *total boundedness* in l_p for a set A satisfying them.

Indeed, by condition 2, for any

$$\forall \varepsilon > 0 \, \exists N \in \mathbb{N} \, \forall x := \{x_n\}_{n=1}^{\infty} \in A : \| \{0, \ldots, 0, x_{N+1}, \ldots\} \|_p^p$$

$$= \sum_{n=N+1}^{\infty} |x_n|^p < \varepsilon^p /2. \tag{2.19}$$

By condition 1, the set

$$A_N := \{\{x_1, \ldots, x_N, 0, \ldots\} \mid x := \{x_n\}_{n=1}^{\infty} \in A\}$$

is *bounded* in l_p.

Exercise 2.91. Explain.

Being *isometric* to the bounded set

$$\tilde{A}_N := \{(x_1, \ldots, x_N) \mid x := \{x_n\}_{n=1}^{\infty} \in A\}$$

in $l_p^{(N)}$, where boundedness is equivalent to total boundedness, A_N is *totally bounded* in l_p.

Therefore, A_N has a finite $\varepsilon/2^{1/p}$-net $\{y_1, \ldots, y_m\} \subseteq A_N$ (see Remarks 2.44) with some $m \in \mathbb{N}$, where

$$y_i := \{y_1^{(i)}, \ldots, y_N^{(i)}, 0, \ldots\}, \ i = 1, \ldots, m,$$

which is an ε-net for A since, for any $x := \{x_n\}_{n=1}^{\infty} \in A$, there is an $i = 1, \ldots, m$ such that

$$\rho_p(\{x_1, \ldots, x_N, 0, \ldots\}, y_i) < \varepsilon/2^{1/p},$$

and hence, in view of (2.19),

$$\rho_p^p(x, y_i) = \sum_{n=1}^{N} |x_n - y_n^{(i)}|^p + \sum_{n=N+1}^{\infty} |x_n|^p = \rho_p^p(\{x_1, \ldots, x_N, 0, \ldots\}, y_i)$$

$$+ \| \{0, \ldots, 0, x_{N+1}, \ldots\} \|_p^p < \varepsilon^p /2 + \varepsilon^p /2 = \varepsilon^p$$

(see Definition 2.4).

Hence, A is *totally bounded* in l_p. $\qquad\qquad\qquad\square$

The following statement can be established to hold by mimicking the proof of the prior proposition.

Theorem 2.38 (Compactness in (c_0, ρ_∞)). *A set A is precompact (a closed set A is compact) in (c_0, ρ_∞) if*

1. *A is bounded;*
2. $\forall \varepsilon > 0 \ \exists N \in \mathbb{N} \ \forall x = \{x_n\}_{n=1}^{\infty} \in A : \ \sup_{n \geq N+1} |x_n| < \varepsilon.$

Exercise 2.92. Prove.

Remark 2.53. In fact, Theorems 2.37 and 2.38 provide not only *sufficient*, but also *necessary* conditions for (pre)compactness in the corresponding spaces, being particular cases of the (pre)compactness characterization in a *Banach*[12] *space* with a *Schauder*[13] *basis* (see, e. g., [31], also Section 3.2.6).

2.15.5 Other Forms of Compactness

Here we are to introduce two other forms of compactness and to prove that, in a metric space setting, they are equivalent to that in the Heine–Borel sense.

2.15.5.1 Sequential Compactness
Definition 2.37 (Sequential Compactness). A *nonempty set A* in a metric space (X, ρ) is said to be *sequentially compact* in (X, ρ) if every sequence $\{x_n\}_{n=1}^{\infty}$ in A has a subsequence $\{x_{n(k)}\}_{k=1}^{\infty}$ convergent to a limit in A.

A metric space (X, ρ) is called *sequentially compact* if the set X is sequentially compact in (X, ρ).

Remarks 2.54.
– Sequential compactness is also called compactness in the *Bolzano*[14]*–Weierstrass*[15] *sense.*
– Formulated entirely in terms of convergence, *sequential compactness* is a topological property, i. e., is preserved by a *homeomorphism*.
– The sequential compactness of a nonempty set A in a metric space (X, ρ) is equivalent to the sequential compactness of (A, ρ) as a subspace of (X, ρ).

Examples 2.27.
1. *Finite sets* are *sequentially compact* in an arbitrary metric space (X, ρ) and only finite sets are sequentially compact in a *discrete* metric space (X, ρ_d).
2. The sets $[0, \infty)$, $(0, 1]$, and $\{1/n\}_{n \in \mathbb{N}}$ are *not sequentially compact* in $\mathbb{R}$ and the set $\{0\} \cup \{1/n\}_{n \in \mathbb{N}}$ is.

---

12 Stefan Banach (1892–1945).
13 Juliusz Schauder (1899–1943).
14 Bernard Bolzano (1781–1848).
15 Karl Weierstrass (1815–1897).

3. The set $E := \{e_n := \{\delta_{nk}\}_{k=1}^{\infty}\}_{n\in\mathbb{N}}$ in (c_{00},ρ_{∞}) is *not sequentially* compact in (c_{00},ρ_{∞}) (also in (c_0,ρ_{∞}), (c,ρ_{∞}), and l_{∞}).
   The same example works in l_p $(1 \le p < \infty)$.

Exercise 2.93. Verify (cf. Examples 2.25).

2.15.5.2 Limit-Point Compactness

Definition 2.38 (Limit-Point Compactness). An *infinite set A* in a metric space (X,ρ) is said to be *limit-point compact* if every infinite subset of A has a limit point in A.

A metric space (X,ρ) is called *limit-point compact* if the set X is limit-point compact in (X,ρ).

Remarks 2.55.
- Being characterized in terms of convergence (see Proposition 2.16), *limit-point compactness* is a topological property, i. e., is preserved by a *homeomorphism*.
- The limit-point compactness of an infinite set A in a metric space (X,ρ) is equivalent to the limit-point compactness of (A,ρ) as a subspace of (X,ρ).

Examples 2.28.
1. No set is limit-point compact in a *discrete metric space* (X,ρ_d).
2. The sets $[0,\infty)$, $(0,1]$, and $\{1/n\}_{n\in\mathbb{N}}$ are *not limit-point compact* in $\mathbb{R}$ and the set $\{0\} \cup \{1/n\}_{n\in\mathbb{N}}$ is.
3. The set $E := \{e_n := \{\delta_{nk}\}_{k=1}^{\infty}\}_{n\in\mathbb{N}}$ in (c_{00},ρ_{∞}) is *not limit-point compact*.
   The same example works in (c_0,ρ_{∞}), (c,ρ_{∞}), and l_p $(1 \le p \le \infty)$.

Exercise 2.94. Verify (cf. Examples 2.25 and 2.27).

2.15.6 Equivalence of Different Forms of Compactness

Theorem 2.39 (Equivalence of Different Forms of Compactness). *For an infinite set A in a metric space (X,ρ), the following statements are equivalent.*
1. *A is compact.*
2. *A is sequentially compact.*
3. *A is limit-point compact.*

Proof. To prove the statement, let us show that the following closed chain of implications

$$1 \Rightarrow 2 \Rightarrow 3 \Rightarrow 1$$

holds true.

$1 \Rightarrow 2$. Suppose that an infinite set A is *compact* and let $\{x_n\}_{n=1}^{\infty}$ be an arbitrary sequence in A.

By the *Characterization of Compactness of a Metric Space* (Corollary 2.7), the subspace (A, ρ) is *complete* and *totally bounded*.

By the *Characterization of Total Boundedness* (Theorem 2.33), the sequence $\{x_n\}_{n=1}^{\infty}$ in (A, ρ) contains a *fundamental* subsequence $\{x_{n(k)}\}_{k=1}^{\infty}$, which, by the completeness of (A, ρ), implies that the latter converges to a limit $x \in A$.

Hence, A is *sequentially compact*.

$2 \Rightarrow 3$. Suppose that an infinite set A is *sequentially compact* and let B be an arbitrary infinite subset of A. Then, by the *Properties of Countable Sets* (Theorem 1.3), we can choose a sequence $\{x_n\}_{n=1}^{\infty}$ in B such that

$$x_m \neq x_n, \ m \neq n. \tag{2.20}$$

By the sequential compactness of A, $\{x_n\}_{n=1}^{\infty}$ contains a subsequence $\{x_{n(k)}\}_{k=1}^{\infty}$, which converges to a limit $x \in A$.

In view of (2.20), each open ball centered at x contains infinitely many elements of B distinct from x.

Exercise 2.95. Explain.

Hence, x is a limit point of B, which proves that A is *limit-point compact*.

$3 \Rightarrow 1$. Suppose that an infinite A is *limit-point compact* and let $\{x_n\}_{n=1}^{\infty}$ be an arbitrary sequence in A, with $B := \{x_n\}_{n \in \mathbb{N}}$ being the set of its values.

If B is *infinite*, then, by the limit-point compactness of A, B has a limit point $x \in A$, and hence, by the *Sequential Characterization of Limit Points* (Theorem 2.16), there is a subsequence $\{x_{n(k)}\}_{k=1}^{\infty}$ convergent to x.

Exercise 2.96. Explain.

If B is *finite*, then there is a *constant* subsequence $\{x_{n(k)}\}_{k=1}^{\infty}$, which is also convergent to a limit in A.

Thus, an arbitrary sequence in A contains a *convergent* to a limit in A (and hence, *fundamental*) subsequence, which, by the *Characterization of Total Boundedness* (Theorem 2.33), implies that the subspace (A, ρ) is *totally bounded* and, by the *Fundamental Sequence with Convergent Subsequence Proposition* (Proposition 2.22) (see Section 2.19, Problem 24), also implies that the subspace (A, ρ) is *complete*.

Exercise 2.97. Explain.

Hence, by the *Characterization of Compactness of a Metric Space* (Corollary 2.7), the subspace (A, ρ) is *compact*, i. e., the set A is *compact* in (X, ρ). □

As an immediate implication we obtain the following corollary.

Corollary 2.8 (Compact Set Consisting of Isolated Points). *A compact set A in a metric space (X,ρ) consisting of isolated points is finite.*

Exercise 2.98. Prove.

Hence, another reason for the noncompactness of the set $\mathbb{Z}$ of the integers in $\mathbb{R}$, besides being *unbounded*, is its being an infinite set consisting of isolated points (cf. Examples 2.25).

Another direct implication of the *Equivalence of Different Forms of Compactness Theorem* (Theorem 2.39) is the following classical result.

Theorem 2.40 (Bolzano–Weierstrass Theorem). *Each bounded sequence of real or complex numbers contains a convergent subsequence.*

Exercise 2.99. Prove.

2.15.7 Compactness and Continuity

In this section, we consider the profound interplay between *compactness* and *continuity*.

Theorem 2.41 (Continuous Image of a Compact Set). *Let (X,ρ) and (Y,σ) be metric spaces and a function $f : X \to Y$ be continuous on X. If A is a compact set in (X,ρ), then $f(A)$ is a compact set in (Y,σ), i. e., a continuous function maps compact sets to compact sets.*

Proof. Let A be an arbitrary compact set in (X,ρ).

For an arbitrary open cover

$$\mathcal{O} = \{O_i\}_{i\in I}$$

of $f(A)$ in (Y,σ), by the *Properties of Inverse Image* (Theorem 1.4),

$$A \subseteq f^{-1}(f(A)) \subseteq f^{-1}\left(\bigcup_{i\in I} O_i\right) = \bigcup_{i\in I} f^{-1}(O_i),$$

and hence, by the *Characterization of Continuity* (Theorem 2.52) (see Section 2.19, Problem 14), the collection

$$f^{-1}(\mathcal{O}) := \left\{f^{-1}(O_i)\right\}_{i\in I}$$

is an *open cover* of A in (X,ρ).

By the compactness of A in (X,ρ), there is a finite subcover $\{f^{-1}(O_{i_1}),\dots,f^{-1}(O_{i_N})\}$ of A with some $N \in \mathbb{N}$.

Then, since image preserves unions (see Exercise 1.4),

$$f(A) \subseteq f\left(\bigcup_{k=1}^{N} f^{-1}(O_{i_k})\right) = \bigcup_{k=1}^{N} f\left(f^{-1}(O_{i_k})\right) \subseteq \bigcup_{k=1}^{N} O_{i_k},$$

and hence, $\{O_{i_1}, \ldots, O_{i_N}\}$ is a finite subcover of $f(A)$, which proves the compactness of $f(A)$ and completes the proof. $\qquad\square$

Exercise 2.100. Prove the prior theorem via the *sequential approach*.

Hint. Use the *Sequential Characterization of Local Continuity* (Theorem 2.10).

Theorem 2.42 (Homeomorphism Test). *Let* (X,ρ), (Y,σ) *be metric spaces, and* $f : X \to Y$ *be a bijective continuous function. If the space* (X,ρ) *is compact, then* f *is a homeomorphism.*

Exercise 2.101. Prove.

Hint. To prove the continuity of the *inverse* f^{-1} apply the *Properties of Compact Sets* (Theorem 2.34), the prior theorem, and *Characterization of Continuity* (Theorem 2.52) and show that the images of *closed sets* in (X,ρ) are closed in (Y,σ).

From the *Continuous Image of a Compact Set Theorem* (Theorem 2.41) and the *Properties of Compact Sets* (Theorem 2.34), we immediately obtain

Corollary 2.9 (Boundedness of Continuous Functions on Compact Sets). *A continuous function from one metric space to another is bounded on compact sets.*

The following is a generalized version of a well-known result from calculus.

Theorem 2.43 (Weierstrass Extreme Value Theorem). *Let* (X,ρ) *be a compact metric space. A continuous real-valued function* $f : X \to \mathbb{R}$ *attains its absolute minimum and maximum values on* X, *i. e., there exist* $x_*, x^* \in X$ *such that*

$$f(x_*) = \inf_{x \in X} f(x) \quad and \quad f(x^*) = \sup_{x \in X} f(x).$$

Proof. As follows from the *Continuous Image of a Compact Set Theorem* (Theorem 2.41), the *image set* $f(X)$ is *compact* in $\mathbb{R}$, and hence, by the *Heine–Borel Theorem* (Theorem 2.36), being *closed* and *bounded* in $\mathbb{R}$, contains

$$\inf_{x \in X} f(x) \quad and \quad \sup_{x \in X} f(x).$$

This implies that there exist $x_*, x^* \in X$ such that

$$f(x_*) = \inf_{x \in X} f(x) \quad and \quad f(x^*) = \sup_{x \in X} f(x),$$

which completes the proof. $\qquad\square$

From the prior theorem and the *Continuity of Composition Theorem* (Theorem 2.12), we obtain

Corollary 2.10 (Extreme Value Theorem for Modulus). *Let* (X,ρ) *be a compact metric space. For a continuous complex-valued function* $f : X \to \mathbb{C}$, *the modulus function* $|f|$ *attains its absolute minimum and maximum values on* X.

We also have the following

Corollary 2.11 (Nearest and Farthest Points Property Relative to Compact Sets). *Let A be a nonempty compact set in a metric space* (X, ρ). *Then, for any* $x \in X$, *in A, there are a nearest point to* x, *i. e.*,

$$\exists y \in A : \rho(x, y) = \inf_{p \in A} \rho(x, p) =: \rho(x, A),$$

and a farthest point from x, *i. e.*,

$$\exists z \in A : \rho(x, z) = \sup_{p \in A} \rho(x, p).$$

Exercise 2.102. Prove.

Hint. Use the *Lipschitz continuity* of the distance function

$$f(y) := \rho(x, y), \ y \in A,$$

on the compact metric space (A, ρ) (see the *Inverse Triangle Inequality* (Theorem 2.7), cf. also Section 2.19, Problem 12).

Remark 2.56. If $x \in A$, then the *unique* nearest to x point in A is, obviously, x itself. In general, the statement says nothing about the *uniqueness* of the nearest and farthest points.

Exercise 2.103.
(a) Give an example showing that nearest and farthest points in a compact need not be unique.
(b) If A is a *nonempty compact set* in a *discrete* metric space (X, ρ_d), which points in A are the *nearest* to and the *farthest* from a point $x \in X$?

The following celebrated result shows that compactness jointly with continuity yields uniform continuity.

Theorem 2.44 (Heine–Cantor Uniform Continuity Theorem). *Let* (X, ρ) *and* (Y, σ) *be metric spaces and* $f \in C(X, Y)$. *If the space* (X, ρ) *is compact, f is uniformly continuous on X.*

Proof. Let us prove the statement *by contradiction.*
Assume that f is *not* uniformly continuous on X (see Definition 2.8). Hence,

$$\exists \varepsilon > 0 \ \forall n \in \mathbb{N} \ \exists x_n', x_n'' \in X \text{ with } \rho(x_n', x_n'') < 1/n :$$
$$\sigma(f(x_n'), f(x_n'')) \geq \varepsilon. \tag{2.21}$$

Since the space (X, ρ) is *compact*, by the *Equivalence of Different Forms of Compactness Theorem* (Theorem 2.39), it is *sequentially compact*, and hence, the sequence $\{x_n'\}_{n=1}^{\infty}$ contains a subsequence $\{x_{n(k)}'\}_{k=1}^{\infty}$ convergent to an element $x \in X$ in (X, ρ), i. e.,

$$\rho(x, x_{n(k)}') \to 0, \ k \to \infty.$$

In respect to (2.21), by the *triangle inequity*, we have:

$$0 \le \rho(x, x''_{n(k)}) \le \rho(x, x'_{n(k)}) + \rho(x'_{n(k)}, x''_{n(k)}) < \rho(x, x'_{n(k)}) + 1/n(k) \to 0, \ k \to \infty,$$

which, by the *Squeeze Theorem*, implies that

$$x''_{n(k)} \to x, \ k \to \infty, \ \text{in} \ (X, \rho).$$

By the *continuity* of f,

$$f(x'_{n(k)}) \to f(x) \ \text{and} \ f(x''_{n(k)}) \to f(x), \ k \to \infty, \ \text{in} \ (Y, \sigma),$$

which *contradicts* (2.21) showing that the assumption is false and completing the proof. □

Exercise 2.104. Give an example showing that the requirement of *compactness* is essential and cannot be dropped.

2.16 Space $(C(X, Y), \rho_\infty)$

Here, we introduce and take a closer look at certain abstract sets of functions that are naturally metrizable.

Theorem 2.45 (Space $(M(X, Y), \rho_\infty)$). *Let X be a nonempty set and (Y, σ) be metric space. The set $M(X, Y)$ of all bounded functions $f : X \to Y$ is a metric space relative to the supremum metric (or uniform metric)*

$$M(X, Y) \ni f, g \mapsto \rho_\infty(f, g) := \sup_{x \in X} \sigma(f(x), g(x)), \tag{2.22}$$

which is complete provided (Y, σ) is complete.

Exercise 2.105. Prove the statement.
(a) Verify that $(M(X, Y), \rho_\infty)$ is a metric space, including the fact that ρ_∞ is well defined by (2.22).
(b) Suppose that the space (Y, σ) is *complete* and, by mimicking the proof of the *Completeness of* $(M(T), \rho_\infty)$ (Theorem 2.24) (see Examples 2.14), prove that the space $(M(X, Y), \rho_\infty)$ is *complete*.

Remarks 2.57.
– In particular, for $Y = \mathbb{R}$ or $Y = \mathbb{C}$, we have the (real/complex) space $(M(X), \rho_\infty)$ (see Examples 2.4) with

$$M(X) \ni f, g \mapsto \rho_\infty(f, g) := \sup_{x \in X} |f(x) - g(x)|,$$

which turns into l_∞ with $X = \mathbb{N}$, and thus, as immediate corollaries, we obtain the *Completeness of* $(M(T), \rho_\infty)$ (Theorem 2.24) and the *Completeness of* l_∞ (Corollary 2.1) (see Examples 2.14).

– Convergence in $(M(X,Y), \rho_\infty)$ is the *uniform convergence* on X, i. e.,

$$f_n \to f, \; n \to \infty, \text{ in } (M(X,Y), \rho_\infty)$$

iff

$$\forall \varepsilon > 0 \; \exists N \in \mathbb{N} \; \forall n \geq N \; \forall x \in X : \; \sigma(f_n(x), f(x)) < \varepsilon.$$

– If (X, ρ) is a *compact* metric space, by the *Boundedness of Continuous Functions on Compact Sets Corollary* (Corollary 2.9), we have the inclusion

$$C(X, Y) \subseteq M(X, Y)$$

and, by the *Weierstrass Extreme Value Theorem* (Corollary 2.43) applied to the continuous real-valued function

$$X \ni x \mapsto \sigma(f(x), g(x)) \in \mathbb{R},$$

for $f, g \in C(X, Y)$, in (2.22), we can use max instead of sup:

$$C(X, Y) \ni f, g \mapsto \rho_\infty(f, g) := \max_{x \in X} \sigma(f(x), g(x)).$$

In particular, for $Y = \mathbb{R}$ or $Y = \mathbb{C}$, we have the (real/complex) space $(C(X), \rho_\infty)$ with

$$C(X) \ni f, g \mapsto \rho_\infty(f, g) = \max_{x \in X} |f(x) - g(x)|,$$

which turns into $(C[a, b], \rho_\infty)$ with $X = [a, b]$ $(-\infty < a < b < \infty)$.

Theorem 2.46 (Completeness of $(C(X, Y), \rho_\infty)$). *Let (X, ρ) and (Y, σ) be metric spaces, the former being compact and the latter being complete. Then the space $(C(X, Y), \rho_\infty)$ is complete.*

In particular, for $Y = \mathbb{R}$ or $Y = \mathbb{C}$, complete is the (real/complex) space $(C(X), \rho_\infty)$.

Proof. By the *Characterization of Completeness* (Proposition 2.4), to prove the completeness of the space $(C(X, Y), \rho_\infty)$, it suffices to prove the *closedness* of the set $C(X, Y)$ in the space $(M(X, Y), \rho_\infty)$, which, by the prior theorem, is *complete* due to the completeness of (Y, σ).

To show that $C(X, Y)$ is *closed* in $(M(X, Y), \rho_\infty)$ (regardless whether (Y, σ) is complete or not), consider an arbitrary sequence $\{f_n\}_{n=1}^\infty$ in $C(X, Y)$ convergent to an f in $(M(X, Y), \rho_\infty)$, i. e.,

$$\forall \varepsilon > 0 \; \exists N \in \mathbb{N} \; \forall n \geq N : \; \rho_\infty(f, f_n) := \sup_{x \in X} \sigma(f(x), f_n(x)) < \varepsilon/3. \qquad (2.23)$$

On the other hand, since f_N is *uniformly continuous* on X, by the *Heine–Cantor Uniform Continuity Theorem* (Theorem 2.44),

$$\exists \delta > 0 \; \forall x', x'' \in X \text{ with } \rho(x',x'') < \delta : \; \sigma(f_N(x'),f_N(x'')) < \varepsilon/3. \qquad (2.24)$$

By (2.23), (2.24), and the *triangle inequality*, we infer that

$$\forall x', x'' \in X \text{ with } \rho(x',x'') < \delta : \; \sigma(f(x'),f(x'')) \leq \sigma(f(x'),f_N(x'))$$
$$+ \sigma(f_N(x'),f_N(x'')) + \sigma(f_N(x''),f(x'')) < \varepsilon/3 + \varepsilon/3 + \varepsilon/3 = \varepsilon,$$

which implies that f is *uniformly continuous* on X, and hence (see Remark 2.15), $f \in C(X,Y)$.

Thus, by the *Sequential Characterization of Closed Sets* (Theorem 2.18), $C(X,Y)$ is *closed* in $(M(X,Y),\rho_\infty)$, which completes the proof. □

Remark 2.58. As a particular case, we obtain the *Completeness of* $(C[a,b],\rho_\infty)$ (Theorem 2.25) (cf. Exercise 2.52).

2.17 Arzelà–Ascoli Theorem

Here, we are to prove *Arzelà*[16]*–Ascoli*[17] *Theorem* characterizing (pre)compactness in the space $(C(X),\rho_\infty)$, where the domain space (X,ρ) is compact.

2.17.1 Uniform Boundedness and Equicontinuity

We first need to introduce the important notions of *uniform boundedness* and *equicontinuity*.

Definition 2.39 (Uniform Boundedness). Let X be a nonempty set. A nonempty collection $F \subseteq M(X)$ is called *uniformly bounded* on X if

$$\exists C > 0 \; \forall f \in F : \; \sup_{x \in X} |f(x)| \leq C.$$

Remark 2.59. The uniform boundedness of $F \subseteq M(X)$ on X is, obviously, equivalent to the boundedness of F in the space $(M(X),\rho_\infty)$.

Example 2.29.
1. The set of all constant functions on X is *not uniformly bounded* on X.
2. The set $\{x^n\}_{n\in\mathbb{N}}$ is *uniformly bounded* on $[0,1]$, but not on $[0,2]$.

---

16 Cesare Arzelà (1847–1912).
17 Giulio Ascoli (1843–1896).

Exercise 2.106. Verify.

Definition 2.40 (Equicontinuity). Let (X,ρ) be a metric space. A nonempty set $F \subseteq C(X)$ is called *equicontinuous* on X if

$$\forall \varepsilon > 0 \; \exists \delta > 0 \; \forall f \in F, \; \forall x', x'' \in X \text{ with } \rho(x', x'') < \delta : \; |f(x') - f(x'')| < \varepsilon.$$

Remark 2.60. Thus, the equicontinuity of F on X is the *uniform continuity* of all functions from F on X in, so to speak, equal extent.

Examples 2.30.
1. If X is *finite*, any nonempty set $F \subseteq C(X)$ is *equicontinuous* on X.
2. The set of all constant functions on X is equicontinuous.
3. More generally, the set of all Lipschitz continuous functions on X (see Definition 2.9) with the same Lipschitz constant L is equicontinuous on X.
4. The set $\{x^n\}_{n \in \mathbb{N}}$ is *not equicontinuous* on $[0, 1]$.

Exercise 2.107. Verify.

Remark 2.61. If (X,ρ) is a *compact* metric space, a set $F \subseteq C(X)$ shares the properties of *uniform boundedness* and *equicontinuity* with its closure $\overline{F}$ in $(C(X),\rho_\infty)$.

Exercise 2.108. Verify.

2.17.2 Arzelà–Ascoli Theorem

Theorem 2.47 (Arzelà–Ascoli Theorem). *Let (X,ρ) be a compact metric space. A set F is precompact (a closed set F is compact) in the space $(C(X),\rho_\infty)$ iff F is uniformly bounded and equicontinuous on X.*

Proof. "Only if" part. Let a set F be *precompact* in the space $(C(X),\rho_\infty)$. In view of the *completeness* of the latter (Theorem 2.46), by the *Hausdorff Criterion* (Theorem 2.35), this is equivalent to the fact that F is *totally bounded* in $(C(X),\rho_\infty)$, and hence (Proposition 2.32), is F *bounded* in $(C(X),\rho_\infty)$, i. e., *uniformly bounded* on X.

By the total boundedness of F, for an arbitrary $\varepsilon > 0$, we can choose a finite $\varepsilon/3$-net $\{f_1, \ldots, f_N\} \subseteq F$ with some $N \in \mathbb{N}$ for F (see Remarks 2.44).

Since, by the *Heine–Cantor Uniform Continuity Theorem* (Theorem 2.44), the functions $f_1, \ldots, f_N$ are *uniformly continuous* on X,

$$\forall i = 1, \ldots, n \; \exists \delta(i) > 0 \; \forall x', x'' \in X \text{ with } \rho(x', x'') < \delta(i) : \; |f_i(x') - f_i(x'')| < \varepsilon/3.$$

Thus, setting $\delta := \min_{1 \le i \le n} \delta(i) > 0$, we have:

$$\forall i = 1, \ldots, n, \; \forall x', x'' \in X \text{ with } \rho(x', x'') < \delta : \; |f_i(x') - f_i(x'')| < \varepsilon/3, \qquad (2.25)$$

i. e., the set $\{f_1, \ldots, f_N\}$ is *equicontinuous* on X.

Since, for each $f \in F$,

$$\exists k = 1, \ldots, n : \rho_\infty(f, f_k) := \max_{x \in X} |f(x) - f_k(x)| < \varepsilon/3,$$

in view of (2.25), we infer that, for all $x', x'' \in X$ with $\rho(x', x'') < \delta$,

$$|f(x') - f(x'')| \le |f(x') - f_k(x')| + |f_k(x') - f_k(x'')| + |f_k(x'') - f(x'')|$$
$$< \varepsilon/3 + \varepsilon/3 + \varepsilon/3 = \varepsilon.$$

Hence, F is *equicontinuous* on X.

   "*If*" part. Suppose that a set F in $(C(X), \rho_\infty)$ is *uniformly bounded* and *equicontinuous* on X. In view of the completeness of $(C(X), \rho_\infty)$ (Theorem 2.46), by the *Hausdorff Criterion* (Theorem 2.35), we are to show the *total boundedness* for F.

   By the *Compactness and Separability Proposition* (Proposition 2.10), there is a *countable everywhere dense set* $\{x_i\}_{i \in I}$, where $I \subseteq \mathbb{N}$, in (X, ρ).

   If the set $\{x_i\}_{i \in I}$ is *finite*, i. e., $I = \{1, \ldots, n\}$ with some $n \in \mathbb{N}$, then

$$X = \overline{\{x_1, \ldots, x_n\}} = \{x_1, \ldots, x_n\}$$

is finite itself, and hence $(C(X), \rho_\infty) = (M(X), \rho_\infty)$, which is *isometric* to $l_\infty^{(n)}$, where, by the *Heine–Borel Theorem* (Theorem 2.36), boundedness alone implies precompactness. Hence, the set F, being *bounded*, is precompact in $(C(X), \rho_\infty)$.

   Suppose now that the set $\{x_i\}_{i \in I}$ is *infinite*, i. e., $I = \mathbb{N}$, and consider an arbitrary sequence $\{f_n\}_{n=1}^\infty \subseteq F$. Since, by the *uniform boundedness* of F, the *numeric sequence* $\{f_n(x_1)\}_{n=1}^\infty$ is *bounded*, by the *Bolzano–Weierstrass Theorem* (Theorem 2.40), the sequence $\{f_n\}_{n=1}^\infty$ contains a subsequence $\{f_{1,n}\}_{n=1}^\infty$ such that

$$f_{1,n}(x_1) \to f(x_1) \in \mathbb{C}, \; n \to \infty.$$

Similarly, since the *numeric sequence* $\{f_{1,n}(x_2)\}_{n=1}^\infty$ is *bounded*, $\{f_{1,n}\}_{n=1}^\infty$ contains a subsequence $\{f_{2,n}\}_{n=1}^\infty$ such that

$$f_{2,n}(x_2) \to f(x_2) \in \mathbb{C}, \; n \to \infty.$$

Continuing inductively, we obtain a countable collection of sequences

$$\{\{f_{m,n}\}_{n=1}^\infty \mid m \in \mathbb{Z}_+\},$$

such that, for each $m \in \mathbb{N}$, $\{f_{m,n}\}_{n=1}^\infty$ is a subsequence of $\{f_{(m-1),n}\}_{n=1}^\infty$, with

$$\{f_{0,n}\}_{n=1}^\infty := \{f_n\}_{n=1}^\infty,$$

and

$$\forall i = 1, \ldots, m : f_{m,n}(x_i) \to f(x_i) \in \mathbb{C}, \; n \to \infty.$$

Hence, for the *"diagonal subsequence"* $\{f_{n,n}\}_{n=1}^{\infty}$ of $\{f_n\}_{n=1}^{\infty}$, we have:

$$\forall i \in \mathbb{N} : f_{n,n}(x_i) \to f(x_i), \ n \to \infty. \tag{2.26}$$

By the *equicontinuity* of F on X,

$$\forall \varepsilon > 0 \ \exists \delta > 0 \ \forall n \in \mathbb{N}, \ \forall x', x'' \in X \text{ with } \rho(x', x'') < \delta :$$
$$|f_{n,n}(x') - f_{n,n}(x'')| < \varepsilon. \tag{2.27}$$

Since, due to its denseness in (X, ρ), the set $\{x_i\}_{i\in\mathbb{N}}$ is a δ-net for X (see Examples 2.22), by the *compactness* of X, there is a *finite* subnet $\{x_{i_1}, \dots, x_{i_N}\} \subseteq \{x_i\}_{i\in\mathbb{N}}$ with some $N \in \mathbb{N}$.

Hence,

$$\forall x \in X \ \exists k = 1, \dots, N : \rho(x, x_{i_k}) < \delta.$$

In view of this and (2.27), for any $m, n \in \mathbb{N}$ and an arbitrary $x \in X$, we have:

$$|f_{m,m}(x) - f_{n,n}(x)| \le |f_{m,m}(x) - f_{m,m}(x_{i_k})| + |f_{m,m}(x_{i_k}) - f_{n,n}(x_{i_k})|$$
$$+ |f_{n,n}(x_{i_k}) - f_{n,n}(x)| < \varepsilon + \sum_{j=1}^{N} |f_{m,m}(x_{i_j}) - f_{n,n}(x_{i_j})| + \varepsilon,$$

and hence,

$$\forall m, n \in \mathbb{N} : \rho_{\infty}(f_{m,m}, f_{n,n}) := \max_{x\in X} |f_{m,m}(x) - f_{n,n}(x)|$$
$$< \varepsilon + \sum_{j=1}^{N} |f_{m,m}(x_{i_j}) - f_{n,n}(x_{i_j})| + \varepsilon.$$

The middle term in the right-hand side *vanishing* as $m, n \to \infty$ by (2.26), we conclude that the subsequence $\{f_{n,n}\}_{n=1}^{\infty}$ is *fundamental* in $(C(X), \rho_{\infty})$. Hence, by the *Characterization of Total Boundedness* (Theorem 2.33), the set F is *totally bounded* in $(C(X), \rho_{\infty})$, which completes the proof. □

Remarks 2.62.

– In view of the *completeness* of the space $(C(X), \rho_{\infty})$ (Theorem 2.46), by the *Hausdorff Criterion* (Theorem 2.35), in the *Arzelà–Ascoli Theorem*, *"precompact"* can be replaced with *"totally bounded"*.

– In particular, if X is finite, as follows from Examples 2.30, and the corresponding segment of the proof of the *"if"* part, the *Arzelà–Ascoli Theorem* is consistent with the *Heine–Borel Theorem* (Theorem 2.36).

Examples 2.31.

1. The set $\{(\frac{x-a}{b-a})^n\}_{n\in\mathbb{N}}$ is uniformly bounded, but not equicontinuous on $[a, b]$ ($-\infty <$ $a < b < \infty$) (cf. Examples 2.30), and hence, by the *Arzelà–Ascoli Theorem*, is *not precompact* (*not totally bounded*) in $(C[a, b], \rho_\infty)$.

2. The set of all Lipschitz continuous functions on $[a, b]$ ($-\infty < a < b < \infty$) with the same Lipschitz constant L is equicontinuous, but, containing all constants, not uniformly bounded on $[a, b]$, and hence, by the *Arzelà–Ascoli Theorem*, is not precompact (not totally bounded) in $(C[a, b], \rho_\infty)$.

2.17.3 Application: Peano's Existence Theorem

As a somewhat unexpected application of the *Arzelà–Ascoli Theorem* (Theorem 2.47), we obtain of the following profound classical result.

Theorem 2.48 (Peano's Existence Theorem). *If a real-valued function $f(\cdot, \cdot)$ is continuous on a closed rectangle*[18]

$$R := [x_0 - a, x_0 + a] \times [y_0 - b, y_0 + b],$$

with some $(x_0, y_0) \in \mathbb{R}^2$ and $a, b > 0$, then, there exists an $h \in (0, a]$ such that the initial-value problem

$$\frac{dy}{dx} = f(x, y), \ y(x_0) = y_0, \tag{2.28}$$

has a solution on the interval $[x_0 - h, x_0 + h]$.

Proof. Since, by the *Heine–Borel Theorem* (Theorem 2.36), the closed rectangle R is compact in $l_\infty^{(2)}$, by the *Weierstrass Extreme Value Theorem* (Theorem 2.43),

$$\exists M > 0 : \max_{(x,y)\in R} |f(x, y)| \le M.$$

Let us fix such an M and set $h := \min[a, \frac{b}{M}] > 0$.

By the *Heine–Cantor Uniform Continuity Theorem* (Theorem 2.44), f is *uniformly continuous* on the *compact* set R and hence,

$$\forall n \in \mathbb{N} \ \exists 0 < \delta(n) \le b \ \forall (x', y'), (x'', y'') \in R, \ |x' - x''| \le \delta(n),$$
$$|y' - y''| \le \delta(n) : \ |f(x', y') - f(x'', y'')| < 1/n. \tag{2.29}$$

For each $n \in N$, let $x_0 = x_0^{(n)} < \cdots < x_{k(n)}^{(n)} = x_0 + h$ ($k(n) \in \mathbb{N}$) be a partition of $[x_0, x_0 + h]$ such that

$$\max_{1 \le i \le k(n)} \left[x_i^{(n)} - x_{i-1}^{(n)} \right] \le \min \left[\delta(n), \frac{\delta(n)}{M} \right]. \tag{2.30}$$

---

18 Giuseppe Peano (1858–1932).

Let us define the *polygonal approximation* $y_n(\cdot)$ for the desired solution $y(\cdot)$ on $[x_0, x_0 + h]$ as follows:

$$y_n(x) = y_0 + f(x_0, y_0)(x - x_0), \quad x_0 \le x \le x_1^{(n)}, \quad n \in \mathbb{N},$$

and, for $n \in \mathbb{N}$, $i = 2, \ldots, k(n)$.

$$y_n(x) = y_n(x_{i-1}^{(n)}) + f(x_{i-1}^{(n)}, y_n(x_{i-1}^{(n)}))(x - x_{i-1}^{(n)}), \quad x_{i-1}^{(n)} \le x \le x_i^{(n)}.$$

Exercise 2.109. Verify that $(x_i^{(n)}, y_n(x_i^{(n)})) \in R$, for all $n \in \mathbb{N}$, $i = 0, \ldots, k(n)$.

Assigning arbitrary real values to $y_n'(x_i^{(n)})$, $n \in \mathbb{N}$, $i = 0, \ldots, k(n)$, we have the following integral representation:

$$y_n(x) = y_0 + \int_{x_0}^{x} y_n'(t)\, dt, \quad n \in \mathbb{N}, \ x \in [x_0, x_0 + h]. \tag{2.31}$$

Whence, we conclude that

$$|y_n(x) - y_0| = \left| \int_{x_0}^{x} y_n'(t)\, dt \right| \le M|x - x_0| \le M \frac{b}{M} = b, \quad n \in \mathbb{N}, \ x \in [x_0, x_0 + h],$$

which implies that, for all $n \in \mathbb{N}$,

$$(x, y_n(x)) \in R, \ x \in [x_0, x_0 + h],$$

and that the set of functions $F := \{y_n\}_{n \in \mathbb{N}}$ is *uniformly bounded* on $[x_0, x_0 + h]$.

Since, for each $n \in \mathbb{N}$, the absolute value of the *slope* of y_n, except, possibly, at the partition points $x_i^{(n)}$, $i = 0, \ldots, k(n)$, does not exceed M, as follows from the *Mean Value Theorem*,

$$\forall n \in \mathbb{N}, \ \forall x', x'' \in [x_0, x_0 + h] : \ |y_n(x') - y_n(x'')| \le M|x' - x''|,$$

which implies that the functions of the set F are Lipschitz continuous on $[x_0, x_0 + h]$ with the same Lipschitz constant M, and hence, F is *equicontinuous* on $[x_0, x_0 + h]$ (see Examples 2.30).

By the *Arzelà–Ascoli Theorem* (Theorem 2.47), the set $\{y_n\}_{n \in \mathbb{N}}$ is *relatively compact* in the space $(C[x_0, x_0 + h], \rho_\infty)$, i.e., its *closure* $\overline{F}$ is compact, and hence, by the *Equivalence of Different Forms of Compactness Theorem* (Theorem 2.39), the sequence $\{y_n\}_{n=1}^{\infty}$ contains a subsequence $\{y_{n(i)}\}_{i=1}^{\infty}$ *uniformly convergent* on $[x_0, x_0 + h]$ to a function $y \in \overline{F} \subseteq C[x_0, x_0 + h]$.

Fixing an arbitrary $n \in \mathbb{N}$, for any $x \in [x_0, x_0 + h]$ distinct from the partition points $x_i^{(n)}$, $n \in \mathbb{N}$, $i = 0, \ldots, k(n)$, choosing a $j = 1, \ldots, k(n)$ such that $x_{j-1}^{(n)} < x < x_j^{(n)}$, by the *Mean Value Theorem* and in view of (2.30), we have:

$$|y_n(x) - y_n(x_{j-1}^{(n)})| \le M|x - x_{j-1}^{(n)}| \le M \frac{\delta(n)}{M} = \delta(n),$$

which, in view of (2.29), implies that

$$|f(x, y_n(x)) - f(x_{j-1}^{(n)}, y_n(x_{j-1}^{(n)}))| < 1/n.$$

Since, on the interval $(x_{j-1}^{(n)}, x_j^{(n)})$, $y_n'(x) = f(x_{j-1}^{(n)}, y_n(x_{j-1}^{(n)}))$, for all $n \in \mathbb{N}$ and $i = 0, \ldots, k(n)$, we have:

$$|f(x, y_n(x)) - y_n'(x)| < 1/n, \ x \in [x_0, x_0 + h], \ x \ne x_i^{(n)}. \tag{2.32}$$

By integral representation (2.31), for all $n \in \mathbb{N}$ and $x \in [x_0, x_0 + h]$ we have:

$$y_n(x) = y_0 + \int_{x_0}^{x} y_n'(t)\, dt = y_0 + \int_{x_0}^{x} f(t, y_n(t))\, dt + \int_{x_0}^{x} [y_n'(t) - f(t, y_n(t))]\, dt. \tag{2.33}$$

Since, by the *uniform continuity* of f on R, $\{f(t, y_{n(i)}(t))\}_{i=1}^{\infty}$ *uniformly converges to* $f(t, y(t))$ on $[x_0, x_0 + h]$ and, due to (2.32),

$$\int_{x_0}^{x} |y_n'(t) - f(t, y_n(t))|\, dt \le h/n,$$

passing to the limit in (2.33) with $n = n(i)$ as $i \to \infty$, we obtain the following integral representation for y:

$$y(x) = y_0 + \int_{x_0}^{x} f(t, y(t))\, dt, \ x \in [x_0, x_0 + h],$$

which, by the *Fundamental Theorem of Calculus*, implies that y is a solution of initial value problem (2.28) on $[x_0, x_0 + h]$.

Using the same construct on $[x_0 - h, x_0]$, we extend the obtained solution to the interval $[x_0 - h, x_0 + h]$. □

Corollary 2.12 (Existence of a Local Solution). *If $f(\cdot, \cdot)$ is a real-valued function continuous on an open subset $D \subseteq \mathbb{R}^2$. Then, for each $(x_0, y_0) \in D$, the initial-value problem (2.28) has a local solution.*

Remark 2.63. *Peano's Existence Theorem* guarantees the *local existence* of the initial-value problem solution only, saying nothing about its *uniqueness* (cf. *Picard's*[19] *Existence and Uniqueness Theorem* (see, e. g., [16])).

Example 2.32. The initial value problem

$$y' = |y|^{1/2}, \ y(0) = 0,$$

---

19 Émile Picard (1856–1941).

whose right-hand side is *continuous* but *not Lipschitz continuous* in y on $\mathbb{R}^2$, has *infinitely many* solutions on $(-\infty, \infty)$: $y = 0$ as well as

$$y = \begin{cases} \frac{(x-C)^2}{4} & \text{for } x \geq C, \\ 0 & \text{for } x < C \end{cases} \quad \text{or} \quad y = \begin{cases} -\frac{(x+C)^2}{4} & \text{for } x \leq -C, \\ 0 & \text{for } x > -C, \end{cases} \quad C \geq 0.$$

Exercise 2.110. Verify.

2.18 Stone–Weierstrass Theorem

The classical *Weierstrass Approximation Theorem* stating that every real-/complex-valued function continuous on an interval $[a, b]$ $(-\infty < a < b < \infty)$ can be *uniformly approximated* on $[a, b]$ arbitrarily close by a polynomial affords an essential generalization due to Marshall H. Stone[20] known as *Stone–Weierstrass Theorem*. The latter allows to consider instead of the interval $[a, b]$ an arbitrary compact metric space (X, ρ) and to use for approximation instead of polynomials functions from more general subsets of $C(X)$ called *algebras*, which encompass the set of polynomials as a particular case.

2.18.1 Weierstrass Approximation Theorem

Theorem 2.49 (Weierstrass Approximation Theorem). *Every real-/complex-valued function continuous on an interval $[a, b]$ $(-\infty < a < b < \infty)$ can be uniformly approximated on $[a, b]$ arbitrarily close by a polynomial, i. e.,*

$$\forall f \in C[a, b], \ \forall \varepsilon > 0 \ \exists p \in P: \ \max_{a \leq x \leq b} |p(x) - f(x)| < \varepsilon.$$

Equivalently, the set P of all polynomials is dense in the (real or complex) space $(C[a, b], \rho_\infty)$.

We are to obtain this classical result, already referred to in Examples 2.11, as a particular case of the *Stone–Weierstrass Theorem*.

2.18.2 Algebras

First, we introduce the concept of an *algebra*.

Definition 2.41 (Algebra/Subalgebra). An *algebra* over a scalar field $\mathbb{F}$ ($\mathbb{F} = \mathbb{R}$ or $\mathbb{F} = \mathbb{C}$) is a *vector space* $\mathcal{A}$ over $\mathbb{F}$ (see Definition 3.1) with an *associative* and *bilinear*

---

[20] Marshall Harvey Stone (1903–1989).

multiplication of elements

$$\mathcal{A} \ni x, y \mapsto xy \in \mathcal{A} :$$

1.	$(xy)z = x(yz)$, $x, y, z \in X$	*Associativity*
2.	$(\lambda x + \mu y)z = \lambda(xz) + \mu(yz)$, $\lambda, \mu \in \mathbb{F}$, $x, y, z \in \mathcal{A}$	*Left Linearity*
3.	$x(\lambda y + \mu z) = \lambda(xy) + \mu(xz)$, $\lambda, \mu \in \mathbb{F}$, $x, y, z \in \mathcal{A}$	*Right Linearity*

A subset $\mathcal{B}$ of an algebra $\mathcal{A}$, which is also an algebra relative to the induced operations, is called a *subalgebra* of $\mathcal{A}$.

Remark 2.64. If $\mathbb{F} = \mathbb{R}$, the algebra is called *real* and, if $\mathbb{F} = \mathbb{C}$, the algebra is called *complex*.

Exercise 2.111. Prove that a subset $\mathcal{B}$ of an algebra $\mathcal{A}$ is *subalgebra* of $\mathcal{A}$ *iff* $\mathcal{B}$ is closed under addition, scalar multiplication, and multiplication.

Examples 2.33.
1. The space $\mathbb{C}$ is a complex algebra and $\mathbb{R}$ is a *real subalgebra* of $\mathbb{C}$.
2. The set M_n of all $n \times n$ matrices ($n \in \mathbb{N}$) with real/complex entries is a real/complex algebra relative to the usual matrix operations, the subspaces U_n, L_n, D_n of all *upper triangular*, *lower triangular*, and *diagonal* $n \times n$ matrices, respectively, being its subalgebras.
3. The (real or complex) space l_∞ is an algebra relative to the *termwise* linear operations and multiplication, the subspaces c, c_0, and c_{00} being its subalgebras.
4. The space $M(X)$ of all real-/complex-valued functions *bounded* on a nonempty set X is an algebra relative to the *pointwise* linear operations and multiplication, the subspace C of all *constant* on X functions being its subalgebra.
5. If (X, ρ) is a *compact metric space*, the space $C(X)$ of all real-/complex-valued functions *continuous* on X is a subalgebra of $M(X)$ and the subspace C of all *constant* on X functions is a subalgebra of $C(X)$.
6. The set P of all *polynomials* with real/complex coefficients is a subalgebra of the real/complex algebra $C[a, b]$ ($-\infty < a < b < \infty$), its subspace

$$P^{(a)} := \{p \in P \mid p(a) = 0\},$$

being in its turn a subalgebra of P.

Exercise 2.112. Verify.

Lemma 2.1 (Intersection Lemma). *The intersection of an arbitrary nonempty collection* $\{\mathcal{B}_i\}_{i \in I}$ *of subalgebras of an algebra $\mathcal{A}$ is a subalgebra of $\mathcal{A}$.*

Exercise 2.113. Prove.

Remark 2.65. As Example 2.34 below shows, the union of subalgebras need not be a subalgebra.

Due to the prior lemma, the following notion is well defined.

Definition 2.42 (Generated Subalgebra). Let C be a nonempty subset of an algebra $\mathcal{A}$, then

$$a(C) := \bigcap_{B \text{ is subalgebra of } \mathcal{A},\ C \subseteq B} B$$

is the *smallest subalgebra of A containing C* called the *subalgebra generated by C*.

Example 2.34. In $M_2(\mathbb{C})$, for the subalgebras

$$D := \left\{ \begin{bmatrix} a & 0 \\ 0 & c \end{bmatrix} \middle| a, c \in \mathbb{C} \right\} \quad \text{and} \quad N := \left\{ \begin{bmatrix} 0 & b \\ 0 & 0 \end{bmatrix} \middle| b \in \mathbb{C} \right\},$$

of $M_2(\mathbb{C})$, $D \cup N$ is not a subalgebra of $M_2(\mathbb{C})$ and

$$a(D \cup N) = D + N = \left\{ \begin{bmatrix} a & b \\ 0 & c \end{bmatrix} \middle| a, b, c \in \mathbb{C} \right\}$$

is the subalgebra of the upper triangular matrices.

Exercise 2.114.
(a) Verify.
(b) Let (X, ρ) be a *compact* metric space and f be a function in the (real or complex) algebra $C(X)$. Describe the subalgebra of $C(X)$ generated by the singleton $\{f\}$.

Proposition 2.14 (Subalgebras of $C(X)$). *Let (X, ρ) be a compact metric space and A be a subalgebra of the (real or complex) algebra $C(X)$. Then*
(1) *provided $1 \in \mathcal{A}$, i.e., A contains constants, for any polynomial p and an arbitrary $f \in \mathcal{A}$,*

$$p \circ f := p(f) \in \mathcal{A};$$

in fact,

$$\{p \circ f := p(f) \mid p \in P\} = a(\{1, f\});$$

(2) *the closure $\overline{A}$ of A in $(C(X), \rho_\infty)$ is also a subalgebra of $C(X)$.*

Exercise 2.115. Prove.

2.18.3 Stone–Weierstrass Theorem

Let us first prove an important lemma and its corollary.

Lemma 2.2 (Polynomial Approximation of Absolute Value). *For any $c > 0$, the absolute-value function $|\cdot|$ can be uniformly approximated on the interval $[-c, c]$ arbitrarily close by a polynomial with real coefficients, i. e.,*

$$\forall \varepsilon > 0 \; \exists p \in P : \; \max_{-c \le x \le c} |p(x) - |x|| < \varepsilon. \tag{2.34}$$

Proof. Without loss of generality, we can regard $c = 1$. Indeed, if (2.34) holds for $c = 1$, and hence,

$$\forall \varepsilon > 0, \; \forall c > 0 \; \exists p \in P : \; \max_{-1 \le t \le 1} |p(t) - |t|| < \varepsilon/c,$$

where $p(t) = a_0 + a_1 t + \cdots + a_n t^n$ with some $n \in \mathbb{Z}_+$ and $a_k \in \mathbb{R}$, $k = 0, \ldots, n$. Then

$$\max_{-1 \le t \le 1} \left| a_0 c + a_1 (ct) + \cdots + \frac{a_n}{c^{n-1}} (ct)^n - |ct| \right| = c \max_{-1 \le t \le 1} |p(t) - |t|| < \varepsilon.$$

Substituting $x := ct$, we arrive at

$$\max_{-c \le x \le c} \left| a_0 c + a_1 x + \cdots + \frac{a_n}{c^{n-1}} x^n - |x| \right| < \varepsilon.$$

As follows from the binomial series decomposition

$$\sqrt{1+x} = \sum_{k=0}^{\infty} \binom{1/2}{k} x^k = 1 + \frac{1}{2}x - \frac{1}{8}x^2 + \frac{1}{16}x^3 + \cdots, \quad -1 \le x \le 1,$$

where

$$\binom{1/2}{0} := 1,$$

$$\binom{1/2}{k} := \frac{(1/2)(1/2 - 1) \cdots (1/2 - k + 1)}{k!} = (-1)^{k-1} \frac{1 \cdot 3 \cdot 5 \cdots (2k - 3)}{2^k k!}, \quad k \in \mathbb{N},$$

uniformly convergent on $[-1, 1]$ (due to *Raabe's*[21] Test and the *Comparison Test*), the function $\sqrt{x}$ can be uniformly approximated on $[0, 1]$ by the polynomial sequence

$$p_n(x) := s_n(x - 1), \; n \in \mathbb{Z}_+, \; x \in [0, 1],$$

where

$$s_n(x) := \sum_{k=0}^{n} \binom{1/2}{k} x^k, \; n \in \mathbb{Z}_+, \; x \in [-1, 1],$$

is the nth partial sum of the binomial series.

---

21 Joseph Ludwig Raabe (1801–1859).

This implies that

$$\max_{-1 \leq x \leq 1} |p_n(x^2) - |x|| = \max_{-1 \leq x \leq 1} |p_n(x^2) - \sqrt{x^2}| = \max_{0 \leq t \leq 1} |p_n(t) - \sqrt{t}| \to 0, \ n \to \infty,$$

i. e., the sequence of polynomials $\{p_n(x^2)\}_{n=0}^{\infty}$ converges uniformly to $|x|$ on $[-1, 1]$, completing the proof. $\square$

Corollary 2.13 (Abs/Min/Max). *Let (X, ρ) be a compact metric space and $\mathcal{A}$ be a subalgebra of the real algebra $C(X, \mathbb{R})$ such that $1 \in \mathcal{A}$, i. e., $\mathcal{A}$ contains constants. Then, for any $f, g \in \mathcal{A}$,*

$$|f|, \min(f, g), \max(f, g) \in \overline{\mathcal{A}},$$

where $\overline{\mathcal{A}}$ is the closure of $\mathcal{A}$ in $(C(X, \mathbb{R}), \rho_{\infty})$.

Proof. Since the space (X, ρ) is compact, by the *Boundedness of Continuous Functions on Compact Sets Corollary* (Corollary 2.9), an arbitrary $f \in \mathcal{A} \subseteq C(X, \mathbb{R})$ is *bounded*, and hence,

$$\exists c > 0 \ \forall x \in X : \ -c \leq f(x) \leq c.$$

By the *Polynomial Approximation of Absolute Value Lemma* (Lemma 2.2),

$$\forall \varepsilon > 0 \ \exists p \in P : \ \max_{-c \leq t \leq c} |p(t) - |t|| < \varepsilon.$$

Then

$$\rho_{\infty}(p \circ f, |f|) = \max_{x \in X} |p(f(x)) - |f(x)|| < \varepsilon.$$

This, since, by the *Subalgebras of $C(X)$ Proposition* (Proposition 2.14), $p \circ f := p(f) \in \mathcal{A}$, implies that

$$|f| \in \overline{\mathcal{A}}$$

Since, by the same proposition, $\overline{\mathcal{A}}$ is also a *subalgebra* of $C(X, \mathbb{R})$, we infer that, for any $f, g \in \mathcal{A}$,

$$\min(f, g) = \frac{f + g}{2} - \frac{|f - g|}{2} \in \overline{\mathcal{A}} \quad \text{and} \quad \max(f, g) = \frac{f + g}{2} + \frac{|f - g|}{2} \in \overline{\mathcal{A}}. \qquad \square$$

Theorem 2.50 (Stone–Weierstrass Theorem (Real Version)). *Let (X, ρ) be a compact metric space and $\mathcal{A}$ be a subalgebra of the real algebra $C(X, \mathbb{R})$ such that*
(1) $1 \in \mathcal{A}$ *(i. e., $\mathcal{A}$ contains constants);*
(2) $\forall x, y \in X, x \neq y \ \exists f \in \mathcal{A} : f(x) \neq f(y)$ *(i. e., $\mathcal{A}$ separates points).*

Then $\mathcal{A}$ is dense in $(C(X, \mathbb{R}), \rho_{\infty})$.

Proof. Let us first prove that, for arbitrary *distinct* $x_1, x_2 \in X$ and any $a, b \in \mathbb{R}$, there exists a function $h \in \mathcal{A}$ such that

$$h(x_1) = a \quad \text{and} \quad h(x_2) = b.$$

Since $\mathcal{A}$ *separates points*,

$$\exists f \in \mathcal{A} : f(x_1) \neq f(x_2).$$

Considering that $\mathcal{A}$ *contains constants*, we obtain the desired function $h \in \mathcal{A}$ as follows:

$$h(x) := a + (b - a)\frac{f(x) - f(x_1)}{f(x_2) - f(x_1)}, \quad x \in X.$$

Let $f \in C(X, \mathbb{R})$ and $\varepsilon > 0$ be arbitrary.
For any $x, y \in X$, there is a function $f_{x,y} \in \mathcal{A}$ such that

$$f_{x,y}(x) = f(x) \quad \text{and} \quad f_{x,y}(y) = f(y).$$

For $x \neq y$, the statement follows from the above and, for $x = y$, it is trivial.
Fixing an arbitrary $x \in X$, since

$$\forall y \in X : f_{x,y}(y) - f(y) = 0,$$

by the *continuity* of the function $f_{x,y}(z) - f(z)$, $z \in X$, at y, we have:

$$\forall y \in X \, \exists r(y) > 0 \, \forall z \in B(y, r(y)) : f_{x,y}(z) < f(z) + \varepsilon/2. \tag{2.35}$$

Since X is *compact*, its open cover

$$\{B(y, r(y)) \mid y \in X\}$$

contains a finite subcover

$$\{B(y_1, r(y_1)), \ldots, B(y_m, r(y_m))\}$$

with some $m = m(x, \varepsilon) \in \mathbb{N}$.
Then, by the *Abs/Min/Max Corollary* (Corollary 2.13) and induction,

$$f_x := \min(f_{x,y_1}, \ldots, f_{x,y_m}) \in \overline{\mathcal{A}}$$

and, by (2.35),

$$f_x(z) < f(z) + \varepsilon/2, \, z \in X. \tag{2.36}$$

Since

$$\forall x \in X : f_x(x) - f(x) = 0,$$

by the *continuity* of the function $f_x(z) - f(z), z \in X$, at x, we have:

$$\forall x \in X \, \exists r(x) > 0 \, \forall z \in B(x, r(x)) : f(z) - \varepsilon/2 < f_x(z). \tag{2.37}$$

Since X is *compact*, its open cover

$$\{B(x, r(x)) \mid x \in X\}$$

contains a finite subcover

$$\{B(x_1, r(x_1)), \dots, B(x_n, r(x_n))\}$$

with some $n = n(\varepsilon) \in \mathbb{N}$.

Then, by the *Abs/Min/Max Corollary* (Corollary 2.13) and induction,

$$g := \max(f_{x_1}, \dots, f_{x_n}) \in \overline{\mathcal{A}}$$

and, by (2.36) and (2.37),

$$f(z) - \varepsilon/2 < g(z) < f(z) + \varepsilon/2, \ z \in X, \tag{2.38}$$

i. e.,

$$\rho_\infty(g, f) < \varepsilon/2.$$

Since $g \in \overline{\mathcal{A}}$, there is an $h \in \mathcal{A}$ such that

$$\rho_\infty(h, g) < \varepsilon/2,$$

and hence, by the *triangle inequality*,

$$\rho_\infty(h, f) \leq \rho_\infty(h, g) + \rho_\infty(g, f) < \varepsilon/2 + \varepsilon/2 = \varepsilon,$$

which completes the proof. □

Remarks 2.66.
- Each condition of the *Stone–Weierstrass Theorem (Real Version)* is essential and cannot be dropped.
- The conditions of *Stone–Weierstrass Theorem (Real Version)* are *insufficient* for the case of the complex algebra $C(X, \mathbb{C})$.

Exercise 2.116. Give the corresponding counterexamples.

Hint. For the latter, consider the subalgebra P of all polynomials of the complex algebra $C(X, \mathbb{C})$, where $X := \{z \mid |z| \leq 1\}$, and show that $\bar{z} \notin \bar{P}$, where $\bar{z} := x - iy$ is the *complex conjugate* of $z = x + iy$ and $\bar{P}$ is the closure of P in $(C(X, \mathbb{C}), \rho_\infty)$.

Thus, the complex version for the *Stone–Weierstrass Theorem* naturally calls for an additional condition.

Theorem 2.51 (Stone–Weierstrass Theorem (Complex Version)). *Let (X, ρ) be a compact metric space and $\mathcal{A}$ be a subalgebra of the complex algebra $C(X, \mathbb{C})$ such that*
(1) $1 \in \mathcal{A}$ (*i. e., $\mathcal{A}$ contains constants*);
(2) $\forall x, y \in X, x \neq y \ \exists f \in \mathcal{A} : f(x) \neq f(y)$ (*i. e., $\mathcal{A}$ separates points*);
(3) $\forall f \in \mathcal{A} : \bar{f} \in \mathcal{A}$ (*i. e., $\mathcal{A}$ self-conjugate*).

Then $\mathcal{A}$ is dense in $(C(X, \mathbb{C}), \rho_\infty)$.

Proof. Let

$$\mathcal{A}_\mathbb{R} := \{f \in \mathcal{A} \mid f(x) \in \mathbb{R}, \ x \in X\}.$$

Obviously, $\mathcal{A}_\mathbb{R}$ is a *subalgebra* of the real algebra $C(X, \mathbb{R})$ *containing the constants*. It also *separates the points*. Indeed, by the same argument as in the proof of the *Stone–Weierstrass Theorem (Real Version)*, since $\mathcal{A}$ *contains constants* and *separates points*,

$$\forall x, y \in X, \ x \neq y \ \exists f \in \mathcal{A} : f(x) = 0, \ f(y) = 1.$$

Since $\mathcal{A}$ is *self-conjugate*,

$$g := f + \bar{f} = 2\,\mathrm{Re}f \in \mathcal{A}_\mathbb{R} \quad \text{and} \quad g(x) = 0 \neq 2 = g(y).$$

Hence, the *Stone–Weierstrass Theorem (Real Version)* applies to $\mathcal{A}_\mathbb{R}$ as a subalgebra of $C(X, \mathbb{R})$.
For any $f \in C(X, \mathbb{C})$,

$$f = \mathrm{Re}f + i\,\mathrm{Im}f$$

with $\mathrm{Re}f, \mathrm{Im}f \in C(X, \mathbb{R})$, by the *Stone–Weierstrass Theorem (Real Version)*,

$$\exists g, h \in \mathcal{A}_\mathbb{R} : \rho_\infty(\mathrm{Re}f, g) < \varepsilon/2 \quad \text{and} \quad \rho_\infty(\mathrm{Im}f, h) < \varepsilon/2.$$

Hence, for $g + ih \in \mathcal{A}$, we have:

$$\rho_\infty(f, g + ih) \leq \rho_\infty(\mathrm{Re}f, g) + \rho_\infty(\mathrm{Im}f, h) < \varepsilon/2 + \varepsilon/2 = \varepsilon,$$

which proves the statement. □

2.18.4 Applications

The *Stone–Weierstrass Theorem* has a number of immediate useful applications.

Remark 2.67. Applying the *Stone–Weierstrass Theorem* to the *subalgebra of polynomials P* in the (real or complex) algebra $C[a, b]$ $(-\infty < a < b < \infty)$, we immediately obtain the *Weierstrass Approximation Theorem* (Theorem 2.49).

Exercise 2.117. Verify that P satisfies the required conditions in both real and complex cases.

Corollary 2.14 (Approximation by Polynomials in z and $\bar{z}$). *Let F be a compact set in $\mathbb{C}$. An arbitrary function $f \in C(F, \mathbb{C})$ can be uniformly approximated on F arbitrarily close by a polynomial in z and $\bar{z}$, i. e., the set of all polynomials in z and $\bar{z}$ is dense in $(C(F, \mathbb{C}), \rho_\infty)$.*

Corollary 2.15 (Approximation by Multivariable Polynomials). *Let F be a compact set in $l_p^{(n)}(\mathbb{R})$ $(n \in \mathbb{N}, 1 \le p \le \infty))$. An arbitrary function f in the (real or complex) algebra $C(F)$ can be uniformly approximated on F arbitrarily close by a polynomial in n variables, i. e., the set of all polynomials in n variables is dense in $(C(F), \rho_\infty)$.*

Exercise 2.118. Prove Corollaries 2.14 and 2.15.

Corollary 2.16 (Approximation by Trigonometric Polynomials). *An arbitrary $2L$-periodic $(L > 0)$ function f in the (real or complex) algebra $C(\mathbb{R})$ can be uniformly approximated on $\mathbb{R}$ arbitrarily close by a trigonometric polynomial, i. e., a function of the form*

$$p_n(x) = a_0 + \sum_{k=1}^{n} \left(a_k \cos \frac{k\pi t}{L} + b_k \sin \frac{k\pi t}{L} \right),$$

where $n \in \mathbb{Z}_+$, $a_i, b_j \in \mathbb{R}$ (or $\mathbb{C}$), $i = 0, 1, \ldots, n$, $j = 1, \ldots, n$.

Exercise 2.119. Prove.

Hint. Apply the *Stone–Weierstrass Theorem* to the algebra T of all trigonometric polynomials in $C(X)$, where (X, ρ) is the *compact* metric space with $X := [-L, L]$, the endpoints considered to be *identical*, and

$$\rho(x, y) := \min \left[|x - y|, 2L - |x - y| \right], \quad x, y \in X.$$

2.19 Problems

1. For which values of $p > 0$ is the mapping

$$\mathbb{R} \ni x, y \mapsto |x - y|^p \in \mathbb{R}$$

   a *metric* on $\mathbb{R}$?

2. Is the mapping

$$x := (x_1,\ldots,x_n), y := (y_1,\ldots,y_n) \mapsto \rho_p(x,y) := \left[\sum_{i=1}^n |x_i - y_i|^p\right]^{1/p}$$

with $0 < p < 1$ a metric on the n-space ($n \in \mathbb{N}$, $n \ge 2$)?

3. Let (X,ρ) be a metric space. Which of the following mappings
    (a) $X \ni x,y \mapsto d(x,y) := \sqrt{\rho(x,y)}$,
    (b) $X \ni x,y \mapsto d(x,y) := \rho^2(x,y)$,
    (c) $X \ni x,y \mapsto d(x,y) := \min(\rho(x,y),1)$,
    (d) $X \ni x,y \mapsto d(x,y) := \frac{\rho(x,y)}{\rho(x,y)+1}$
    are *metrics* on X?

4. Let $1 \le p \le \infty$. Show that $C[a,b]$ ($-\infty < a < b < \infty$) is a *metric space* relative to the *p-metric*

$$C[a,b] \ni f,g \mapsto \rho_p(f,g) := \begin{cases} \left[\int_a^b |f(t) - g(t)|^p \, dt\right]^{1/p} & \text{if } 1 \le p < \infty, \\ \max_{a \le t \le b} |f(t) - g(t)| & \text{if } p = \infty. \end{cases}$$

Hint. The cases of $p = 1$ and $p = \infty$ are considered in Examples 2.4. To consider the case of $1 < p < \infty$, first prove *Hölder's inequality* for functions:

$$\int_a^b |f(t)g(t)| \, dt \le \|f\|_p \|g\|_q, \quad f,g \in C[a,b],$$

where $1 \le p,q \le \infty$ are the *conjugate indices* and $\|f\|_p$ is the *p-norm* of f, i. e., the distance of f from the *zero function*:

$$C[a,b] \ni f \mapsto \|f\|_p := \rho_p(f,0) = \begin{cases} \left[\int_a^b |f(t)|^p \, dt\right]^{1/p} & \text{if } 1 \le p < \infty, \\ \max_{a \le t \le b} |f(t)| & \text{if } p = \infty, \end{cases}$$

which is obvious for the symmetric pairs $p = 1$, $q = \infty$ and $p = \infty$, $q = 1$ and follows from *Young's Inequality* (Theorem 2.1) for $1 < p,q < \infty$ in the same manner as *Minkowski's Inequality for n-Tuples* (Theorem 2.2). Then deduce *Minkowski's inequality* for functions:

$$\|f + g\|_p \le \|f\|_p + \|g\|_p, \quad f,g \in C[a,b],$$

with $1 < p < \infty$ similarly to how it is done for n-tuples.

5.

Definition 2.43 (Cartesian Product of Metric Spaces). Let (X_1,ρ_1), (X_2,ρ_2) be metric spaces. The *Cartesian product* $X = X_1 \times X_2$ is a *metric space* relative to the *product metric*

$$X \ni (x_1,x_2), (y_1,y_2) \mapsto \rho((x_1,x_2),(y_1,y_2)) := \sqrt{\rho_1^2(x_1,y_1) + \rho_2^2(x_2,y_2)}.$$

The product space (X, ρ) is naturally called the *Cartesian product* of the spaces (X_1, ρ_1) and (X_2, ρ_2).

Verify that (X, ρ) is a *metric space*.

Remark 2.68. The definition can be naturally extended to arbitrary finite products.

6. * Prove

Proposition 2.15 (Characterization of Convergence in (c_0, ρ_∞)).
In (c_0, ρ_∞), $\{x_k^{(n)}\}_{k=1}^\infty =: x^{(n)} \to x := \{x_k\}_{k=1}^\infty$, $n \to \infty$, iff
(1) $\forall k \in \mathbb{N}: x_k^{(n)} \to x_k$, $n \to \infty$, *and*
(2) $\forall \varepsilon > 0 \, \exists N \in \mathbb{N} \, \forall n \in \mathbb{N}: \sup_{k \geq N+1} |x_k^{(n)}| < \varepsilon.$

7. * Prove

Proposition 2.16 (Characterization of Convergence in l_p $(1 \leq p < \infty)$). *In l_p $(1 \leq p < \infty)$, $\{x_k^{(n)}\}_{k=1}^\infty =: x^{(n)} \to x := \{x_k\}_{k=1}^\infty$, $n \to \infty$, iff*
(1) $\forall k \in \mathbb{N}: x_k^{(n)} \to x_k$, $n \to \infty$, *and*
(2) $\forall \varepsilon > 0 \, \exists N \in \mathbb{N} \, \forall n \in \mathbb{N}: \sum_{k=N+1}^\infty |x_k^{(n)}|^p < \varepsilon.$

8. Show that the power sequence $\{t^n\}_{n=0}^\infty$ converges in $(C[0, 1/2], \rho_\infty)$, but does not converge in $(C[0, 1], \rho_\infty)$.

9. Prove

Proposition 2.17 (Characterization of Convergence in Product Space). *Let (X, ρ) be the Cartesian product of metric spaces (X_1, ρ_1) and (X_2, ρ_2) (see Problem 5). Then*

$$(x_1^{(n)}, x_2^{(n)}) \to (x_1, x_2), \ n \to \infty, \ in \ (X, \rho)$$

iff

$$x_i^{(n)} \to x_i, \ n \to \infty, \ in \ (X_i, \rho_i), \ i = 1, 2,$$

i.e., the convergence of a sequence in the Cartesian product of metric spaces of (X_1, ρ_1) and (X_2, ρ_2) is equivalent to the componentwise convergence in the corresponding spaces.

Remark 2.69. The statement can be naturally extended to arbitrary finite products.

10. Prove

Proposition 2.18 (Joint Continuity of Metric). *If $x_n \to x$ and $y_n \to y$, $n \to \infty$, in a metric space (X, ρ),*

$$\rho(x_n, y_n) \to \rho(x, y), \ n \to \infty.$$

Remark 2.70. Joint continuity of metric immediately implies its continuity in each argument:

$$\rho(x_n, y) \to \rho(x, y) \quad \text{and} \quad \rho(x, y_n) \to \rho(x, y), \ n \to \infty.$$

11. Let (X, ρ) be a metric space and $f : X \to l_p^{(n)}$ ($n \in \mathbb{N}$, $1 \le p \le \infty$):

$$X \ni x \mapsto f(x) := (f_1(x), \dots, f_n(x)) \in l_p^{(n)}$$

Prove that $f(\cdot)$ is *continuous* at a point $x_0 \in X$ iff each scalar component function $f_i(\cdot)$, $i = 1, \dots, n$, is continuous at x_0.

12.

Definition 2.44 (Distance to a Set). Let A be a nonempty set in a metric space (X, ρ). For any $x \in X$, the nonnegative number

$$\rho(x, A) := \inf_{y \in A} \rho(x, y)$$

is called the *distance from the point x to the set A*.

Prove that the *distance-to-a-set function $f(x) := \rho(x, A)$, $x \in X$*, is *Lipschitz continuous* on X.

13. Prove that, for a set A in a metric space (X, ρ), the *interior* of A is the largest open set contained in A, i. e.,

$$\text{int}(A) = \bigcup_{O \in \mathscr{G}, O \subseteq A} O,$$

where $\mathscr{G}$ is the metric topology of (X, ρ)

Hint. Show first that $\text{int}(A)$ is an open set.

14. Prove

Theorem 2.52 (Characterization of Continuity). *Let (X, ρ) and (Y, σ) be metric spaces.*
A function $f : X \to Y$ is continuous iff, for each open set O (closed set C) in (Y, σ), the inverse image $f^{-1}(O)$ ($f^{-1}(C)$) is an open (respectively, closed) set in (X, ρ).

15. Prove that, for a set A in a metric space (X, ρ), the *closure* of A is the smallest closed set containing A, i. e.,

$$\overline{A} = \bigcap_{C \in \mathscr{C}, A \subseteq C} C,$$

where $\mathscr{C}$ is the collection of all closed sets in (X, ρ).

Hint. Show first that $\overline{A}$ is a closed set.

16. Prove that, for a nonempty set A in a metric space (X, ρ),

$$\overline{A} = \{x \in X \mid \rho(x, A) = 0\}$$

(see Definition 2.44).

17. Prove

Proposition 2.19 (Subspace of Separable Metric Space). *A subspace of a separable metric space (X, ρ) is separable.*

18. Prove

Proposition 2.20 (Characterization of the Separability of Product Space). *The Cartesian product (X, ρ) of metric spaces (X_1, ρ_1) and (X_2, ρ_2) (see Problem 5) is separable iff each metric space (X_i, ρ_i), $i = 1, 2$, is separable.*

19. Prove

Proposition 2.21 (Continuous Image of Separable Metric Space). *Let (X, ρ) and (Y, σ) be metric spaces, the former one being separable and $f \in C(X, Y)$. Then the subspace $(f(X), \sigma)$ of (Y, σ) is separable, i. e., the image of a separable space under a continuous mapping is separable.*

20. Prove that the space $(M(X), \rho_\infty)$, where X is an infinite set, is *not separable*.
21. Prove that, for a set A in a metric space (X, ρ),

$$\overline{A} = X \setminus \text{ext}(A), \text{ i. e., } \overline{A} = \text{int}(A) \cup \partial A.$$

22. On $C[a, b]$ $(-\infty < a < b < \infty)$, are the metrics

$$\rho_1(f, g) := \int_a^b |f(t) - g(t)| \, dt, \ f, g \in C[a, b],$$

and

$$\rho_\infty(f, g) := \max_{a \le t \le b} |f(t) - g(t)|, \ f, g \in C[a, b],$$

equivalent? Explain.

23. Show in *two different ways* that the real line $\mathbb{R}$ with the regular distance is not isometric to plane $\mathbb{R}^2$ with the Euclidean distance.

24. Prove

Proposition 2.22 (Fundamental Sequence with Convergent Subsequence). *If a fundamental sequence $\{x_n\}_{n=1}^\infty$ in a metric space (X, ρ) contains a subsequence $\{x_{n(k)}\}_{k=1}^\infty$ such that*

$$\exists x \in X : x_{n(k)} \to x, \ k \to \infty, \text{ in } (X, \rho),$$

then

$$x_n \to x, \ n \to \infty, \ in \ (X,\rho).$$

25. Prove

Proposition 2.23 (Characterization of Completeness of Product Space). *The Cartesian product* (X,ρ) *of metric spaces* (X_1,ρ_1) *and* (X_2,ρ_2) *(see Problem 5) is complete iff each metric space* (X_i,ρ_i), $i = 1,2$, *is complete.*

26. Let s be the set of all real/complex *sequences*.
    (a) Prove that s is a metric space relative to the mapping

$$s \ni x := \{x_k\}_{k=1}^{\infty}, y := \{y_k\}_{k=1}^{\infty} \mapsto \rho(x,y) := \sum_{k=1}^{\infty} \frac{1}{2^k} \frac{|x_k - y_k|}{|x_k - y_k| + 1}.$$

    (b) Describe *convergence* in (s,ρ).
    (c) Prove that the space (s,ρ) is *complete*.
27. Let (X,ρ) and (Y,σ) be *isometric* metric spaces. Prove that
    (a) (X,ρ) is *separable* iff (Y,σ) is *separable*;
    (b) (X,ρ) is *complete* iff (Y,σ) is *complete*.
28. Consider on $\mathbb{R}$ the following metric:

$$d(x,y) := |\arctan x - \arctan y|, \ x,y \in \mathbb{R}.$$

    (a) Show that the metric space $(\mathbb{R},d)$ is *incomplete*.
    (b) Construct a *completion* of $(\mathbb{R},d)$.
29. (a) Prove

Proposition 2.24 (Finite Intersections of Open Dense Sets). *In a metric space* (X,ρ), *any finite intersection of open dense sets is dense.*

    (b) Give an example showing that the condition of *openness* is essential and cannot be dropped.
    (c) Give an example showing that an infinite intersection of open dense sets need not be dense.
30. Show that the set P of all *polynomials* with real/complex coefficients is of the *first category* in $(C[a,b],\rho_\infty)$ $(-\infty < a < b < \infty)$.
31. (a) Prove that, in a *complete* metric space (X,ρ), for any *first-category* set A, the complement A^c is a *second-category* set.
    (b) Is the *converse* true?
32.

Definition 2.45 (Centered Collections of Sets). A collection $\mathscr{C}$ of subsets of a set X is said to have the *finite intersection property*, or to be *centered*, if the intersection of any finite subcollection of $\mathscr{C}$ is nonempty.

Examples 2.35.
1. The collection $\mathscr{C} = \{[n, \infty)\}_{n \in \mathbb{N}}$ is *centered*.
2. The collection $\mathscr{C} = \{[n, n+1)\}_{n \in \mathbb{N}}$ is *not centered*.
3. The collection of all open/closed balls in a metric space (X, ρ) centered at a fixed point $x \in X$ is *centered*.

(a) Verify the examples.
(b) Prove

> **Theorem 2.53** (Centered-Collection Characterization of Compactness).
> *A metric space (X, ρ) is compact iff every centered collection $\mathscr{C}$ of closed subsets of X has a nonempty intersection.*

> and

> **Corollary 2.17** (Nested Sequences of Closed Sets in Compact Spaces).
> *If $\{C_n\}_{n=1}^{\infty}$ is a sequence of nonempty closed sets in a compact metric space (X, ρ) such that*

$$C_n \supseteq C_{n+1}, \quad n \in \mathbb{N},$$

> *then*

$$\bigcap_{n=1}^{\infty} C_n \neq \emptyset.$$

33. Prove

> **Theorem 2.54** (Centered Collections of Compact Sets). *In a metric space (X, ρ), every centered collection of compact sets has a nonempty intersection*

> **Hint.** For an arbitrary centered collection $\mathscr{C} = \{C_i\}_{i \in I}$ of compact sets in (X, ρ), fix a $j \in I$ and, in the *compact subspace* (C_j, ρ), consider the centered collection of closed subsets $\{C_i \cap C_j\}_{i \in I}$.

34. Use the prior theorem to prove

> **Theorem 2.55** (Cantor's Intersection Theorem). *If $\{C_n\}_{n=1}^{\infty}$ is a sequence of non-empty compact sets in a metric space (X, ρ) such that*

$$C_n \supseteq C_{n+1}, \quad n \in \mathbb{N},$$

> *then*

$$\bigcap_{n=1}^{\infty} C_n \neq \emptyset.$$

35. Which of the following sets are compact in $l_p^{(2)}$ $(1 \leq p \leq \infty)$

(a) $\{(x,y) \mid x = 0\}$,

(b) $\{(x,y) \mid 0 < x^2 + y^2 \le 1\}$,

(c) $\{(x,y) \mid |y| \le 1,\ x^2 + y^2 \le 4\}$,

(d) $\{(x,y) \mid y = \sin\frac{1}{x},\ 0 < x \le 1\} \cup \{(x,y) \mid x = 0,\ -1 \le y \le 1\}$?

36. Show that the set

$$\{\{x_n\}_{n=1}^{\infty} \mid |x_n| \le 1/n\}_{n \in \mathbb{N}}$$

is *compact* in l_p $(1 < p < \infty)$ and (c_0, ρ_∞).

37. Prove

Proposition 2.25 (Characterization of Compactness of Product Space). *The Cartesian product* (X, ρ) *of metric spaces* (X_1, ρ_1) *and* (X_2, ρ_2) *(see Problem 5) is compact iff each metric space* (X_i, ρ_i), $i = 1, 2$, *is compact.*

Hint. Use the *sequential approach.*

38. Let (X_1, ρ_1) and (X_2, ρ_2) be metric spaces and $T : X_1 \to X_2$ be a *continuous transformation.* Prove that, if (X_1, ρ_1) is *compact*, the graph of T

$$G_T := \{(x, Tx) \in X_1 \times X_2 \mid x \in X_1\}$$

is a *compact set* in the product space (X, ρ) of (X_1, ρ_1) and (X_2, ρ_2) (see Problem 5).

39.

Definition 2.46 (Distance Between Sets). For nonempty sets A and B in a metric space (X, ρ), the nonnegative number

$$\rho(A, B) := \inf_{x \in A, y \in B} \rho(x, y)$$

is called the *distance between the sets A and B.*

Prove that, for nonempty compact sets A and B in a metric space (X, ρ), the *distance* between A and B is attained, i. e.,

$$\exists x_0 \in A, y_0 \in B : \rho(x_0, y_0) = \rho(A, B),$$

which, provided $A \cap B = \emptyset$, implies that

$$\rho(A, B) > 0.$$

Hint. Use the results of Problems 10 and 37.

40. Prove

Proposition 2.26 (Nearest Point Property in the n-Space). *Let* A *be a nonempty closed set in the (real or complex) space* $l_p^{(n)}$ $(n \in \mathbb{N}, 1 \le p \le \infty)$. *Then, for any* $x \in X$, *in* A, *there is a nearest point to* x, *i. e.,*

$$\exists y \in A : \rho(x, y) = \inf_{p \in A} \rho(x, p) =: \rho(x, A).$$

Give an example showing that a *farthest point* from x in A need not exist.

41. Let (X,ρ) be a metric space and $\{f_n\}_{n=1}^\infty$ be an equicontinuous on X sequence in $C(X)$, i. e.,

$$\forall \varepsilon > 0 \ \exists \delta > 0 \ \forall n \in \mathbb{N}, \ \forall x', x'' \in X, \ \rho(x', x'') < \delta : \ |f_n(x') - f_n(x'')| < \varepsilon.$$

Show that, if $\{f_n\}_{n=1}^\infty$ *converges pointwise* on X to a function f, i. e.,

$$\forall x \in X : f_n(x) \to f(x), \ n \to \infty,$$

then f is *uniformly continuous* on X.

42. Show in *two different ways* that a nontrivial sphere/closed ball is *not* a compact set in $(C[a,b], \rho_\infty)$ $(-\infty < a < b < \infty)$.

Hint. Consider the unit sphere $S(0,1)$ in $(C[0,1], \rho_\infty)$.

43. Describe the subalgebra $\mathcal{A}$ of $C[a,b]$ $(-\infty < a < b < \infty)$ generated by $\{1, t\}$ and its closure $\overline{\mathcal{A}}$ in $(C[a,b], \rho_\infty)$.

44. Describe the subalgebra $\mathcal{A}$ of $C[a,b]$ $(0 \le a < b < \infty)$ generated by $\{1, t^2\}$ and its closure $\overline{\mathcal{A}}$ in $(C[a,b], \rho_\infty)$.

45. Let (X,ρ) be the *Cartesian product* of *compact* metric spaces (X_1, ρ_1) and (X_2, ρ_2) (see Problem 5). Prove that,

$$\forall f \in C(X, \mathbb{R}), \ \forall \varepsilon > 0 \ \exists n \in \mathbb{N} \ \exists g_{i1}, \ldots, g_{in} \in C(X_i, \mathbb{R}), \ i = 1, 2 :$$

$$\max_{(x_1, x_2) \in X} \left| f(x_1, x_2) - \sum_{k=1}^n g_{1k}(x_1) g_{2k}(x_2) \right| < \varepsilon.$$

Is the *complex version* of the statement valid?

Hint. Use the result of Problem 37.

3 Normed Vector and Banach Spaces

In this chapter, we introduce *normed vector spaces* and study their properties emerging from the remarkable interplay between their linear and topological structures.

3.1 Vector Spaces

First, we introduce and study *vector spaces* endowed with linear structure alone.

3.1.1 Definition, Examples, Properties

Definition 3.1 (Vector Space). A *vector space* (also a *linear space*) over a scalar field $\mathbb{F}$ ($\mathbb{F} = \mathbb{R}$ or $\mathbb{F} = \mathbb{C}$) is a set X of elements, also called *vectors*, equipped with the two *linear operations* for

- *vector addition:* $X \ni x, y \mapsto x + y \in X$ and
- *scalar multiplication:* $\mathbb{F} \ni \lambda, X \ni x \mapsto \lambda x \in X$

subject to the following *vector space axioms*:

1. $x + y = y + x$, $x, y \in X$.   *Commutativity*
2. $(x + y) + z = x + (y + z)$, $x, y, z \in X$.   *Associativity*
3. $\exists\, 0 \in X\colon x + 0 = x$, $x \in X$.   Existence of *additive identity* (*zero vector*)
4. $\forall x \in X \,\exists\, -x \in X : x + (-x) = 0$. Existence of *additive inverses* or *opposite vectors*
5. $\lambda(\mu x) = (\lambda\mu)x$, $\lambda, \mu \in \mathbb{F}$, $x \in X$.   *Associativity* of *Scalar Multiplication*
6. $1x = x$, $x \in X$.   *Neutrality of Scalar Identity*
7. $\lambda(x + y) = \lambda x + \lambda y$, $\lambda \in \mathbb{F}$, $x, y \in X$.   *Right Distributivity*
8. $(\lambda + \mu)x = \lambda x + \mu x$, $\lambda, \mu \in \mathbb{F}$, $x \in X$.   *Left Distributivity*

Remarks 3.1.
- Henceforth, without further specifying, we understand that $\mathbb{F}$ stands for $\mathbb{R}$ or $\mathbb{C}$ and the underlying vector space is called *real* in the former case and *complex* in the latter.
- For any complex vector space X, the *associated real space* $X_{\mathbb{R}}$ is obtained by restricting the scalars to $\mathbb{R}$.

Examples 3.1.
1. The sets of all *bound vectors* (directed segments with a fixed origin) on a line, in a plane, or in the 3-space with the ordinary addition (by the *parallelogram law*) and the usual scalar multiplication are *real vector spaces*.
2. The sets of all *free vectors* (directed segments) on a line, in a plane, or in the 3-space (any two vectors of the same direction and length considered identical)

https://doi.org/10.1515/9783110614039-003

with the ordinary addition (by the *triangle* or *parallelogram law*) and the usual scalar multiplication are *real vector spaces*.

3. The set $\mathbb{R}$ is a *real vector space* and the set $\mathbb{C}$ is a *complex vector space*.

4. The *n-space* $\mathbb{F}^n$ of all ordered *n-tuples* ($n \in \mathbb{N}$) of numbers with the *componentwise linear operations*

$$(x_1,\ldots,x_n) + (y_1,\ldots,y_n) := (x_1 + y_1,\ldots,x_n + y_n),$$
$$\lambda(x_1,\ldots,x_n) := (\lambda x_1,\ldots,\lambda x_n),$$

is a vector space over $\mathbb{F}$.

5. The set $M_{m \times n}$ of all $m \times n$ ($m, n \in \mathbb{N}$) *matrices* with entries from $\mathbb{F}$ and the *entrywise linear operations* of matrix addition and scalar multiplication is a vector space over $\mathbb{F}$.

6. The set s of all $\mathbb{F}$-valued sequences with the *termwise linear operations* is a vector space over $\mathbb{F}$.

7. Due to *Minkowski's Inequality for Sequences* (Theorem 2.4), the set $l_p(\mathbb{F})$ ($1 \le p \le \infty$) with the *termwise linear operations* is a vector space over $\mathbb{F}$.

8. The set $F(T)$ of all $\mathbb{F}$-valued functions on a nonempty set T with the *pointwise linear operations*

$$(f + g)(t) := f(t) + g(t),\ t \in T,$$
$$(\lambda f)(t) := \lambda f(t),\ t \in T,$$

is a vector space over $\mathbb{F}$.

However, the set $F_+(T)$ of all *nonnegative* functions on a nonempty set T with the *pointwise linear operations* is *not* a real vector space.

9. The set $M(T)$ of all $\mathbb{F}$-valued functions *bounded* on a nonempty set T with the pointwise linear operations is a vector space over $\mathbb{F}$.

However, the set $U(T)$ of all $\mathbb{F}$-valued functions *unbounded* on an *infinite* set T with the pointwise linear operations is *not* a vector space over $\mathbb{F}$.

10. The set $R[a, b]$ ($-\infty < a < b < \infty$) of all $\mathbb{F}$-valued functions *Riemann integrable* on $[a, b]$ with the pointwise linear operations is a vector space over $\mathbb{F}$.

11. Let (X, ρ) be a *metric space*. The set $C(X)$ of all $\mathbb{F}$-valued functions *continuous* on X with the pointwise linear operations is a vector space over $\mathbb{F}$.

In particular, for $X := [a, b]$ ($-\infty < a < b < \infty$), $C[a, b]$ is a vector space over $\mathbb{F}$.

12. The set P of all *polynomials* with coefficients from $\mathbb{F}$ and pointwise linear operations is a vector space over $\mathbb{F}$.

13. The set P_n of all polynomials of degree *at most n* ($n \in \mathbb{Z}_+$) with coefficients from $\mathbb{F}$ and the pointwise linear operations is also a vector space over $\mathbb{F}$.

However, the set $\hat{P}_n$ of all polynomials of degree n ($n \in \mathbb{N}$) with coefficients from $\mathbb{F}$ and the pointwise linear operations is *not* a vector space over $\mathbb{F}$.

Exercise 3.1. Verify.

Theorem 3.1 (Properties of Vector Spaces). *In a vector space X over* $\mathbb{F}$,
(1) *the zero vector 0 is unique;*
(2) *for each vector $x \in X$, the opposite vector $-x$ is unique;*
(3) $\forall x \in X : -(-x) = x;$
(4) *for $\lambda \in \mathbb{F}$ and $x \in X$, $\lambda x = 0$ iff $\lambda = 0$ or $x = 0$;* (Zero Product Rule)
(5) $(-1)x = -x.$

Remark 3.2. Observe that the same notation 0 is used to designate both the *scalar zero* and the *zero vector*, such an economy of symbols being rather common.

Proof.
(1) Assume that a vector $0'$ is also an additive identity. Then, by the *vector space axioms*,

$$0' = 0' + 0 = 0.$$

(2) Assume that y and z are *additive inverses* of x, i. e., $x + y = x + z = 0$. Then, by the *vector space axioms*,

$$y = y + 0 = y + (x + z) = (y + x) + z = 0 + z = z.$$

(3) Immediately follows from (2).
(4) "*If*" part.
If $\lambda = 0$, for any $x \in X$, by the *vector space axioms*,

$$0x = (0 + 0)x = 0x + 0x$$
$$0x + (-0x) = [0x + 0x] + (-0x)$$
$$0 = 0x + [0x + (-0x)]$$
$$0 = 0x + 0$$
$$0 = 0x.$$

The case of $x = 0$ is considered similarly.

Exercise 3.2. Verify.

"*Only if*" part. Let us prove this part *by contradiction*, assuming that, for some $\lambda \neq 0$ and $x \neq 0$,

$$\lambda x = 0.$$

Multiplying through by $1/\lambda$, by the *vector space axioms* and what is proved in the "*if*" part, we arrive at

$$(1/\lambda)(\lambda x) = (1/\lambda)0$$
$$((1/\lambda)\lambda) x = 0$$

$$1x = 0$$
$$x = 0,$$

which is a *contradiction* proving the statement.

(5) By the *vector space axioms* and (4),

$$x + (-1)x = 1x + (-1)x = [1 + (-1)]x = 0x = 0.$$

Whence, by (2), $(-1)x = -x$. □

3.1.2 Homomorphisms and Isomorphisms

Important in the theory of vector spaces are the following notions of *homomorphism* and *isomorphism*.

Definition 3.2 (Homomorphism of Vector Spaces). Let X and Y be vector spaces over $\mathbb{F}$.

A *homomorphism of X to Y* is a mapping $T : X \to Y$ preserving linear operations:

$$T(\lambda x + \mu y) = \lambda Tx + \mu Ty, \quad \lambda, \mu \in \mathbb{F}, \ x, y \in X.$$

When $Y = X$, a homomorphism $T : X \to X$ is called an *endomorphism of X*.

Examples 3.2.
1. Multiplication by a number $\lambda \in \mathbb{F}$ in a vector space X is an *endomorphism* of X.
2. Multiplication by an $m \times n$ ($m, n \in \mathbb{N}$) matrix $[t_{ij}]$ with entries from $\mathbb{F}$ is a *homomorphism* of $\mathbb{F}^n$ to $\mathbb{F}^m$ and, provided $m = n$, is an *endomorphism* of $\mathbb{F}^n$.

Definition 3.3 (Isomorphism of Vector Spaces). Let X and Y be vector spaces over $\mathbb{F}$.

An *isomorphism of X to Y* is a *one-to-one homomorphism $T : X \to Y$*. It is said to *isomorphically embed X in Y*.

If an isomorphism $T : X \to Y$ is onto (i. e., *surjective*), it is called an *isomorphism between X and Y* and the spaces are called *isomorphic*.

An isomorphism between X and itself is called an *automorphism of X*.

Examples 3.3.
1. Multiplication by a *nonzero* number $\lambda \in \mathbb{F} \setminus \{0\}$ in a vector space X is an *automorphism* of X.
2. Multiplication by a *nonsingular* $n \times n$ ($n \in \mathbb{N}$) matrix $[t_{ij}]$ with entries from $\mathbb{F}$ is an *automorphism* of $\mathbb{F}^n$.
3. For $n \in \mathbb{Z}_+$, the mapping

$$\mathbb{F}^{n+1} \ni (a_0, a_1, \ldots, a_n) \mapsto \sum_{k=0}^{n} a_k t^k$$

is an *isomorphism* between $\mathbb{F}^{n+1}$ and P_n, the space of all polynomials of degree at most n with coefficients from $\mathbb{F}$.

Remarks 3.3.
-   Isomorphism is an *equivalence relation* on the set of all vector spaces.

    Exercise 3.3. Verify the former.

-   Isomorphic vector spaces are *linearly indistinguishable*, i. e., identical as vector spaces.

3.1.3 Subspaces

As well as for metric spaces (cf. Section 2.3), we can consider *subspaces* of vector spaces.

Definition 3.4 (Subspace of a Vector Space). A subset Y of a vector space X over $\mathbb{F}$ that is itself a vector space over $\mathbb{F}$ relative to the induced linear operations is called a *subspace* of X.

Remarks 3.4.
-   Thus, Y is a *subspace* of a vector space X over $\mathbb{F}$ *iff* Y is *closed under the linear operations*, i. e.,
    (a) $Y + Y \subseteq Y$ and
    (b) $\lambda Y \subseteq Y, \lambda \in \mathbb{F}$.
-   Each nontrivial vector space has at least two subspaces: the *trivial* one $\{0\}$ and the whole X.
-   Each subspace of a vector space always contains at least one element, 0, and hence, cannot be empty.
-   A subspace Y of X such that $Y \neq X$ is called a *proper subspace* of X.

Examples 3.4.
1. The set $\mathbb{R}$ is a *subspace* of the associated real space $\mathbb{C}_\mathbb{R}$ (see Remarks 3.1).
2. Due to the set-theoretic inclusions

    $$c_{00} \subset l_p \subset l_q \subset c_0 \subset c \subset l_\infty,$$

    where $1 \leq p < q < \infty$, (cf. Examples 2.3) and the closedness under the termwise linear operations, each of the above sequence spaces is a *proper subspace* of the next one.
3. Due to the set-theoretic inclusion

    $$M(T) \subseteq F(T)$$

and the closedness under the pointwise linear operations, $M(T)$ is a subspace of $F(T)$.

Exercise 3.4. When is it a *proper subspace*?

4. Due to the set-theoretic inclusions

$$P_n \subset P \subset C[a,b] \subset M[a,b] \subset F[a,b]$$

$(n \in \mathbb{Z}_+, -\infty < a < b < \infty)$, and the closedness under the pointwise linear operations, each of the above function spaces is a *proper subspace* of the next one.

5. The set

$$Y = \{(x,y) \mid xy \geq 0\}$$

is *not* a subspace of $\mathbb{R}^2$.

Exercise 3.5.
(a) Verify.
(b) Describe all subspaces in $\mathbb{R}$, $\mathbb{R}^2$ and $\mathbb{R}^3$.

Exercise 3.6. Let Y be a *subspace* in a vector space X over $\mathbb{F}$. Show that
(a) $Y + Y = Y$;
(b) $\lambda Y = Y, \lambda \in \mathbb{F} \setminus \{0\}$.

Proposition 3.1 (Sum and Intersection of Subspaces). *In a vector space X,*
(1) *the sum of a finite number of subspaces is a subspace;*
(2) *an arbitrary intersection of subspaces is a subspace.*

Exercise 3.7.
(a) Prove.
(b) Give an example showing that the *union* of subspaces need not be a subspace.

The following statement gives conditions *necessary and sufficient* for the union of two subspaces of a vector space to be a subspace.

Proposition 3.2 (Union of Subspaces). *In a vector space X, the union $Y \cup Z$ of subspaces Y and Z is a subspace iff $Y \subseteq Z$ or $Z \subseteq Y$.*

Exercise 3.8. Prove.

Hint. Prove the *"only if"* part *by contradiction* or *by contrapositive*.

3.1.4 Spans and Linear Combinations

By the *Sum and Intersection of Subspaces Proposition* (Proposition 3.1), the following notion is well defined.

Definition 3.5 (Linear Span). Let S be a subset of a vector space X, then

$$\text{span}(S) := \bigcap_{Y \text{ is a subspace of } X,\ S \subseteq Y} Y$$

is the *smallest subspace of X containing S* called the *linear span* (also *span* or *linear hull*) of S.

The *span* of S, also called the *subspace generated by S*

Exercise 3.9. Show that, in a vector space X,
(a) $\text{span}(\emptyset) = \{0\}$;
(b) $\text{span}(X) = X$;
(c) for subspaces $Y_1, \ldots, Y_n$ ($n \in \mathbb{N}$) of X,

$$\text{span}\left(\bigcup_{k=1}^{n} Y_k\right) = \sum_{k=1}^{n} Y_k := \left\{\sum_{k=1}^{n} y_k \,\middle|\, y_k \in Y_k,\ k = 1, \ldots, n\right\}.$$

Definition 3.6 (Linear Combination). Let X be a vector space over $\mathbb{F}$. A *linear combination* of vectors $x_1, \ldots, x_n \in X$ ($n \in \mathbb{N}$) with coefficients $\lambda_1, \ldots, \lambda_n \in \mathbb{F}$ is the sum

$$\sum_{k=1}^{n} \lambda_k x_k.$$

Remark 3.5. The linear combination with $\lambda_1 = \cdots = \lambda_n = 0$ is called *trivial*. Obviously, any *trivial* linear combination is the *zero vector*.

Exercise 3.10. Give an example showing that the *converse* is not true.

Proposition 3.3 (Span's Structure). *For a nonempty subset S of a vector space X over $\mathbb{F}$, $\text{span}(S)$ is the set of all linear combinations of its elements:*

$$\text{span}(S) = \left\{\sum_{k=1}^{n} \lambda_k x_k \,\middle|\, x_1, \ldots, x_n \in S, \lambda_1, \ldots, \lambda_n \in \mathbb{F},\ n \in \mathbb{N}\right\}.$$

Exercise 3.11. Prove.

Exercise 3.12.
(a) In c_0, describe $\text{span}(\{e_n := \{\delta_{nk}\}_{k=1}^{\infty}\}_{n\in\mathbb{N}})$, where δ_{nk} is the *Kronecker delta*.
(b) In $C[a, b]$ ($-\infty < a < b < \infty$), describe $\text{span}(\{t^n\}_{n\in\mathbb{Z}_+})$.

3.1.5 Linear Independence, Hamel Bases, Dimension

Fundamental to vector spaces is the concept of *linear (in)dependence*.

3.1.5.1 Linear Independence/Dependence

Definition 3.7 (Linearly Independent/Dependent Set). A nonempty subset S of a vector space X is called *linearly independent* if none of its vectors is spanned by the other vectors of S, i. e.,

$$\forall x \in S : x \notin \text{span}(S \setminus \{x\}),$$

and is said to be *linearly dependent* otherwise.

We also say that the vectors of S are *linearly independent/dependent*.

Remark 3.6. A linearly independent set S in a vector space X cannot contain the *zero vector* 0. Thus, the notion of linear independence is well-defined only in a nontrivial vector space.

Examples 3.5.
1. A singleton $\{x\}$ is a linearly dependent set in a vector space X iff $x = 0$.
2. A two-vector set $\{x, y\}$ is a linearly dependent in a nontrivial vector space X iff $x = \lambda y$ or $y = \lambda x$ with some $\lambda \in \mathbb{F}$.
3. The set $\{(1, 0), (1, 1)\}$ is linearly independent and the set $\{(1, 0), (2, 0)\}$ is linearly dependent in $\mathbb{F}^2$.

Exercise 3.13. Verify the prior remark and examples.

Proposition 3.4 (Characterization of Linear Independence). *A nonempty subset S of a nontrivial vector space X is linearly independent iff only the trivial linear combinations of its vectors equal 0, i. e.,*

$$\forall \{x_1, \ldots, x_n\} \subseteq S \ (n \in \mathbb{N}) : \sum_{k=1}^{n} \lambda_k x_k = 0 \iff \lambda_1 = \cdots = \lambda_n = 0.$$

Exercise 3.14.
(a) Prove.

   Hint. Reason *by contrapositive*.

(b) Show that any nonempty subset of a linearly independent set is linearly independent.

(c) Show that a set of a vector space containing a linearly dependent subset, in particular the *zero vector*, is linearly dependent.

Remark 3.7. The prior characterization is often used as a definition of linear independence, especially for finite sets of vectors.

3.1.5.2 Hamel Bases

Definition 3.8 (Basis of a Vector Space). A *basis* B (also *Hamel*[1] *basis* or *algebraic basis*) of a nontrivial vector space X is a *maximal linearly independent* subset of X, or equivalently, a linearly independent set of X spanning the entire X.

Remark 3.8. The *"maximality"* is understood relative to the set-theoretic inclusion $\subseteq$ (see Sections 1.1.1 and A.2).

Exercise 3.15. Prove the equivalence of the two definitions.

Examples 3.6.
1. The singleton $\{1\}$ is a basis for both $\mathbb{R}$ and $\mathbb{C}$, but not for the associated real space $\mathbb{C}_\mathbb{R}$ (see Remarks 3.1), whose bases coincide with those of $\mathbb{R}^2$.
2. The set of n ($n \in \mathbb{N}$) ordered n-tuples

$$\{e_1 := (1,0,\ldots,0), e_2 := (0,1,\ldots,0),\ldots,e_n := (0,0,\ldots,1)\}$$

 is a basis for the n-space $\mathbb{F}^n$ called the *standard basis*.
3. The sets $\{(1,0),(0,-1)\}$ and $\{(1,0),(1,1)\}$ are both bases for $\mathbb{F}^2$.
4. The set $E := \{e_n := \{\delta_{nk}\}_{k=1}^\infty\}_{n\in\mathbb{N}}$, where δ_{nk} is the *Kronecker delta*, is a basis for c_{00}, but not for c_0.
5. The set $\{1,t,\ldots,t^n\}$ is a basis of P_n ($n \in \mathbb{Z}_+$).
6. The set $\{1,t,\ldots,t^n,\ldots\}$ is a basis for P, but not for $C[a,b]$ ($-\infty < a < b < \infty$).

Exercise 3.16. Verify.

Theorem 3.2 (Basis Theorem). *Each linearly independent set S in a nontrivial vector space X can be extended to a basis B of X.*

Proof. Let S be an arbitrary linearly independent set in a nontrivial vector space X.

Let $\mathscr{L}$ be the collection of all linearly independent sets in X *partially ordered* by the set-theoretic inclusion $\subseteq$ (see Section A.2). For an arbitrary *chain* $\mathscr{C}$ in $(\mathscr{L},\subseteq)$, the set

$$L := \bigcup_{C\in\mathscr{C}} C \subseteq X,$$

is *linearly independent* and is an *upper bound* of $\mathscr{C}$ in $(\mathscr{L},\subseteq)$.

Exercise 3.17. Verify.

Hence, by *Zorn's Lemma (Precise Version)* (Theorem A.6), there is a *maximal element* B in $(\mathscr{L},\subseteq)$, i.e., a *maximal linearly independent* subset in X, such that $S \subseteq B$, which completes the proof. □

---

[1] Georg Karl Wilhelm Hamel (1877–1954).

Remark 3.9. A basis of a nontrivial vector space is never unique. One can always produce a new basis via, e. g., multiplying each basis vector by a nonzero scalar.

Theorem 3.3 (Representation Theorem). *A nonempty subset $B := \{x_i\}_{i \in I}$ of a nontrivial vector space X over $\mathbb{F}$ is a basis for X iff any element $x \in X$ can be uniquely represented as a sum*

$$\sum_{i \in I} c_i x_i,$$

in which only a finite number of the coefficients $c_i \in \mathbb{F}, i \in I$, are nonzero.

Exercise 3.18. Prove.

Remark 3.10. The prior representation is called the *representation* of x relative to the basis B and the numbers $c_i, i \in I$, are called the *coordinates* of x relative to B.

Corollary 3.1 (Representation of Nonzero Elements). *Each nonzero element x of a nontrivial vector space X with a basis B allows a unique representation as a linear combination of elements of B with nonzero coefficients.*

3.1.5.3 Dimension
Theorem 3.4 (Dimension Theorem). *All bases of a nontrivial vector space have equally many elements.*

Proof. Let A and B be two arbitrary bases in X.

We are to show that their *cardinalities* $|A|$ and $|B|$ are equal.

The case when A or B is a *finite set* is considered in *linear algebra* (see, e. g., [40]) by reducing it to the question of the existence of nontrivial solutions of a homogeneous linear system, which has more unknowns than equations.

Exercise 3.19. Fill in the details.

Suppose that both A or B are *infinite*. By the *Representation of Nonzero Elements Corollary* (Corollary 3.1), each $y \in B$ is a linear combination with nonzero coefficients of some elements $x_1, \ldots, x_n \in A$ ($n \in \mathbb{N}$), and at most n elements of B can be associated with the set $\{x_1, \ldots, x_n\}$ or its subset in such a way.

Exercise 3.20. Explain.

Since A is infinite, by the *Cardinality of the Collection of Finite Subsets* (Proposition 1.2), the cardinality of the collection of all its finite subsets is $|A|$. Hence, considering that $\aleph_0 \le |A|$, where $\aleph_0$ is the cardinality of $\mathbb{N}$ (see Examples 1.1), by the arithmetic of cardinals (see, e. g., [21, 26, 38]),

$$|B| \le \aleph_0 |A| = |A|$$

Similarly, $|A| \le |B|$ and thus, $|A| = |B|$, which completes the proof. $\square$

By the *Dimension Theorem*, the following fundamental notion is well-defined.

Definition 3.9 (Dimension of a Vector Space). The *dimension* of a nontrivial vector space X is the *common cardinality* of all bases of X.

Notation. $\dim X$.

The dimension of a trivial space is naturally defined to be 0.

We call a vector space *finite dimensional* if its dimension is a finite number and *infinite dimensional* otherwise.

If $\dim X = n$ with some $n \in \mathbb{Z}_+$, the vector space is called n-*dimensional*.

Examples 3.7.
1. $\dim X = 0$ *iff* $X = \{0\}$.
2. $\dim \mathbb{C} = \begin{cases} 1 & \text{over } \mathbb{C}, \\ 2 & \text{over } \mathbb{R}. \end{cases}$
3. $\dim F^n = n$ over $\mathbb{F}$ $(n \in \mathbb{N})$.
4. $\dim P_n = n + 1$ $(n \in \mathbb{Z}_+)$.
5. $\dim P = \dim c_{00} = \aleph_0$.
6. $\dim C[a, b] \geq \mathfrak{c}$ $(-\infty < a < b < \infty)$, where $\mathfrak{c}$ is the *cardinality of the continuum*, i. e., $\mathfrak{c} = |\mathbb{R}|$ (see Examples 1.1).

Exercise 3.21. Explain. For 6, give a corresponding example.

As the following theorem shows, all vector spaces of the same dimension are *linearly indistinguishable*.

Theorem 3.5 (Isomorphism Theorem). *Two nontrivial vector spaces X and Y are isomorphic iff* $\dim X = \dim Y$.

Proof. "Only if" part. Suppose that X and Y are isomorphic and let $T : X \rightarrow Y$ be an *isomorphism* between X and Y.

Then the set $B \subseteq X$ is a *basis* of X iff $T(B)$ is a *basis* in Y

Exercise 3.22. Verify.

This, considering that T is a *bijection*, implies that $\dim X = \dim Y$.

"*If*" part. Suppose $\dim X = \dim Y$.

Then, we can choose bases $B_X = \{x_i\}_{i \in I}$ of X and $B_Y = \{y_i\}_{i \in I}$ of Y sharing the *indexing set* I whose *cardinality* $|I| = \dim X = \dim Y$ and establish an *isomorphism* T between X and Y by matching the vectors with the *identical basis representations* relative to B_X and B_Y, respectively (see the *Representation Theorem* (Theorem 3.3)):

$$X \ni x = \sum_{i \in I} c_i x_i \mapsto Tx := \sum_{i \in I} c_i y_i \in Y,$$

in particular, $Tx_i = y_i$, $i \in I$.

Exercise 3.23. Verify that T is an *isomorphism* between X and Y. □

Corollary 3.2 (*n*-Dimensional Vector Spaces). *Each n-dimensional vector space (n* $\in$ $\mathbb{N}$) *over* $\mathbb{F}$ *is isomorphic to* $\mathbb{F}^n$.

3.1.6 New Spaces from Old

Here, we discuss several ways of generating new vector spaces.

3.1.6.1 Direct Products and Sums

Definition 3.10 (Direct Product). Let $\{X_i\}_{i \in I}$ be a nonempty collection of nonempty sets.

The *direct product* (also *Cartesian product*) of the sets X_i, $i \in I$, is the set of all choice functions on the indexing set I:

$$\prod_{i \in I} X_i := \left\{ x : I \to \bigcup_{i \in I} X_i \,\middle|\, x(i) = x_i \in X_i,\ i \in I \right\}.$$

For each $i \in I$, the value $x(i)$ of a function $x \in \prod_{i \in I} X_i$ at i is also called the *ith coordinate* of x and denoted x_i, the function itself is also called an *I-tuple* and denoted $(x_i)_{i \in I}$, the set X_i is called the *ith factor space*, and, for each $j \in I$, the mapping

$$\pi_j : \prod_{i \in I} X_i \to X_j$$

assigning to each *I*-tuple its *j*th coordinate:

$$\pi_j((x_i)_{i \in I}) := x_j, \quad (x_i)_{i \in I} \in \prod_{i \in I} X_i,$$

is called the *projection mapping* of $\prod_{i \in I} X_i$ on X_j, or simply, the *j*th *projection mapping*.

Each set X_i, $i \in I$, being a *vector space* over $\mathbb{F}$, the product space $\prod_{i \in I} X_i$ is also a vector space over $\mathbb{F}$ relative to the *coordinatewise linear operations*:

$$(x_i)_{i \in I} + (y_i)_{i \in I} := (x_i + y_i)_{i \in I}, \quad (x_i)_{i \in I}, (y_i)_{i \in I} \in X,$$
$$\lambda(x_i)_{i \in I} := (\lambda x_i)_{i \in I}, \quad \lambda \in \mathbb{F}, (x_i)_{i \in I} \in X.$$

Remarks 3.11.

- The *nonemptiness* of $\prod_{i \in I} X_i$ is guaranteed by the *Axiom of Choice*.
- In particular, if $X_i = X$, $i \in I$, $\prod_{i \in I} X_i$ is the set of all X-valued functions on I:

$$x : I \to X$$

and we use the notation X^I.

Examples 3.8.

1. For $I = \{1, \ldots, n\}$ $(n \in \mathbb{N})$,

$$\prod_{i=1}^{n} X_i = X_1 \times \cdots \times X_n := \{(x_1, \ldots, x_n) \mid x_i \in X_i, \ i = 1, \ldots, n\}$$

is the set of all *choice n-tuples*.

In particular, if $X_i = X$, $i = 1, \ldots, n$, we use the notation X^n, which includes the case of the *n*-space $\mathbb{F}^n$.

2. For $I = \mathbb{N}$,

$$\prod_{i=1}^{\infty} X_i = \{(x_i)_{i \in \mathbb{N}} \mid x_i \in X_i, \ i \in \mathbb{N}\}$$

is the set of all *choice sequences*.

In particular, if $X_i = X$, $i \in \mathbb{N}$, we obtain $X^{\mathbb{N}}$, the set of all X-valued sequences, which includes the case of the space $s := \mathbb{F}^{\mathbb{N}}$ of all $\mathbb{F}$-valued sequences.

3. For $I = [0, 1]$ and $X_i = \mathbb{F}$, $i \in [0, 1]$,

$$\prod_{i \in [0,1]} X_i = \mathbb{F}^{[0,1]} = F[0, 1],$$

where $F[0, 1]$ is the set of all $\mathbb{F}$-valued functions on $[0, 1]$.

Definition 3.11 (Direct Sum). Let $\{X_i\}_{i \in I}$ be a nonempty collection of vector spaces.

The *direct sum* of X_i, $i \in I$, is a *subspace* of the direct product $\prod_{i \in I} X_i$ defined as follows:

$$\bigoplus_{i \in I} X_i := \left\{ (x_i)_{i \in I} \in \prod_{i \in I} X_i \ \middle| \ x_i = 0 \text{ for all but a finite number of } i \in I \right\}$$

Remark 3.12. As is easily seen,

$$\bigoplus_{i \in I} X_i = \prod_{i \in I} X_i \ \Leftrightarrow \ I \text{ is a } finite \text{ set.}$$

Examples 3.9.

1. For $I = \mathbb{N}$ and $X_i = \mathbb{F}$, $i \in \mathbb{N}$,

$$\prod_{i=1}^{\infty} X_i = s \quad \text{and} \quad \bigoplus_{i=1}^{\infty} X_i = c_{00}.$$

2. For $I = [0, 1]$ and $X_i = \mathbb{F}$, $i \in [0, 1]$, $\prod_{i=1}^{\infty} X_i = F[0, 1]$, whereas $\bigoplus_{i=1}^{\infty} X_i$ is the subset of all $\mathbb{F}$-valued functions on $[0, 1]$ equal to zero for all but a finite number of values $t \in [0, 1]$.

3.1.6.2 Quotient Spaces

Let Y be a subspace of a vector space X.

The binary relation on X defined as follows:

$$x \sim y \iff y - x \in Y$$

is an *equivalence relation* called the *equivalence modulo Y*.

Exercise 3.24.

(a) Verify.

(b) Show that the equivalence class modulo Y represented by an element $x \in X$ is of the form

$$[x] := x + Y,$$

i. e., the equivalence classes modulo Y are the *translations* of Y.

Definition 3.12 (Quotient Space). For a subspace Y of a vector space X over $\mathbb{F}$, the set X/Y of all equivalence classes modulo Y, called the *cosets modulo Y*, is a vector space over $\mathbb{F}$ relative to the linear operations

$$[x] + [y] := [x + y] = (x + y) + Y \quad \text{and} \quad \lambda[x] := [\lambda x] = \lambda x + Y, \ x, y \in X, \ \lambda \in \mathbb{F},$$

called the *quotient space of X modulo Y*.

Exercise 3.25.

(a) Verify that X/Y is a vector space.

(b) What is the *zero element* of X/Y?

(c) Describe $X/\{0\}$ and X/X.

(d) Describe $\mathbb{R}^2/\{(x, y) \in \mathbb{R}^2 \mid y = mx$ with some $m \in \mathbb{R}\}$.

Definition 3.13 (Canonical Homomorphism). The homomorphism

$$X \ni x \mapsto Tx := [x] = x + Y \in X/Y$$

of X onto X/Y is called the *canonical homomorphism* of X onto X/Y.

Exercise 3.26.

(a) Verify that $T : X \to X/Y$ so defined is a *homomorphism* of X onto X/Y.

(b) Determine the *kernel* (the *null space*) of the *canonical homomorphism* T:

$$\ker T := \{x \in X \mid Tx = 0\}.$$

(c) Describe the *canonical homomorphism* of $\mathbb{R}^2$ onto

$$\mathbb{R}^2/\{(x, y) \in \mathbb{R}^2 \mid y = mx \text{ with some } m \in \mathbb{R}\}.$$

3.1.7 Disjoint and Complementary Subspaces, Direct Sum Decompositions, Deficiency and Codimension

Definition 3.14 (Disjoint and Complementary Subspaces). Two subspaces Y and Z of a vector space X are called *disjoint* if

$$Y \cap Z = \{0\}.$$

Two disjoint subspaces Y and Z of a vector space X are called *complementary* if every $x \in X$ allows a *unique decomposition*

$$x = y + z$$

with some $y \in Y$ and $z \in Z$.

Remarks 3.13.
- For complementary subspaces, nontrivial is only the *existence* of the decomposition, the *uniqueness* immediately following from the disjointness.
- The definitions naturally extend to any finite number of subspaces.

Examples 3.10.
1. The subspaces

$$\left\{(x,y,z) \in \mathbb{R}^3 \mid z = 0\right\} \quad \text{and} \quad \left\{(x,y,z) \in \mathbb{R}^3 \mid x = y = 0\right\}$$

   are *complementary* in $\mathbb{R}^3$.
2. The subspaces

$$\left\{(x,y,z) \in \mathbb{R}^3 \mid y = z = 0\right\} \quad \text{and} \quad \left\{(x,y,z) \in \mathbb{R}^3 \mid x = z = 0\right\}$$

   are *disjoint*, but *not complementary* in $\mathbb{R}^3$.
3. The subspaces

$$\left\{(x,y,z) \in \mathbb{R}^3 \mid z = 0\right\} \quad \text{and} \quad \left\{(x,y,z) \in \mathbb{R}^3 \mid y = 0\right\}$$

   are *not disjoint* in $\mathbb{R}^3$.
4. The subspaces c_0 and $\text{span}(\{(1,1,1,\dots)\}) = \{(\lambda,\lambda,\lambda,\dots) \mid \lambda \in \mathbb{F}\}$ are *complementary* in c.

Theorem 3.6 (Existence of a Complementary Subspace). *Every subspace of a vector space X has a complementary subspace.*

Proof. Let Y be a subspace of X. Consider the collection $\mathscr{S}$ of all subspaces in X disjoint from Y partially ordered by the set-theoretic inclusion $\subseteq$.

Exercise 3.27. Why is $\mathscr{S}$ *nonempty?*

Let $\mathscr{C}$ be an arbitrary *chain* in $(\mathscr{S}, \subseteq)$. Then

$$L := \bigcup_{C \in \mathscr{C}} C,$$

is also a *subspace* of X *disjoint* from Y.

Exercise 3.28. Verify.

Clearly, L is an *upper bound* of $\mathscr{C}$ in $(\mathscr{D}, \subseteq)$.

By *Zorn's Lemma* (Theorem A.5), there is a *maximal element* Z in $(\mathscr{S}, \subseteq)$, i. e., a *maximal* disjoint from Y subspace of X.

Let us show that Z is *complementary* to Y by *contradiction* assuming that

$$\exists x \in X \text{ such that } x \notin Y + Z.$$

Then

$$Z' := \text{span}\,(Z \cup \{x\}) = Z + \text{span}\,(\{x\})$$

is also a *subspace* of X disjoint from Y.

Exercise 3.29. Verify.

Since $Z \subset Z'$, this *contradicts* the *maximality* of Z in $(\mathscr{S}, \subseteq)$ showing that Z is *complementary* to Y. □

Remark 3.14. The complementary subspace need not be unique.

Exercise 3.30. Give a corresponding example.

However, as we see below, all complementary subspaces of a given subspace are *isomorphic*.

Proposition 3.5 (Direct Sum Decompositions). *Let Y be a subspace of a vector space X. Then, for any subspace Z complementary to Y, X is isomorphic to the direct sum $Y \oplus Z$.*

Exercise 3.31. Prove.

Remark 3.15. We say that X is the direct sum of Y and Z and write

$$X = Y \oplus Z,$$

calling the latter a *direct sum decomposition* of X and immediately obtain the following corollary.

Corollary 3.3 (Sum of Dimensions). *Let Y be a subspace of a vector space X. Then, for any subspace Z complementary to Y,*

$$\dim X = \dim Y + \dim Z.$$

Examples 3.11.
1. $\mathbb{R}^2 = \{(x,y) \in \mathbb{R}^2 \mid y = 0\} \oplus \{(x,y) \in \mathbb{R}^2 \mid x = 0\}$.
2. $\mathbb{R}^3 = \{(x,y,z) \in \mathbb{R}^3 \mid z = 0\} \oplus \{(x,y,z) \in \mathbb{R}^3 \mid x = y = 0\}$.
3. $\mathbb{R}^3 = \{(x,y,z) \in \mathbb{R}^3 \mid y = z = 0\} \oplus \{(x,y,z) \in \mathbb{R}^3 \mid x = z = 0\} \oplus \{(x,y,z) \in \mathbb{R}^3 \mid x = y = 0\}$.
4. $c = c_0 \oplus \mathrm{span}(\{(1,1,1,\dots)\})$.

Definition 3.15 (Deficiency). The *deficiency* of a subspace Y in a vector space X is the *dimension* $\dim(X/Y)$ of the *quotient space* X/Y.

Exercise 3.32. What is the *deficiency*
(a) of $\{(x,y) \in \mathbb{R}^2 \mid y = 0\}$ in $\mathbb{R}^2$;
(b) of $\{(x,y,z) \in \mathbb{R}^3 \mid x = y = 0\}$ in $\mathbb{R}^3$;
(c) of P_n $(n \in \mathbb{Z}_+)$ in P?

Definition 3.16 (Hyperplane). A subspace Y in a vector space X of deficiency 1 is called a *hyperplane*.

Examples 3.12.
1. If $B = \{x_i\}_{i \in I}$ is a basis of X, then, for any $j \in I$, $Y := \mathrm{span}\{B \setminus \{x_j\}\}$ is a hyperplane in X.
2. c_0 is a hyperplane in c.

Exercise 3.33.
(a) Verify.
(b) Describe the quotient space c/c_0.

Proposition 3.6 (Dimension of Complementary Subspaces). *Let Y be a subspace of a vector space X. Then, each complementary subspace Z of Y is isomorphic to the quotient space X/Y, and hence,*

$$\dim Z = \dim(X/Y).$$

Exercise 3.34. Prove.

Remark 3.16. Thus, all complementary subspaces of a subspace Y of a vector space X are *isomorphic* and, by the *Isomorphism Theorem* (Theorem 3.5), we obtain

Corollary 3.4 (Codimension of a Subspace). *All complementary subspaces of a subspace Y of a vector space X have the same dimension, which is called the* codimension *of Y and coincides with its deficiency:*

$$\mathrm{codim}\, Y = \dim(X/Y)$$

and we have:

$$\dim X = \dim Y + \operatorname{codim} Y = \dim Y + \dim(X/Y).$$

Remark 3.17. Thus, a hyperplane can be equivalently defined as a subspace of codimension 1, and hence, can be described as in Examples 3.12.

3.2 Normed Vector and Banach Spaces

Here, we introduce the notion of *norm* on a vector space, combining its linear structure with topology, and study the surprising and beautiful results of the profound interplay between the two.

3.2.1 Definitions and Examples

Recall that $\mathbb{F}$ stands for either $\mathbb{R}$ or $\mathbb{C}$.

Definition 3.17 (Normed Vector Space). A *normed vector space* over $\mathbb{F}$ is a *vector space* X over $\mathbb{F}$ equipped with a *norm*, i. e., a mapping

$$\|\cdot\| : X \to \mathbb{R}$$

subject to the following *norm axioms:*

1. $\|x\| \geq 0$, $x \in X$.   *Nonnegativity*
2. $\|x\| = 0$ iff $x = 0$.   *Separation*
3. $\|\lambda x\| = |\lambda|\|x\|$, $\lambda \in \mathbb{F}$, $x \in X$.   *Absolute Homogeneity/Scalability*
4. $\|x + y\| \leq \|x\| + \|y\|$, $x, y \in X$.   *Subadditivity/Triangle Inequality*

The space is said to be *real* if $\mathbb{F} = \mathbb{R}$ and *complex* if $\mathbb{F} = \mathbb{C}$.

Notation. $(X, \|\cdot\|)$.

Remarks 3.18.

- A function $\|\cdot\| : X \to \mathbb{R}$ satisfying the norm axioms of *absolute scalability* and *subadditivity* only, which immediately implies the following weaker version of the *separation axiom:*

  2w. $\|x\| = 0$ if $x = 0$,

  and hence, also the axiom of *nonnegativity*, is called a *seminorm* on X and $(X, \|\cdot\|)$ is called a *seminormed vector space* (see the examples to follow).
- A norm $\|\cdot\|$ on a vector space X generates a metric on X, called the *norm metric*, as follows:

$$X \times X \ni (x, y) \mapsto \rho(x, y) := \|x - y\|, \tag{3.1}$$

which turns X into a *metric space*, endows it with the *norm metric topology*, and brings to life all the relevant concepts: *openness, closedness, denseness, category, boundedness, total boundedness*, and *compactness* for sets, various forms of *continuity* for functions, *fundamentality* and *convergence* for sequences, *and separability* and *completeness* for spaces.

If $\| \cdot \|$ is a *seminorm*, (3.1) defines a *semimetric* (or *pseudometric*) (see Remark 2.1).

‒ Due to the *axioms* of *subadditivity* and *absolute scalability*, the linear operations of *vector addition*

$$X \times X \ni (x, y) \mapsto x + y \in X$$

and *scalar multiplication*

$$\mathbb{F} \times X \ni (\lambda, x) \mapsto \lambda x \in X$$

are *jointly continuous*.

‒ The following immediate implication of the *subadditivity*

$$|\|x\| - \|y\|| \le \|x - y\|, \ x, y \in X, \tag{3.2}$$

showing that the norm is *Lipschitz continuous* on X holds. Observe that the inequality applies to *seminorms*, as well.

Exercise 3.35. Verify.

Definition 3.18 (Subspace of a Normed Vector Space). If $(X, \| \cdot \|)$ is a *normed vector space* and $Y \subseteq X$ is a *linear subspace* of X, then the restriction of the norm $\| \cdot \|$ to Y is a norm on Y and the normed vector space $(Y, \| \cdot \|)$ is called a *subspace* of $(X, \| \cdot \|)$.

Exercise. Prove that, if Y is a *subspace* of a normed vector space $(X, \| \cdot \|)$, then its closure $\overline{Y}$ is also a *subspace* of $(X, \| \cdot \|)$.

Definition 3.19 (Banach Space). A *Banach space* is a normed vector space $(X, \|\cdot\|)$ *complete* relative to the norm metric, i. e., such that every *Cauchy sequence* (or *fundamental sequence*)

$$\{x_n\}_{n=1}^{\infty} \subseteq X : \rho(x_n, x_m) = \|x_n - x_m\| \to 0, \ n, m \to \infty,$$

converges to an element $x \in X$:

$$\lim_{n \to \infty} x_n = x \Leftrightarrow \rho(x_n, x) = \|x_n - x\| \to 0, \ n \to \infty.$$

Examples 3.13.
1. The vector space $\mathbb{F}$ (i. e., $\mathbb{R}$ or $\mathbb{C}$) is a *Banach space* relative to the *absolute-value norm*:

$$\mathbb{F} \ni x \mapsto |x|,$$

which generates the usual metric (see Examples 2.14).

2. By the *Completeness of the n-Space* (Theorem 2.22), the vector space $l_p^{(n)}$ ($n \in \mathbb{N}$, $1 \le p \le \infty$) is a *Banach space* relative to the *p-norm*

$$\mathbb{F}^n \ni x = (x_1, \dots, x_n) \mapsto \|x\|_p := \begin{cases} \left[\sum_{k=1}^n |x_k|^p\right]^{1/p} & \text{if } 1 \le p < \infty, \\ \max_{1 \le k \le n} |x_k| & \text{if } p = \infty, \end{cases}$$

which generates the *p-metric*.

The *nonnegativity*, *separation*, and *absolute scalability axioms* are trivially verified. The *subadditivity axiom* is satisfied based on *Minkowski's Inequality for n-Tuples* (Theorem 2.3).

3. By the *completeness of l_p* ($1 \le p \le \infty$) (Theorem 2.23, Corollary 2.1), the vector space l_p ($1 \le p \le \infty$) is a *Banach space* relative to the *p-norm*

$$l_p \ni x = \{x_k\}_{k=1}^\infty \mapsto \|x\|_p := \begin{cases} \left[\sum_{k=1}^\infty |x_k|^p\right]^{1/p} & \text{if } 1 \le p < \infty, \\ \sup_{k \ge 1} |x_k| & \text{if } p = \infty, \end{cases}$$

which generates the *p-metric*.

Verifying the *nonnegativity*, *separation*, and *absolute scalability axioms* is trivial. The *subadditivity axiom* follows from *Minkowski's Inequality for Sequences* (Theorem 2.4).

4. The normed vector spaces $(c_0, \|\cdot\|_\infty)$ and $(c, \|\cdot\|_\infty)$ are *Banach spaces*, being *closed proper subspaces* of l_∞ (see Exercises 2.56).

Remark 3.19. By the *Nowhere Denseness of Closed Proper Subspace Proposition* (Proposition 3.16) (see Section 3.6, Problem 9), the subspace c_0 is *nowhere dense* in $(c, \|\cdot\|_\infty)$ and, in its turn, the subspace c is nowhere dense in l_∞. Both claims can also be verified directly.

Exercise 3.36. Verify directly.

5. The normed vector space $(c_{00}, \|\cdot\|_\infty)$ is *incomplete* since it is a *subspace* of the Banach space $(c_0, \|\cdot\|_\infty)$ that is *not closed* (see Exercises 2.56).

6. Let T be a nonempty set. By the *completeness of* $(M(T), \rho_\infty)$ (Theorem 2.24), the vector space $M(T)$ (see Examples 3.1) is a *Banach space* relative to the *supremum norm*

$$M(T) \ni f \mapsto \|f\|_\infty := \sup_{t \in T} |f(t)|,$$

which generates the *supremum metric*, the norm axioms being readily verified.

Exercise 3.37. Verify.

7. Let (X, ρ) be a *compact metric space*. By the *completeness of $C(X, Y)$* (Theorem 2.46), the vector space $C(X)$ (see Examples 3.1) is a *Banach space* relative

to the *maximum norm*:

$$C(X) \ni f \mapsto \|f\|_\infty := \max_{x \in X} |f(x)|,$$

which generates the *maximum metric*.

In particular, this includes the case of $(C[a,b], \|\cdot\|_\infty)$ $(-\infty < a < b < \infty)$.

8. The vector space P of *all polynomials* is an *incomplete* normed vector space relative to the *maximum norm*

$$\|f\|_\infty := \max_{a \le t \le b} |f(t)|$$

$(-\infty < a < b < \infty)$ since it is a *subspace* of the Banach space $(C[a,b], \|\cdot\|_\infty)$ that is *not closed* (see Exercises 2.56).

9. For each $n \in \mathbb{Z}_+$, the *finite-dimensional subspace* P_n of all polynomial of degree at most n is a *Banach space* relative to the *maximum norm*, which, as follows from the *Completeness of Finite-Dimensional Spaces Theorem* (Theorem 3.12) (see Section 3.3), is true in a more general context.

10. By the *incompleteness of* $(C[a,b], \rho_p)$ $(1 \le p < \infty)$ (Proposition 2.3), the vector space $C[a,b]$ $(-\infty < a < b < \infty)$ is an *incomplete* normed vector space relative to the *p-norm*

$$C[a,b] \ni f \mapsto \|f\|_p := \left[\int_a^b |f(t)|^p \, dt \right]^{1/p} \quad (1 \le p < \infty).$$

11. The vector space $R[a,b]$ $(-\infty < a < b < \infty)$ of all $\mathbb{F}$-valued functions *Riemann integrable* on $[a,b]$ is a *seminormed vector space* relative to the *integral seminorm*

$$R[a,b] \ni f \mapsto \|f\|_1 := \int_a^b |f(t)| \, dt.$$

12. The set $BV[a,b]$ of all $\mathbb{F}$-valued functions of *bounded variation* on $[a,b]$ $(-\infty < a < b < \infty)$ (see Examples 2.4), with pointwise linear operations is *seminormed vector space* relative to the *total-variation seminorm*

$$BV[a,b] \ni f \mapsto \|f\| := V_a^b(f)$$

(see Exercise 2.14) and is a Banach space relative to the *total-variation norm*

$$BV[a,b] \ni f \mapsto \|f\| := |f(a)| + V_a^b(f)$$

(see Proposition 3.14, Section 3.6, Problem 7).

3.2.2 Series and Completeness Characterization

Unlike in a metric space, which is void of addition, in a normed vector space, one can consider convergence not only for sequences, but also for series and even characterize the completeness of the space in those terms.

Definition 3.20 (Convergence of Series in Normed Vector Space). Let $(X, \| \cdot \|)$ be a normed vector space and $\{x_n\}_{n=1}^{\infty}$ be a sequence of its elements.

The *series* $\sum_{k=1}^{\infty} x_k$ is said to *converge* (or to be *convergent*) in $(X, \|\cdot\|)$ if the sequence of its partial sums $\{s_n := \sum_{k=1}^{n} x_k\}_{n=1}^{\infty}$ converges to an element $x \in X$, i. e.,

$$\lim_{n \to \infty} \sum_{k=1}^{n} x_k = x.$$

We call x the *sum* of the series and write

$$\sum_{k=1}^{\infty} x_k = x.$$

The *series* $\sum_{k=1}^{\infty} x_k$ is said to *absolutely converge* (or to be *absolutely convergent*) in $(X, \| \cdot \|)$ if the numeric series $\sum_{k=1}^{\infty} \|x_k\|$ converges.

A series that does not converge in $(X, \| \cdot \|)$ is called *divergent*.

The following three statements are the generalization of their well-known counterparts from the classical analysis.

Proposition 3.7 (Necessary Condition of Convergence for Series). *If a series $\sum_{k=1}^{\infty} x_k$ converges in a normed vector space $(X, \| \cdot \|)$, then*

$$x_n \to 0, \; n \to \infty, \; in \; (X, \| \cdot \|).$$

Exercise 3.38. Prove.

The equivalent *contrapositive* statement is as follows.

Proposition 3.8 (Divergence Test for Series). *If, for a series $\sum_{k=1}^{\infty} x_k$ in a normed vector space $(X, \| \cdot \|)$*

$$x_n \not\to 0, \; n \to \infty,$$

it diverges.

Theorem 3.7 (Cauchy's Convergence Test for Series). *For a series $\sum_{k=1}^{\infty} x_k$ in a normed vector space $(X, \|\cdot\|)$ to converge, it is necessary and, provided $(X, \|\cdot\|)$ is a Banach space, sufficient that*

$$\forall \varepsilon > 0 \; \exists N \in \mathbb{N} \; \forall n \geq N, \; \forall p \in \mathbb{N} : \left\| \sum_{k=n+1}^{n+p} x_k \right\| < \varepsilon.$$

Exercise 3.39. Prove.

It appears that one can characterize the completeness of the space in terms of series' convergence.

Theorem 3.8 (Series Characterization of a Banach Space). *A normed vector space $(X, \| \cdot \|)$ is a Banach space iff every absolutely convergent series of its elements converges.*

Proof. "*Only if*" part. This part follows directly from the *Cauchy's Convergence Test for Series* (Theorem 3.7) by *subadditivity* of norm.

Exercise 3.40. Fill in the details.

"*If*" part. Let $\{x_n\}_{n=1}^{\infty}$ be an arbitrary *fundamental* sequence in $(X, \| \cdot \|)$. Then it contains a subsequence $\{x_{n(k)}\}_{k=1}^{\infty}$ such that

$$\|x_{n(k+1)} - x_{n(k)}\| \le \frac{1}{2^k}, \quad k \in \mathbb{N}.$$

Exercise 3.41. Explain.

This, by the *Comparison Test*, implies that the telescoping series

$$x_{n(1)} + \left[x_{n(2)} - x_{n(1)}\right] + \left[x_{n(3)} - x_{n(2)}\right] + \cdots$$

absolutely converges and hence, by the premise, *converges* in $(X, \| \cdot \|)$, i. e.,

$$\exists x \in X : s_m := x_{n(1)} + \left[x_{n(2)} - x_{n(1)}\right] + \left[x_{n(3)} - x_{n(2)}\right] + \cdots + \left[x_{n(m)} - x_{n(m-1)}\right]$$
$$= x_{n(m)} \to x, \; n \to \infty, \text{ in } (X, \| \cdot \|).$$

Therefore, the subsequence $\{x_{n(k)}\}_{k=1}^{\infty}$ converges to x in $(X, \| \cdot \|)$, which, by the *Fundamental Sequence with Convergent Subsequence Proposition* (Proposition 2.22) (see Section 2.19, Problem 24), implies that the fundamental sequence $\{x_n\}_{n=1}^{\infty}$ itself converges to x in $(X, \| \cdot \|)$, completing the proof. $\square$

3.2.3 Comparing Norms, Equivalent Norms

Norms on a vector space can be naturally compared by their strength, which reflects the *strength* of the corresponding norm metric topology (see, e. g., [38, 41]).

Definition 3.21 (Comparing Norms). Let $\| \cdot \|_1$ and $\| \cdot \|_2$ be norms on a vector space X. The norm $\| \cdot \|_1$ is said to be *stronger* than $\| \cdot \|_2$, or $\| \cdot \|_2$ is said to be *weaker* than $\| \cdot \|_1$, if

$$\exists C > 0 : \|x\|_2 \le C\|x\|_1, \; x \in X.$$

Definition 3.22 (Equivalent Norms). Two norm $\| \cdot \|_1$ and $\| \cdot \|_2$ on a vector space X are called *equivalent* if

$$\exists\, c, C > 0 : \quad c\|x\|_1 \le \|x\|_2 \le C\|x\|_1, \quad x \in X,$$

i. e., each norm is both stronger and weaker than the other.

Remarks 3.20.

– Fundamentality/convergence of a sequence relative to the norm metric generated by a stronger norm is preserved relative the norm metric generated by a weaker one, but not vice versa.

– The equivalence of norms on a vector space X is an *equivalence relation* on the set of all norms on X.

– Equivalent norms generate equivalent norm metrics in the sense of (2.5) (see Section 2.12.1), which define the same topology on X.

– Two norms $\|\cdot\|_1$ and $\|\cdot\|_2$ on a vector space X are equivalent *iff* the identity mapping $I : X \to X$

$$(X, \| \cdot \|_1) \ni x \mapsto Ix := x \in (X, \| \cdot \|_2)$$

is a *bi-Lipschitzian isomorphism*, i. e., both the mapping I and its inverse

$$(X, \| \cdot \|_2) \ni x \mapsto I^{-1}x := x \in (X, \| \cdot \|_1)$$

are *Lipschitz continuous*.

Exercise 3.42.

(a) Verify.

(b) Show that all p-norms ($1 \le p \le \infty$) on $\mathbb{F}^n$ ($n \in \mathbb{N}$) are equivalent (see Exercise 2.43).

(c) Show that, on $C[a, b]$, the *maximum norm*

$$C[a, b] \ni f \mapsto \|f\|_\infty := \max_{a \le t \le b} |f(t)|$$

is *stronger* than the *integral norm*

$$C[a, b] \ni f \mapsto \|f\|_1 := \int_a^b |f(t)|\, dt,$$

but the two norms are *not equivalent*.

3.2.4 Isometric Isomorphisms

Combining the notions of isometry and isomorphism, one can define *isometric isomorphism*.

Definition 3.23 (Isometric Isomorphism of Normed Vector Spaces). Let $(X, \|\cdot\|_X)$ and $(Y, \|\cdot\|_Y)$ be normed vector spaces. A mapping $T : X \to Y$ that is *simultaneously* an *isometry* and an *isomorphism* is called an *isometric isomorphism from X to Y*. It is said to *isometrically embed X in Y*.

If the isometric isomorphism T is *onto* (i. e., *surjective*), it is called an *isometric isomorphism between X and Y* are called *isometrically isomorphic* or *linearly isometric*.

Examples 3.14.
1. For $n \in \mathbb{N}$, $1 \leq p \leq \infty$, the mapping

$$l_p^{(n)} \ni x = (x_1, \ldots, x_n) \mapsto Tx := \{x_1, \ldots, x_n, 0, 0, \ldots\} \in l_p$$

is an *isometric embedding* of $l_p^{(n)}$ in l_p.
2. Let X be an *n-dimensional* vector space ($n \in \mathbb{N}$) with a *basis* $B := \{x_1, \ldots, x_n\}$ and $1 \leq p \leq \infty$. The mapping

$$X \ni x = \sum_{k=1}^{n} c_k x_k \mapsto Tx := (c_1, \ldots, c_n) \in l_p^{(n)}$$

is an *isomorphism* between the spaces X and $l_p^{(n)}$, which also becomes an *isometry* when the former is equipped with the norm

$$X \ni x = \sum_{k=1}^{n} c_k x_k \mapsto \|x\|_X := \|Tx\|_p.$$

This follows from a more general construct (see Section 3.6, Problem 11).

Remarks 3.21.
– *Isometric isomorphism* is an *equivalence relation* on the set of all normed vector spaces.
– Isometrically isomorphic metric spaces are both *metrically* and *linearly indistinguishable*, in particular, they have the *same dimension* and are *separable* or *complete* only simultaneously.

3.2.5 Completion

A completion procedure similar to that for metric spaces is in place for normed vector spaces.

Theorem 3.9 (Completion Theorem for Normed Vector Spaces). *An arbitrary normed vector space $(X, \|\cdot\|_X)$ over $\mathbb{F}$ can be isometrically embedded, as a dense subspace, in a Banach space $(\tilde{X}, \|\cdot\|_{\tilde{X}})$ over $\mathbb{F}$ called a completion of $(X, \|\cdot\|_X)$.*
Any two completions of $(X, \|\cdot\|_X)$ are isometrically isomorphic.

Proof. The statement largely follows directly from the *Completion Theorem for Metric Spaces* (Theorem 2.28), considering that the constructed completion space $\tilde{X}$ of the equivalence classes of the *asymptotically equivalent* fundamental sequences of elements of X is a Banach space relative to the linear operations defined as follows:

$$\tilde{X} \ni [x], [y] \mapsto [x] + [y] := [x + y] \quad \text{and} \quad \lambda[x] := [\lambda x], \ x, y \in X, \ \lambda \in \mathbb{F},$$

where $x := \{x_n\}_{n=1}^{\infty}$ and $y := \{y_n\}_{n=1}^{\infty}$ are *arbitrary* representatives of the classes $[x]$ and $[y]$, respectively, and the norm

$$\tilde{X} \ni [x] \mapsto \|[x]\|_{\tilde{X}} := \lim_{n \to \infty} \|x_n\|_X,$$

which generates precisely the *completion metric* $\tilde{\rho}$ on $\tilde{X}$ defined by (2.9).

Exercise 3.43. Verify that the linear operations and the mapping $\| \cdot \|_{\tilde{X}}$ are *well defined* and that the latter is a *norm on* $\tilde{X}$.

Hint. The existence of the limit and its independence of the choice of the representative sequence follow from inequality (3.2) applied to norm $\| \cdot \|_X$.

The fact that any two completions of $(X, \| \cdot \|_X)$ are *isometric* is proved in *Completion Theorem for Metric Spaces* (Theorem 2.28).

Exercise 3.44. Prove that any two completions of $(X, \| \cdot \|_X)$ are also *isomorphic*, and hence, they are *isometrically isomorphic*. □

Remark 3.22. It follows immediately that, if $(X_0, \| \cdot \|)$ is a *dense subspace* of a Banach space $(X, \| \cdot \|)$, the latter is a *completion* of the *former*.

Examples 3.15.
1. $(c_0, \| \cdot \|_\infty)$ is a *completion* of $(c_{00}, \| \cdot \|_\infty)$ (cf. Examples 2.15).
2. l_p $(1 \le p < \infty)$ is a *completion* of $(c_{00}, \| \cdot \|_p)$ (cf. Examples 2.15).
3. $(C[a, b], \| \cdot \|_\infty)$ $(-\infty < a < b < \infty)$ is a *completion* of $(P, \| \cdot \|_\infty)$ (cf. Examples 2.15).

3.2.6 Topological and Schauder Bases

In a Banach space setting, the notion of *basis* acquires two other meanings discussed below.

Definition 3.24 (Topological Basis of a Normed Vector Space). A *topological basis* of a *normed vector space* $(X, \|\cdot\|)$ is a *linearly independent* subset B of X, whose *span* is *dense* in X:

$$\overline{\text{span}(B)} = X.$$

Remarks 3.23.
- We require substantially less from a *topological basis* than from its *algebraic basis*:

$$\overline{\text{span}(B)} = X \text{ as opposed to } \text{span}(B) = X.$$

- Clearly, every *algebraic basis* of a normed vector space is also its *topological basis*, and hence, the *existence* of a topological basis in any normed vector space immediately follows from the *Basis Theorem* (Theorem 3.2), which also implies that a topological basis of a normed vector space is never unique (Remark 3.9).
- As the following examples show, a topological basis of a normed vector space need not be its algebraic basis.

Examples 3.16.
1. A subset B is a *topological basis* of a finite-dimensional normed vector space $(X, \| \cdot \|)$ *iff* it is an *algebraic basis* of X, which follows directly from Theorem 3.13 (see Section 3.3.2) stating that each *finite-dimensional subspace* of a normed vector space is *closed*.
2. The set $E := \{e_n := \{\delta_{nk}\}_{k=1}^{\infty}\}_{n \in \mathbb{N}}$, where δ_{nk} is the *Kronecker delta*, is
   - an *algebraic basis* for $(c_{00}, \| \cdot \|_{\infty})$;
   - a *topological basis*, but not an *algebraic basis* for $(c_0, \| \cdot \|_{\infty})$ and l_p $(1 \leq p < \infty)$, and
   - not even a *topological basis* for $(c, \| \cdot \|_{\infty})$ and l_{∞}.
3. The set $\{t^n\}_{n \in \mathbb{Z}_+}$ is
   - an *algebraic basis* for P;
   - a *topological basis*, but not an *algebraic basis* for $(C[a, b], \| \cdot \|_{\infty})$ $(-\infty < a < b < \infty)$, and
   - not even a *topological basis* for $(M[a, b], \| \cdot \|_{\infty})$.

Exercise 3.45.
(a) Verify.

Hint. Show that, in $(c_0, \| \cdot \|_{\infty})$ and l_p, any element $x := \{x_k\}_{k=1}^{\infty}$ can be represented as the sum of the following series:

$$x = \sum_{k=0}^{\infty} x_k e_k.$$

(b) Show that by appending the sequence $e_0 = \{1, 1, 1, \ldots\}$ to the set $E := \{e_n\}_{n \in \mathbb{N}}$, we obtain the set $E' := \{e_n\}_{n \in \mathbb{Z}_+}$, which is a *topological basis* for $(c, \| \cdot \|_{\infty})$, with each element $x := \{x_k\}_{k=1}^{\infty} \in c$ being represented as the sum of the following series:

$$x = (\lim_{n \to \infty} x_n) e_0 + \sum_{k=1}^{\infty} (x_k - \lim_{n \to \infty} x_n) e_k.$$

Exercise 3.46. Does, for normed vector spaces, the analogue of the *Dimension Theo-rem* (Theorem 3.4) hold for topological bases? In other words, is the notion of "*topological dimension*" well-defined?

Hint. In $(C[a, b], \|\cdot\|_\infty)$, consider the topological basis $\{t^n\}_{n\in\mathbb{Z}_+} \cup \{e^{ct}\}_{c\in\mathbb{R}}$ and compare its *cardinality* to that of $\{t^n\}_{n\in\mathbb{Z}_+}$.

Definition 3.25 (Schauder Basis of a Banach Space). A *Schauder basis* (also a *countable basis*) of a Banach space $(X, \|\cdot\|)$ over $\mathbb{F}$ is a *countably infinite set* of elements $\{e_n\}_{n\in\mathbb{N}}$ in X such that

$$\forall x \in X \; \exists! \; \{c_k\}_{k=1}^\infty \subseteq \mathbb{F} : \; x = \sum_{k=1}^\infty c_k e_k.$$

The series is called the *Schauder expansion* of x and the numbers $c_k \in \mathbb{F}$, $k \in \mathbb{N}$, uniquely determined by x, are called the *coordinates* of x relative to the Schauder basis $\{e_n\}_{n\in\mathbb{N}}$.

Remarks 3.24.

- Because of the uniqueness of the foregoing series representation, a *Schauder basis* of a Banach space, when it exists, is automatically *linearly independent*. It is also *not unique*.

  Exercise 3.47. Explain.

- A *Schauder basis* of a Banach space, when it exists, is also its *topological basis*. However, as the following examples show, a *topological basis* of a Banach need not be its *Schauder basis*.
- The *order* of vectors in a *Schauder basis* is *important*. A permutation of infinitely many elements may transform a *Schauder basis* into a set, which fails to be one (see, e. g., [31]).

Examples 3.17.

1. The set $E := \{e_n := \{\delta_{nk}\}_{k=1}^\infty\}_{n\in\mathbb{N}}$ (see Examples 3.16) is
   - a (standard) *Schauder basis* for $(c_0, \|\cdot\|_\infty)$ and l_p $(1 \le p < \infty)$, but
   - not a *Schauder basis* for c and l_∞.
2. The set $\{t^n\}_{n\in\mathbb{Z}_+}$ is a *topological basis*, but not a *Schauder basis* for the space $(C[a, b], \|\cdot\|_\infty)$ $(-\infty < a < b < \infty)$, which does have Schauder bases of a more intricate structure (see, e. g., [31]).

Exercise 3.48.

(a) Verify.

(b) Show that the set $E' := \{e_n\}_{n\in\mathbb{Z}_+}$ (see Exercise 3.45) is a *Schauder basis* for c.

Proposition 3.9 (Separability of Banach Space with a Schauder Basis). *A Banach space X with a Schauder basis $\{e_n\}_{n \in \mathbb{N}}$ is separable.*

Exercise 3.49. Prove.

Hint. Consider the set

$$C := \left\{ \sum_{k=1}^{n} c_k e_k \,\middle|\, n \in \mathbb{N} \right\},$$

where c_k, $k = 1, \ldots, n$, are arbitrary rationals/complex rationals.

Remark 3.25. The converse statement is not true. The *basis problem* on whether every separable Banach space has a Schauder basis posed by Stefan Banach was negatively answered by Per Enflo (1944–) in [14].

3.3 Finite-Dimensional Spaces and Related Topics

In this section, we analyze certain features inherent to the important class of finite-dimensional normed vector spaces.

3.3.1 Norm Equivalence and Completeness

The following statement shows that the equivalence of all p-norms ($1 \le p \le \infty$) on $\mathbb{F}^n$ ($n \in \mathbb{N}$) (see Exercise 3.42) is not coincidental.

Theorem 3.10 (Norm Equivalence Theorem). *All norms on a finite-dimensional vector space are equivalent.*

Proof. Let X be a finite-dimensional vector space over $\mathbb{F}$ and $\| \cdot \|$ be an arbitrary norm on X. The case of $\dim X = 0$ being vacuous, suppose that with $\dim X = n$ with some $n \in \mathbb{N}$, and let $B := \{x_1, \ldots, x_n\}$ be a *basis* of X. Then, by the *Representation Theorem* (Theorem 3.3),

$$\forall x \in X \; \exists! \, (c_1, \ldots, c_n) \in \mathbb{F}^n : x = \sum_{k=1}^{n} c_k x_k,$$

and hence, the mapping

$$X \ni x = \sum_{k=1}^{n} c_k x_k \mapsto Tx := (c_1, \ldots, c_n) \in \mathbb{F}^n$$

is an *isomorphism* between X and $\mathbb{F}^n$ and

$$X \ni x = \sum_{k=1}^{n} c_k x_k \mapsto \|x\|_2 := \|(c_1, \ldots, c_n)\|_2 = \left[\sum_{k=1}^{n} |c_k|^2 \right]^{1/2}$$

is a norm on X (see Examples 3.14 and Section 3.6, Problem 11), which is *stronger* than $\|\cdot\|$, since, for any $x \in X$,

$$\|x\| = \left\|\sum_{k=1}^{n} c_k x_k\right\| \qquad \text{by } subadditivity \text{ and } absolute\ scalability \text{ of norm;}$$

$$\leq \sum_{k=1}^{n} |c_k| \|x_k\| \qquad \text{by the } Cauchy\text{-}Schwarz\ inequality \text{ (see (2.2));}$$

$$\leq \left[\sum_{k=1}^{n} \|x_k\|^2\right]^{1/2} \left[\sum_{k=1}^{n} |c_k|^2\right]^{1/2} = C\|x\|_2, \qquad (3.3)$$

where $C := [\sum_{k=1}^{n} \|x_k\|^2]^{1/2} > 0$.

Whence, in view of inequality (3.2),

$$|\|x\| - \|y\|| \leq \|x - y\| \leq C\|x - y\|_2, \quad x, y \in X,$$

which implies that the norm $\|\cdot\|$ is a *Lipschitz continuous* function on $(X, \|\cdot\|_2)$.

The spaces $(X, \|\cdot\|_2)$ and $l_2^{(n)} := (\mathbb{F}^n, \|\cdot\|_2)$ being *isometrically isomorphic*, by the *Heine–Borel Theorem* (Theorem 2.36), the *unit sphere* $S(0,1)$ in $(X, \|\cdot\|_2)$ is *compact*. Hence, by the *Weierstrass Extreme Value Theorem* (Theorem 2.43), the "old" norm $\|\cdot\|$ attains on $S(0,1)$ its *absolute minimum value* $c > 0$.

Exercise 3.50. Explain why $c > 0$.

For each $x \in X \setminus \{0\}$, $x/\|x\|_2 \in S(0,1)$ in $(X, \|\cdot\|_2)$, and hence, by *absolute scalability* of norm, we have:

$$c \leq \left\|\frac{1}{\|x\|_2} x\right\| = \frac{1}{\|x\|_2} \|x\|.$$

Combining the latter with (3.3), we have:

$$c\|x\|_2 \leq \|x\| \leq C\|x\|_2, \quad x \in X,$$

which implies that the norms $\|\cdot\|$ and $\|\cdot\|_2$ are *equivalent*.

Since the norm equivalence is an *equivalence relation*, we conclude that all norms on X are equivalent. □

Remark 3.26. As is seen from the proof of the prior theorem, an n-dimensional ($n \in \mathbb{N}$) normed vector space $(X, \|\cdot\|)$ with a basis $B := \{x_1, \ldots, x_n\}$ can be *equivalently renormed* as follows:

$$X \ni x = \sum_{k=1}^{n} c_k x_k \mapsto \|x\|_2 := \|(c_1, \ldots, c_n)\|_2 = \left[\sum_{k=1}^{n} |c_k|^2\right]^{1/2},$$

which makes the spaces $(X, \|\cdot\|)$ and $l_2^{(n)}$ isometrically isomorphic.

Whence, we obtain immediately the following generalization of the *Heine-Borel Theorem* (Theorem 2.36).

Theorem 3.11 (Generalized Heine-Borel Theorem). *Let $(X, \| \cdot \|)$ be a (real or complex) finite-dimensional normed vector space.*
(1) *A set A is precompact in $(X, \| \cdot \|)$ iff it is bounded.*
(2) *A set A is compact in $(X, \| \cdot \|)$ iff it is closed and bounded.*

Exercise 3.51. Prove.

In the same manner based on the *Norm Equivalence Theorem* (Theorem 3.10) and the renorming procedure as described in Remark 3.26, one can prove the following important statement.

Theorem 3.12 (Completeness of Finite-Dimensional Spaces). *Every finite-dimensional normed vector space is a Banach space.*

Proof. Let $(X, \| \cdot \|)$ be a finite-dimensional normed vector space. The case of $\dim X = 0$ being vacuous, suppose that with $\dim X = n$ with some $n \in \mathbb{N}$.

By the *Norm Equivalence Theorem* (Theorem 3.10), equivalently renorming X as in Remark 3.26, we obtain a normed vector space $(X, \| \cdot \|_2)$, which is *isometrically isomorphic* to the Banach space $l_2^{(n)}$, and hence, *complete*.

The norms $\| \cdot \|$ and $\| \cdot \|_2$ being *equivalent*, the space $(X, \| \cdot \|)$ is also *complete* (see Remarks 3.21). □

3.3.2 Finite-Dimensional Subspaces and Bases of Banach Spaces

3.3.2.1 Finite-Dimensional Subspaces
From the *Completeness of Finite-Dimensional Spaces Theorem* (Theorem 3.12), we immediately obtain the following corollary.

Theorem 3.13 (Closedness of Finite-Dimensional Subspaces). *Every finite-dimensional subspace of a normed vector space is closed.*

Exercise 3.52. Prove.

Theorem 3.14 (Nearest Point Property Relative to Finite-Dimensional Subspaces). *Let Y be a finite-dimensional subspace of a normed vector space $(X, \| \cdot \|)$. Then, for each $x \in X$, there is a nearest point to x in Y, i. e.,*

$$\forall x \in X \, \exists y \in Y : \|x - y\| = \rho(x, Y) := \inf_{u \in Y} \|x - u\|.$$

Proof. If $x \in Y$, $\rho(x, Y) = 0$ and $y = x$.

Suppose that $x \in Y^c$. Then, since Y is *closed*, $\rho(x, Y) > 0$ (cf. Section 2.19, Problem 16).

There is a sequence $\{y_n\}_{n=1}^{\infty} \subset Y$ such that

$$\lim_{n \to \infty} \|x - y_n\| = \rho(x, Y), \tag{3.4}$$

which implies that $\{y_n\}_{n=1}^{\infty}$ is *bounded* in $(Y, \|\cdot\|)$.

Exercise 3.53. Explain.

Hence, in view of the *finite dimensionality* of Y, by the *Generalized Heine-Borel Theorem* (Theorem 3.11) and the *Equivalence of Different Forms of Compactness Theorem* (Theorem 2.39), there is a subsequence $\{y_{n(k)}\}_{k=1}^{\infty}$ convergent to an element $y \in Y$. The latter, in respect to (3.4), by the *continuity* of norm, implies that

$$\|x - y\| = \lim_{k \to \infty} \|x - y_{n(k)}\| = \rho(x, Y)$$

completing the proof. $\qquad\qquad\qquad\qquad\qquad\qquad\qquad\qquad\qquad\qquad\qquad\quad$ □

Remark 3.27. The condition of the *finite dimensionality* of a subspace is essential and cannot be dropped. Indeed, as Examples 4.5 and 4.7 below demonstrate, in the space $(c_{00}, \|\cdot\|_2)$, there exists a *proper closed infinite-dimensional* subspace Y such that to no point $x \in Y^c$, is there a nearest point in Y.

Applying the prior statement to $(C[a, b], \|\cdot\|_{\infty})$ $(-\infty < a < b < \infty)$, we obtain the following corollary.

Corollary 3.5 (Best Approximation Polynomial in $(C[a, b], \|\cdot\|_{\infty})$). *For each* $f \in C[a, b]$ $(-\infty < a < b < \infty)$ *and any* $n \in \mathbb{Z}_+$, *in* $(C[a, b], \|\cdot\|_{\infty})$, *there is a best approximation polynomial* $p_n \in P_n$:

$$\|f - p_n\|_{\infty} := \max_{a \le t \le b} |f(t) - p_n(t)| = \rho(x, P_n) := \inf_{u \in P_n} \|f - u\|_{\infty},$$

i. e., a polynomial nearest to f *in the* $(n + 1)$-*dimensional subspace* P_n *of all polynomial of degree at most* n *in* $(C[a, b], \|\cdot\|_{\infty})$.

Remark 3.28. A nearest point relative to a finite-dimensional subspace need not be unique.

Example 3.18. In $X = l_{\infty}^{(2)}(\mathbb{R})$, i. e., in $\mathbb{R}^2$ with the norm

$$\|(x_1, x_2)\|_{\infty} := \max[|x_1|, |x_2|], \quad (x_1, x_2) \in \mathbb{R}^2,$$

for the point $x := (0, 1)$ and the *subspace* $Y := \text{span}(\{(1, 0)\}) = \{(\lambda, 0) \mid \lambda \in \mathbb{R}\}$,

$$\rho(x, Y) = \inf_{\lambda \in \mathbb{R}} \max[|0 - \lambda|, |1 - 0|] = \inf_{\lambda \in \mathbb{R}} \max[|\lambda|, 1] = \begin{cases} 1 & \text{for } |\lambda| \le 1, \\ > 1 & \text{for } |\lambda| > 1. \end{cases}$$

Hence, in Y, there are *infinitely many* nearest points to x of the form $y = (\lambda, 0)$ with $-1 \le \lambda \le 1$.

3.3.2.2 Bases of Banach Spaces

As is known (see Examples 3.16), a *topological basis* of a Banach space can be *countably infinite*. This, however, is not true for an *algebraic basis* of such a space. The closedness of finite dimensional subspaces (Theorem 3.13) along with the nowhere denseness of closed proper subspaces (Proposition 3.16, see Section 3.6, Problem 9) enable proving the following interesting and profound fact.

Theorem 3.15 (Basis of a Banach Space Theorem). *A (Hamel) basis of a Banach space is either finite or uncountable.*

Proof. Let us prove the statement *by contradiction* and assume that a Banach space $(X, \| \cdot \|)$ has a *countably infinite* Hamel basis $B := \{x_n\}_{n \in \mathbb{N}}$.
   Then

$$X = \bigcup_{n=1}^{\infty} X_n, \tag{3.5}$$

where

$$X_n := \operatorname{span}\left(\{x_1, x_2, \ldots, x_n\}\right), \ n \in \mathbb{N},$$

is an n-dimensional subspace of X.
   By the *Closedness of Finite-Dimensional Subspaces Theorem* (Theorem 3.13), each subspace X_n, $n \in \mathbb{N}$, is *closed*. Hence, being a closed proper subspace of X, by the *Nowhere Denseness of Closed Proper Subspace Proposition* (Proposition 3.16) (see Section 3.6, Problem 9), each subspace X_n, $n \in \mathbb{N}$, is *nowhere dense* in $(X, \| \cdot \|)$.
   By the *Baire Category Theorem* (Theorem 2.31), representation (3.5), *contradicts* the completeness of $(X, \| \cdot \|)$, which proves the statement. ☐

Remark 3.29. Hence, for a Banach space $(X, \| \cdot \|)$,

$$\text{either } \dim X = n, \text{ with some } n \in \mathbb{Z}_+, \text{ or } \dim X \geq \mathfrak{c},$$

where $\mathfrak{c}$ is the cardinality of the continuum, i. e., $\mathfrak{c} = |\mathbb{R}|$ (see Examples 1.1). In the latter case, by the *Dimension-Cardinality Connection Proposition* (Proposition 3.13) (see Section 3.6, Problem 3), the dimension of X is equal to its *cardinality*:

$$\dim X = |X|.$$

In particular,

$$\dim l_p^{(n)} = n \ (n \in \mathbb{N}, \ 1 \leq p \leq \infty), \quad \dim P_n = n + 1 \ (n \in \mathbb{Z}_+)$$

and, by the arithmetic of cardinals (see, e. g., [21, 26, 38]),

$$\dim l_p = \dim c_0 = \dim c = \dim C[a, b] = \mathfrak{c} \ (1 \leq p \leq \infty).$$

3.4 Riesz's Lemma and Implications

The following celebrated statement has a number of profound implications.

Theorem 3.16 (Riesz's Lemma). *If Y is a closed proper subspace of a normed vector space $(X, \| \cdot \|)$, then*[2]

$$\forall \varepsilon \in (0,1) \, \exists x_\varepsilon \in Y^c \text{ with } \|x_\varepsilon\| = 1 : \, \rho(x_\varepsilon, Y) := \inf_{y \in Y} \|x_\varepsilon - y\| > 1 - \varepsilon.$$

Proof. Let $x \in Y^c$ be arbitrary. Then, by the *closedness* of Y, $\rho(x, Y) > 0$, and hence,

$$\forall \varepsilon \in (0,1) \, \exists y_\varepsilon \in Y : \, \rho(x, Y) \le \|x - y_\varepsilon\| < \frac{\rho(x, Y)}{1 - \varepsilon}. \tag{3.6}$$

Setting

$$x_\varepsilon := \frac{1}{\|x - y_\varepsilon\|} (x - y_\varepsilon),$$

we obtain an element $x_\varepsilon \notin Y$ with $\|x_\varepsilon\| = 1$.

Exercise 3.54. Explain.

By *absolute scalability* of norm, we have:

$$\forall y \in Y : \, \|x_\varepsilon - y\| = \left\| \frac{1}{\|x - y_\varepsilon\|} (x - y_\varepsilon) - y \right\|$$

$$= \frac{1}{\|x - y_\varepsilon\|} \|x - (y_\varepsilon + \|x - y_\varepsilon\| y)\|$$

since $y_\varepsilon + \|x - y_\varepsilon\| y \in Y$ and in view of (3.6);

$$> \frac{1 - \varepsilon}{\rho(x, Y)} \rho(x, Y) = 1 - \varepsilon,$$

which completes the proof. ☐

Remark 3.30. The *closedness* condition is essential and cannot be dropped.

Exercise 3.55. Give a corresponding example.

Remark 3.31. As is known (see Remark 2.50), the *unit sphere* $S(0,1)$, although *closed* and *bounded*, is *not compact* in $(c_{00}, \| \cdot \|_\infty)$ (also in $(c_0, \| \cdot \|_\infty)$, $(c, \| \cdot \|_\infty)$, and l_p ($1 \le p \le \infty$)). All the above spaces being *infinite dimensional*, *Riesz's Lemma* (Theorem 3.16) is instrumental for showing that this fact is not coincidental.

---

2 Frigyes Riesz (1880–1956).

Corollary 3.6 (Characterization of Finite-Dimensional Normed Vector Spaces). *A normed vector space* $(X, \| \cdot \|)$ *over* $\mathbb{F}$ *is finite dimensional iff the unit sphere*

$$S(0,1) := \{x \in X \mid \|x\| = 1\}$$

is compact in $(X, \| \cdot \|)$.

Proof. "*Only if*" part. Immediately follows from the *Generalized Heine-Borel Theorem* (Theorem 3.11).

Exercise 3.56. Explain.

"*If*" part. Let us prove this part *by contrapositive* assuming that X is *infinite dimensional*.

Let us fix an element $x_1 \in X$ with $\|x_1\| = 1$. Then the *one-dimensional subspace*

$$X_1 := \text{span}(\{x_1\})$$

is a *closed proper subspace* of X, which, by *Riesz's Lemma* (Theorem 3.16) with $\varepsilon = 1/2$, implies that

$$\exists x_2 \in X_1^c \text{ with } \|x_1\| = 1 : \|x_2 - x_1\| \geq \rho(x_2, X_1) > \frac{1}{2}.$$

Similarly, since the *two-dimensional subspace*

$$X_2 := \text{span}\{x_1, x_2\}$$

is a *closed proper subspace* of X, by *Riesz's Lemma* with $\varepsilon = 1/2$,

$$\exists x_3 \in X_2^c \text{ with } \|x_3\| = 1 : \|x_3 - x_i\| \geq \rho(x_3, X_2) > \frac{1}{2}, \ i = 1, 2.$$

Continuing inductively, we obtain a sequence

$$\{x_n\}_{n=1}^{\infty} \subseteq S(0,1)$$

with the following property:

$$\|x_m - x_n\| > \frac{1}{2}, \ m, n \in \mathbb{N}, \ m \neq n,$$

and thus, not containing a convergent subsequence, which, by the *Equivalence of Different Forms of Compactness* (Theorem 2.39), implies that $S(0,1)$ is not compact in $(X, \| \cdot \|)$ unit sphere and completes the proof. $\square$

An immediate corollary is as follows.

Corollary 3.7 (Noncompactness of Spheres/Balls). *A nontrivial sphere/ball in an infinite-dimensional normed vector space is not compact.*

Exercise 3.57. Prove.

Based on this fact, it is not difficult to prove the following quite counterintuitive statement.

Proposition 3.10 (Nowhere Denseness of Compact Sets). *A compact set in an infinite-dimensional normed vector space is nowhere dense.*

Exercise 3.58. Prove.

Hint. Prove *by contradiction.*

3.5 Convexity, Strictly Convex Normed Vector Spaces

We conclude this chapter with a discussion of the purely linear property of *convexity* and an important class of normed vector spaces called *strictly convex.*

3.5.1 Convexity

Definition 3.26 (Convex Set). A nonempty set C in a vector space X is called *convex* if

$$\forall x, y \in C, \ \forall 0 \le \lambda \le 1: \ \lambda x + (1 - \lambda)y \in C,$$

i. e., for each pair x, y of its points, C contains the *line segment*

$$\{\lambda x + (1 - \lambda)y \mid 0 \le \lambda \le 1\}$$

connecting them.

Examples 3.19.
1. In a vector space X,
   (a) every *singleton* $\{x_0\}$ is *convex*;
   (b) every *subspace* Y is *convex*.
2. In $\mathbb{R}^2$, the sets $\{(x, y) \in \mathbb{R}^2 \mid x \ge 0\}$ and $\{(x, y) \in \mathbb{R}^2 \mid y < 0\}$ are *convex*, but the set $\{(x, y) \in \mathbb{R}^2 \mid xy \ge 0\}$ is *not*.
3. In a normed vector space $(X, \|\cdot\|)$, every (open or closed) *nontrivial ball* is a *convex set*.
   This simple fact, in particular, precludes the mapping

$$\mathbb{F}^n \ni x = (x_1, \ldots, x_m) \mapsto \|x\|_p := \left[\sum_{k=1}^{n} |x_k|^p \right]^{1/p}$$

from being a norm on $\mathbb{F}^n$ ($n = 2, 3, \ldots$) for $0 < p < 1$.

Exercise 3.59.

(a) Verify.

(b) Show that the only convex sets in $\mathbb{R}$ are the *intervals*.

(c) Give more examples of convex sets and of sets, which are not convex, in $\mathbb{R}^2$.

Theorem 3.17 (Properties of Convex Sets). *In a vector space X over $\mathbb{F}$,*

(1) *if C is a convex set, then, for each $x \in X$, the translation $x + C$ is convex;*

(2) *if C is a convex set, then, for each $\mu \in \mathbb{F}$, the product μC is convex;*

(3) *if $\{C_i\}_{i \in I}$ is a nonempty collection of convex sets, then the intersection $\bigcap_{i \in I} C_i$, if nonempty, is convex;*

(4) *provided $(X, \| \cdot \|)$ is a normed vector space, if C is a convex set, then the closure $\overline{C}$ is convex.*

Exercise 3.60.

(a) Prove.

(b) Give an example showing that the union of convex sets need not be convex.

(c) Give an example showing that the converse to (4) is *not* true.

By properties (3) and (4) in the prior theorem, the following notions are well defined.

Definition 3.27 (Convex Hull and Closed Convex Hull). Let S be a nonempty subset of a vector space X, then

$$\mathrm{conv}(S) := \bigcap_{C \text{ is convex, } S \subseteq C} C$$

is the *smallest convex set of X containing S* called the *convex hull* of S.

In a normed vector space $(X, \| \cdot \|)$, the *closed convex hull* of a nonempty set S is the smallest closed convex set containing S, i. e., the closure $\overline{\mathrm{conv}(S)}$ of its convex hull.

Remark 3.32. A nonempty set S in a vector space X is convex *iff* $S = \mathrm{conv}(S)$.

Exercise 3.61.

(a) Verify.

(b) Describe the convex hull of a two-point set $\{x, y\}$ in a vector space X.

(c) Describe the convex hull of a finite set in $\mathbb{R}^2$ (cf. the *Convex Hull's Structure Proposition* (Proposition 3.19), Section 3.6, Problem 17).

3.5.2 Strictly Convex Normed Vector Spaces

The following is an important class of normed vector spaces.

Definition 3.28 (Strictly Convex Normed Vector Space). A strictly convex normed vector space is a normed vector space $(X, \| \cdot \|)$ such that, for arbitrary $x, y \in X, x \neq y$, with $\|x\| = \|y\| = 1$,

$$\forall \lambda \in (0, 1) : \|\lambda x + (1 - \lambda)y\| < 1,$$

i. e., the unit sphere $S(0, 1)$ contains no points of the line segments connecting pairs of its points, except the endpoints.

Examples 3.20.
1. The space $\mathbb{F}$ (i. e., $\mathbb{R}$ or $\mathbb{C}$) is *strictly normed/convex* relative to the *absolute-value norm*.
2. The spaces $l_p^{(n)}$ ($n = 2, 3, \dots$) and l_p ($1 < p < \infty$) are *strictly convex*, whereas the spaces $l_1^{(n)}, l_\infty^{(n)}$ ($n = 2, 3, \dots$) and l_1, l_∞ are *not strictly convex* (see Figure 3.1).

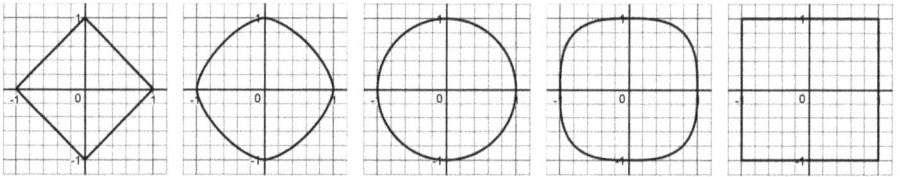

Figure 3.1: The unit sphere in $l_p^{(2)}(\mathbb{R})$ for $p = 1$, $p = 3/2$, $p = 2$, $p = 3$, and $p = \infty$.

3. The space $(C[a, b], \| \cdot \|_\infty)$ $(-\infty < a < b < \infty)$ is not *strictly convex*.
4. As is shown below (see Proposition 4.3), any *inner-product space* is *strictly convex* relative to the inner-product norm.

Exercise 3.62. Verify 1, 3, and 4.

Proposition 3.11 (Characterization of Strict Convexity). *A normed vector space $(X, \| \cdot \|)$ is strictly convex iff for arbitrary nonzero elements $x, y \in X \setminus \{0\}$, the equality*

$$\|x + y\| = \|x\| + \|y\|$$

implies that

$$\exists \lambda > 0 : y = \lambda x.$$

Proof. *"Only if"* part. Let us prove this part *by contrapositive*, assuming that, in a vector space $(X, \| \cdot \|)$,

$$\exists x, y \in X \setminus \{0\}, \ y \neq \lambda x \text{ for any } \lambda > 0 : \|x + y\| = \|x\| + \|y\|. \qquad (3.7)$$

Since $y \neq \lambda x$ for any $\lambda > 0$, we infer that

$$S(0,1) \ni x_1 := \frac{1}{\|x\|} x \neq y_1 := \frac{1}{\|y\|} y \in S(0,1).$$

Setting $\lambda := \frac{\|x\|}{\|x\|+\|y\|} \in (0,1)$, by *absolute scalability* of norm and in view of (3.7), we have:

$$\|\lambda x_1 + (1-\lambda)y_1\| = \left\| \frac{\|x\|}{\|x\| + \|y\|} \frac{1}{\|x\|} x + \frac{\|y\|}{\|x\| + \|y\|} \frac{1}{\|y\|} y \right\|$$

$$= \frac{1}{\|x\| + \|y\|} \|x + y\| = 1,$$

which implies that $(X, \|\cdot\|)$ is *not strictly convex* and concludes the proof of the *"only if"* part.

"If" part. Let us prove the *"if"* part *by contrapositive* as well, assuming that $(X, \|\cdot\|)$ is *not strictly convex*, and hence, in view of *absolute scalability* of norm,

$$\exists x, y \in S(0,1), \ x \neq y, \ \exists \lambda \in (0,1): \ \|\lambda x + (1-\lambda)y\| = 1 = \lambda + (1-\lambda)$$
$$= \|\lambda x\| + \|(1-\lambda)y\|.$$

Observe that

$$\forall \alpha > 0: \ \lambda x \neq \alpha(1-\lambda)y,$$

since otherwise, by *absolute scalability* of norm,

$$\lambda = \|\lambda x\| = \|\alpha(1-\lambda)y\| = \alpha(1-\lambda).$$

The latter implies that $\alpha = \frac{\lambda}{1-\lambda}$, and hence, $x = y$, which is a *contradiction*. This completes the proof of the *"if"* part and the entire statement. $\square$

The following statement justifies the importance of strict convexity.

Theorem 3.18 (Nearest Point Property for Strictly Convex Spaces). *Let C be a convex set in a strictly convex normed vector space $(X, \|\cdot\|)$. Then, for each $x \in X$, there exists at most one nearest point to x in C.*

Proof. If, for an $x \in X$,

$$\rho(x, C) = 0$$

the statement follows immediately.

Exercise 3.63. Explain.

Assume that, for an $x \in X$ with

$$\rho(x, C) > 0,$$

there are two *distinct* nearest points $y_1, y_2 \in C, y_1 \neq y_2$, to x in C.

In view of the *convexity* of C, by the *Convexity of the Set of the Nearest Points* (Proposition 3.20) (see Section 3.6, Problem 18), for every $0 < \lambda < 1$,

$$\lambda y_1 + (1 - \lambda) y_2$$

is also a nearest point to x in C, i. e.,

$$\forall \lambda \in (0, 1) : \|x - [\lambda y_1 + (1 - \lambda) y_2]\| = \rho(x, Y).$$

Whence, by *absolute scalability* and *subadditivity* of norm, for the *distinct* points

$$x_1 := (x - y_1)/\rho(x, C), x_2 := (x - y_2)/\rho(x, C) \in S(0, 1),$$

we have:

$$\forall \lambda \in (0, 1) : 1 = \lambda + (1 - \lambda) = \lambda \|x_1\| + (1 - \lambda) \|x_2\| \geq \|\lambda x_1 + (1 - \lambda) x_2\|$$
$$= \|\lambda(x - y_1)/\rho(x, C) + (1 - \lambda)(x - y_2)/\rho(x, C)\|$$
$$= \frac{1}{\rho(x, C)} \|\lambda(x - y_1) + (1 - \lambda)(x - y_2)\|$$
$$= \frac{1}{\rho(x, C)} \|\lambda x + (1 - \lambda)x - [\lambda y_1 + (1 - \lambda)y_2]\|$$
$$= \frac{1}{\rho(x, C)} \|x - [\lambda y_1 + (1 - \lambda)y_2]\| = \frac{1}{\rho(x, C)} \rho(x, C) = 1,$$

which *contradicts* the strict convexity of $(X, \| \cdot \|)$, completing the proof. ☐

From the prior theorem and the *Nearest Point Property Relative to Finite-Dimensional Subspaces* (Theorem 3.14), we obtain the following corollary.

Corollary 3.8 (Nearest Point Property for Strictly Convex Spaces Relative to Finite-Dimensional Subspaces). *Let Y be a finite-dimensional subspace of a strictly convex normed vector space $(X, \| \cdot \|)$. Then, for each $x \in X$, there is a unique nearest point to x in Y, i. e.,*

$$\forall x \in X \; \exists! y \in Y : \|x - y\| = \rho(x, Y) := \inf_{u \in Y} \|x - u\|.$$

3.6 Problems

1. Prove

   Proposition 3.12 (Characterization of Isomorphisms). *Let X and Y be vector spaces and a mapping $T : X \to Y$ be a homomorphism. Then T is an isomorphism iff its kernel (null space) $\ker T$ is trivial, i. e.,*

   $$\ker T := \{x \in X \mid Tx = 0\} = \{0\}.$$

2. Show that, if Y is a nontrivial *subspace* of a nontrivial vector space X, each basis B_Y of Y is contained in a basis B_X of X.

3. Prove

   Proposition 3.13 (Dimension-Cardinality Connection). *If for a vector space X, $\dim(X) \geq \mathfrak{c}\ (\mathfrak{c} := |\mathbb{R}|)$, then $\dim X = |X|$.*

   Hint. Use the idea of the proof of the *Dimension Theorem* (Theorem 3.4).

4. Show that the metric

   $$s \ni x = \{x_k\}_{k=1}^{\infty}, \{y_k\}_{k=1}^{\infty} \mapsto \rho(x,y) := \sum_{k=1}^{\infty} \frac{1}{2^k} \frac{|x_k - y_k|}{1 + |x_k - y_k|}$$

   (see Section 2.19, Problem 26) is not a *norm metric*.

5. (Cartesian Product of Normed Vector Spaces). Let $(X_1, \|\cdot\|_1)$ and $(X_2, \|\cdot\|_2)$ be normed vector spaces over $\mathbb{F}$.
   (a) Show that the Cartesian product $X = X_1 \times X_2$ is a normed vector space relative to the *product norm*

   $$X_1 \times X_2 \ni x = (x_1, x_2) \mapsto \|x\|_{X_1 \times X_2} = \sqrt{\|x_1\|_1^2 + \|x_2\|_2^2}.$$

   (b) Describe convergence in $(X_1 \times X_2, \|\cdot\|_{X_1 \times X_2})$.
   (c) Show that the product space $(X_1 \times X_2, \|\cdot\|_{X_1 \times X_2})$ is Banach space *iff* each space $(X_i, \|\cdot\|_i)$, $i = 1, 2$, is a Banach space.

6. Show that if $(X, \|\cdot\|)$ is a *seminormed* vector space, then
   (a) $Y := \{x \in X \mid \|x\| = 0\}$ is a subspace of X and
   (b) the quotient space X/Y is a normed vector space relative to the norm

   $$X/Y \ni [x] := x + Y \mapsto \|[x]\|_1 = \|x\|.$$

   Hint. To show that $\|\cdot\|_1$ is well defined on X/Y, i. e., is independent of the choice of the representative of the coset $[x]$, use inequality (3.2).

7. Prove

   Proposition 3.14 (Completeness of $BV[a,b]$). *The vector space* $BV[a,b]$ $(-\infty < a < b < \infty)$ *is a Banach space relative to the total-variation norm*

   $$BV[a,b] \ni f \mapsto \|f\| := |f(a)| + V_a^b(f).$$

   Hint. Show first that

   $$\forall f \in BV[a,b]: \sup_{a \le x \le b} |f(x)| \le |f(a)| + V_a^b(f).$$

8. Prove

   Proposition 3.15 (Completeness of $C^n[a,b]$). *The vector space* $C^n[a,b]$ $(n \in \mathbb{N},$ $-\infty < a < b < \infty)$ *of n-times continuously differentiable on* $[a,b]$ *functions with pointwise linear operations is a Banach space relative to the norm*

   $$C^n[a,b] \ni f \mapsto \|f\|_n := \max_{0 \le k \le n}\left[\max_{a \le t \le b} |f^{(k)}(t)|\right].$$

   Hint. Use the *Total Change Formula* representation:

   $$f^{(k-1)}(t) := f^{(k-1)}(a) + \int_a^t f^{(k)}(s)\,ds, \ k = 1,\ldots,n, \ t \in [a,b].$$

9. Prove

   Proposition 3.16 (Nowhere Denseness of Closed Proper Subspace).
   A closed proper subspace Y *of a normed vector space* $(X, \|\cdot\|)$ *is nowhere dense.*

   Give an example showing the *closedness requirement* cannot be dropped.

10. Prove

    Theorem 3.19 (Quotient Space Norm). *Let* Y *be a closed subspace of a normed vector space* $(X, \|\cdot\|)$. *Then*
    (1)

    $$X/Y \ni [x] := x + Y \mapsto \|[x]\|_1 = \inf_{y \in Y} \|x + y\| = \rho(x, Y)$$

    is a norm on X/Y *and*
    (2) *if* $(X, \|\cdot\|)$ *is a Banach space, then* $(X/Y, \|\cdot\|_1)$ *is a Banach space.*

    Hint. To prove part (2), first describe the convergence in $(X/Y, \|\cdot\|_1)$.

11. (Norm Defining Procedure). Let X be a vector space, $(Y, \|\cdot\|_Y)$ be a *normed vector space*, and $T : X \to Y$ be an *isomorphic embedding* of X in Y.
    Show that the mapping

    $$X \ni x \mapsto \|x\|_X := \|Tx\|_Y$$

    is a *norm* on X, relative to which the isomorphism T is also isometric.

12. Show that the mapping

$$P_n \ni p(t) = \sum_{k=0}^{n} a_k t^k \mapsto \|p\| := \sum_{k=0}^{n} |a_k|$$

is a *norm* on the space P_n of all polynomials of degree at most n ($n \in \mathbb{Z}_+$) with real/complex coefficients.

13. (The 60th W. L. Putnam Mathematical Competition, 1999). Prove that there is a constant $C > 0$ such that, for any polynomial $p(x)$ of degree 1999,

$$|p(0)| \le C \int_{-1}^{1} |p(x)| \, dx.$$

14. In $X = l_1^{(2)}(\mathbb{R})$, i. e., in $\mathbb{R}^2$ with the norm

$$\|(x_1, x_2)\|_1 = |x_1| + |x_2|, \ (x_1, x_2) \in \mathbb{R}^2,$$

determine the *set of all nearest points* to $x := (1, -1)$ in the subspace $Y :=$ span$(\{(1, 1)\}) = \{(\lambda, \lambda) \mid \lambda \in \mathbb{R}\}$.

15. Prove

Proposition 3.17 (Characterization of Finite Dimensionality). *A Banach space* $(X, \| \cdot \|)$ *is finite-dimensional iff every subspace in X is closed.*

Hint. Apply the *Basis of a Banach Space Theorem* (Theorem 3.15).

16. Prove

Proposition 3.18 (Riesz's Lemma in Finite-Dimensional Spaces). *If Y is a proper subspace of a finite-dimensional normed vector space* $(X, \| \cdot \|)$, *then*

$$\exists x \in Y^c : \|x\| = 1 = \rho(x, Y).$$

17. Prove

Proposition 3.19 (Convex Hull's Structure). *For a nonempty set S in a vector space X over* $\mathbb{F}$, conv(S) *is the set of all convex combinations of its elements:*

$$\text{conv}(S) = \left\{ \sum_{k=1}^{n} \lambda_k x_k \ \middle|\ x_1, \dots, x_n \in S, \lambda_1, \dots, \lambda_n \in [0,1] \text{ with } \sum_{k=1}^{n} \lambda_k = 1, n \in \mathbb{N} \right\}.$$

18. Prove

Proposition 3.20 (Convexity of the Set of the Nearest Points). *Let* $(X, \| \cdot \|)$ *be a normed vector space. Then for an arbitrary* $x \in X$, *the set* N_x *of all nearest to x points in a convex set* $C \subseteq X$, *if nonempty, is convex.*

and

Corollary 3.9 (Cardinality of the Set of Nearest Points). *Let* $(X, \| \cdot \|)$ *be a normed vector space. Then for an arbitrary* $x \in X$, *the set* N_x *of all nearest to* x *points in a convex subset* $C \subseteq X$ *is either empty, a singleton, or uncountably infinite.*

19. Prove

Proposition 3.21 (Midpoint Characterization of Strictly Convexity). *A normed vector space* $(X, \| \cdot \|)$ *is strictly convex iff for all* $x, y \in X$, $x \neq y$, *with* $\|x\| = \|y\| = 1$,

$$\left\| \frac{1}{2}(x + y) \right\| < 1,$$

i. e., the unit sphere $S(0, 1)$ *contains no midpoints of the line segments connecting pairs of its distinct points.*

Hint. Prove the *"if"* part *by contradiction* using the *Characterization of Strict Convexity* (Proposition 3.11).

4 Inner Product and Hilbert Spaces

In this chapter, we discuss an important class of abstract spaces whose structure makes possible defining the notions of *orthogonality*, the angle between vectors, and hence, even further warrants the use of the conventional geometric terms and intuition.

4.1 Definitions and Examples

Recall that $\mathbb{F}$ stands for the *scalar field* of $\mathbb{R}$ or $\mathbb{C}$.

Definition 4.1 (Inner Product Space). An *inner product space* (or a *pre-Hilbert*[1] *space*) over $\mathbb{F}$ is a *vector space* X over $\mathbb{F}$ equipped with an *inner product* (or a *scalar product*), i. e., a mapping

$$(\cdot,\cdot) : X \times X \to \mathbb{F}$$

subject to the following *inner product axioms*:

1. $(x,x) \geq 0$, $x \in X$, and $(x,x) = 0$ iff $x = 0$.   *Positive Definiteness*
2. $(x,y) = \overline{(y,x)}$, $x,y \in X$.   *Conjugate Symmetry*
3. $(\lambda x + \mu y, z) = \lambda(x,z) + \mu(y,z)$, $\lambda, \mu \in \mathbb{F}$, $x,y,z \in X$.   *Linearity in the First Argument*

The space is said to be *real* if $\mathbb{F} = \mathbb{R}$ and *complex* if $\mathbb{F} = \mathbb{C}$.

Notation. $(X, (\cdot,\cdot))$.

Remarks 4.1.
- For a *real* inner product space, the axiom of *conjugate symmetry* turns into
  2R. $(x,y) = (y,x)$, $x,y \in X$ *Symmetry*
- From *conjugate symmetry* and *linearity in the first argument*, we immediately derive the following property:
  4. $(x, \lambda y + \mu z) = \overline{\lambda}(x,y) + \overline{\mu}(x,z)$, $\lambda, \mu \in \mathbb{F}$, $x,y,z \in X$
       Conjugate Linearity in the Second Argument
  also called *antilinearity* or *semilinearity*.
  For a *real* inner product space, the latter turns into
  4R. $(x, \lambda y + \mu z) = \lambda(x,y) + \mu(x,z)$, $\lambda, \mu \in \mathbb{R}$, $x,y,z \in X$
       Linearity in the Second Argument,
- From *linearity* and *conjugate symmetry* of an inner product, we immediately infer that

$$\forall x \in X : \ (0,x) = (x,0) = 0.$$

---

[1] David Hilbert (1862–1943).

https://doi.org/10.1515/9783110614039-004

Exercise 4.1. Verify.

Examples 4.1.

1. On $l_2^{(n)}(\mathbb{F})$ $(n \in \mathbb{N})$, the mapping

$$l_2^{(n)} \ni x := (x_1, \ldots, x_n), y := (y_1, \ldots, y_n) \mapsto (x,y) := \sum_{k=1}^{n} x_k \bar{y}_k \in \mathbb{F}$$

is an *inner product*. When the space is real, the conjugation is superfluous.

2. On $l_2(\mathbb{F})$, the mapping

$$l_2 \ni x := \{x_k\}_{k=1}^{\infty}, y := \{y_k\}_{k=1}^{\infty} \mapsto (x,y) := \sum_{k=1}^{\infty} x_k \bar{y}_k \in \mathbb{F}$$

is an *inner product*. When the space is real, the conjugation is superfluous.

Remark 4.2. For $l_2(\mathbb{F})$, the inner product is well-defined due to the *Cauchy–Schwarz inequality* (see (2.2)).

3. On $C[a,b]$ $(-\infty < a < b < \infty)$, the mapping

$$C[a,b] \ni f, g \mapsto (f,g) := \int_a^b f(t)\overline{g(t)}\, dt \in \mathbb{F}$$

is an *inner product*. When the space is real, the conjugation is superfluous.

4. When integration is understood in the *Lebesgue*[2] *sense* (see, e. g., [16, 33, 43, 44, 47]), the mapping

$$f, g \mapsto (f,g) := \int_a^b f(t)\overline{g(t)}\, dt \in \mathbb{F}$$

is an inner product on the space $L_2(a,b)$ $(-\infty \leq a < b \leq \infty)$ of all equivalence classes of equal almost everywhere relative to the *Lebesgue measure* square integrable on (a,b) functions:

$$\int_a^b |f(t)|^2\, dt < \infty,$$

$f(\cdot)$ and $g(\cdot)$ being arbitrary representatives of the equivalence classes f and g, respectively.

Exercise 4.2. Verify 1–3.

---

2 Henri Lebesgue (1875–1941).

Definition 4.2 (Subspace of an Inner Product Space). If $(X, (\cdot, \cdot))$ is a *inner product space* and $Y \subseteq X$ is a *linear subspace* of X, then the restriction of the inner product $(\cdot, \cdot)$ to $Y \times Y$ is an inner product on Y and the inner product space $(Y, (\cdot, \cdot))$ is called a *subspace* of $(X, (\cdot, \cdot))$.

Examples 4.2.
1. The space c_{00} is a subspace of l_2 and c_0 is not.
2. The space $C^1[a, b]$ $(-\infty < a < b < \infty)$ is a subspace of $(C[a, b], (\cdot, \cdot))$ and $R[a, b]$ is not.

4.2 Inner Product Norm, Cauchy–Schwarz Inequality

An inner product generates a norm with certain distinctive properties.

Theorem 4.1 (Inner Product Norm). *On an inner product space $(X, (\cdot, \cdot))$, the mapping*

$$X \ni x \mapsto \|x\| := (x, x)^{1/2} \in \mathbb{R}$$

is a norm, called the inner product norm.

One can easily verify that the norm axioms of *nonnegativity*, *separation*, and *absolute scalability* hold.

Exercise 4.3. Verify.

To show that *subadditivity* holds as well, we are to prove the following important inequality generalizing the *Cauchy–Schwarz inequalities* for n-tuples and sequences (see (2.2)).

Theorem 4.2 (Cauchy–Schwarz Inequality). *In an inner product space $(X, (\cdot, \cdot))$,*

$$\forall x, y \in X : |(x, y)| \le \|x\| \|y\|,$$

with

$$|(x, y)| = \|x\| \|y\| \iff x \text{ and } y \text{ are linearly dependent.}$$

Proof. If $y = 0$, we have:

$$|(x, y)| = 0 = \|x\| \|y\| \tag{4.1}$$

(see Remarks 4.1), and hence, the inequality is trivially true.

Suppose that $y \neq 0$. Then, for every $\lambda \in \mathbb{F}$, by *linearity* and *conjugate linearity* of an inner product and the definition of $\| \cdot \|$,

$$0 \le (x + \lambda y, x + \lambda y) = (x, x + \lambda y) + \lambda(y, x + \lambda y) = (x, x) + \bar{\lambda}(x, y) + \lambda(y, x) + \lambda\bar{\lambda}(y, y)$$
$$= \|x\|^2 + \bar{\lambda}(x, y) + \lambda(y, x) + |\lambda|^2 \|y\|^2.$$

Setting $\lambda = -\frac{(x,y)}{\|y\|^2}$, in view of *conjugate symmetry*, we arrive at

$$0 \le \|x\|^2 - \frac{|(x,y)|^2}{\|y\|^2} - \frac{|(x,y)|^2}{\|y\|^2} + \frac{|(x,y)|^2}{\|y\|^2}.$$

Whence, the *Cauchy–Schwarz inequality* follows immediately.

If the vectors x and y are *linearly dependent*, then either at least one of them is *zero*, in which case equality (4.1) (see Remarks 4.1) trivially holds, or

$$x, y \ne 0 \text{ and } y = \lambda x$$

with some $\lambda \in \mathbb{F} \setminus \{0\}$, in which case, by *conjugate linearity* and the definition of $\|\cdot\|$,

$$|(x,y)| = |(x, \lambda x)| = |\bar{\lambda}| \|x\|^2 = |\lambda| \|x\|^2 = \|x\| \|\lambda x\| = \|x\| \|y\|.$$

Conversely, if for $x, y \in X$,

$$|(x,y)| = \|x\| \|y\|,$$

then, by the definition of $\|\cdot\|$ and *linearity* and *conjugate linearity* of an inner product,

$$\|\|y\|^2 x - (x,y)y\|^2 = (\|y\|^2 x - (x,y)y, \|y\|^2 x - (x,y)y) = \|y\|^2(x, \|y\|^2 x - (x,y)y)$$
$$- (x,y)(y, \|y\|^2 x - (x,y)y) = \|y\|^4 \|x\|^2 - \|y\|^2 |(x,y)|^2 - \|y\|^2 |(x,y)|^2 + |(x,y)|^2 \|y\|^2$$
$$= 2\|y\|^4 \|x\|^2 - 2\|y\|^4 \|x\|^2 = 0.$$

Whence, by *positive definiteness*,

$$\|y\|^2 x - (x,y)y = 0,$$

which proves the linear dependence of x and y.

Exercise 4.4. Explain. □

Remark 4.3. The *Cauchy–Schwarz inequality* (Theorem 4.2) in the spaces $l_2^{(n)}$ ($n \in \mathbb{N}$) and l_2 yields the *Cauchy–Schwarz inequalities* for n-tuples and sequences (see (2.2)) as particular cases.

Now, let us prove the *subadditivity* of $\|\cdot\|$ in Theorem 4.1.

Proof. For every $x, y \in X$, by *linearity* and *conjugate linearity* of an inner product and the *Cauchy–Schwarz Inequality* (Theorem 4.2),

$$\|x + y\|^2 = (x + y, x + y) = (x, x + y) + (y, x + y) = (x,x) + (x,y) + (y,x) + (y,y)$$
$$= \|x\|^2 + (x,y) + \overline{(x,y)} + \|y\|^2 = \|x\|^2 + 2\operatorname{Re}(x,y) + \|y\|^2 \le \|x\|^2 + 2|(x,y)| + \|y\|^2$$
$$\le \|x\|^2 + 2\|x\| \|y\| + \|y\|^2 = (\|x\| + \|y\|)^2.$$

Whence the *subadditivity* of $\|\cdot\|$ follows immediately, and hence, the latter is a norm on X. □

Remarks 4.4.

– While proving the *Inner Product Norm Theorem* (Theorem 4.1), we have come across the following identity

$$\|x \pm y\|^2 = \|x\|^2 \pm 2\operatorname{Re}(x,y) + \|y\|^2, \quad x,y \in X. \tag{4.2}$$

For the case of "–", using the similarity of the above to the geometric *Law of Cosines*, for any vectors $x,y \in X$ with $(x,y) \neq 0$, which, in particular, implies, that $\|x\|, \|y\| \neq 0$, one can define the *angle* between them as follows:

$$\theta := \arccos \frac{\operatorname{Re}(x,y)}{\|x\|\|y\|} \in [0,\pi].$$

The notion is well-defined due to the *Cauchy–Schwarz inequality* (Theorem 4.2) since

$$-1 \le \frac{\operatorname{Re}(x,y)}{\|x\|\|y\|} \le 1.$$

For vectors $x,y \in X$ with $(x,y) = 0$, i. e., "*orthogonal*" (see Sec, 4.4.3), $\theta = \pi/2$.

– The triple $(X,(\cdot,\cdot),\|\cdot\|)$ is used to designate an inner product space $(X,(\cdot,\cdot))$ with the *inner product norm* $\|\cdot\|$.

The *Cauchy–Schwarz Inequality* (Theorem 4.2) has the following immediate important implication.

Proposition 4.1 (Joint Continuity of Inner Product). *On an inner product space $(X,(\cdot,\cdot), \|\cdot\|)$, an inner product is jointly continuous, i. e., if*

$$X \ni x_n \to x \in X \quad and \quad X \ni y_n \to y \in X, \ n \to \infty, \ in \ (X,(\cdot,\cdot),\|\cdot\|).$$

then

$$(x_n,y_n) \to (x,y), \ n \to \infty.$$

Exercise 4.5. Prove.

4.3 Hilbert Spaces

Definition 4.3 (Hilbert Space). A *Hilbert space* is an inner product space $(X,(\cdot,\cdot),\|\cdot\|)$ that is a Banach space relative to the *inner product norm* $\|\cdot\|$.

Examples 4.3.

1. On $l_2^{(n)}(\mathbb{F})$ $(n \in \mathbb{N})$, the inner product

$$l_2^{(n)} \ni x := (x_1, \ldots, x_n), y := (y_1, \ldots, y_n) \mapsto (x, y) := \sum_{k=1}^{n} x_k \bar{y}_k \in \mathbb{F},$$

generates the *Euclidean norm* (the 2-*norm*)

$$l_2^{(n)} \ni x := (x_1, \ldots, x_n) \mapsto \|x\|_2 := \left[\sum_{k=1}^{n} |x_k|^2 \right]^{1/2}.$$

And hence, $l_2^{(n)}$ is a *Hilbert space* (see Examples 3.13).
The real $l_2^{(n)}$ is called the *Euclidean n-space* and the complex $l_2^{(n)}$ is called the *unitary n-space*.

2. On $l_2(\mathbb{F})$ $(n \in \mathbb{N})$, the inner product

$$l_2 \ni x := \{x_k\}_{k=1}^{\infty}, y := \{y_k\}_{k=1}^{\infty} \mapsto (x, y) := \sum_{k=1}^{\infty} x_k \bar{y}_k \in \mathbb{F},$$

generates the 2-*norm*

$$l_2 \ni x := \{x_k\}_{k=1}^{\infty} \mapsto \|x\|_2 := \left[\sum_{k=1}^{\infty} |x_k|^2 \right]^{1/2}.$$

And hence, l_2 is a *Hilbert space* (see Examples 3.13).

3. The space $(c_{00}, (\cdot, \cdot), \| \cdot \|_2)$, considered as a subspace of l_2, is an *incomplete inner product space*.

   Exercise 4.6. Verify.

4. The space $C[a, b]$ $(-\infty < a < b < \infty)$ is an *incomplete inner product space* relative to the *integral inner product*

$$C[a, b] \ni f, g \mapsto (f, g) := \int_a^b f(t)\overline{g(t)} \, dt \in \mathbb{F}$$

generating the 2-*norm*

$$C[a, b] \ni f \mapsto \|f\|_2 := \left[\int_a^b |f(t)|^2 \, dt \right]^{1/2}$$

(see Examples 3.13).

5. The space $L_2(a, b)$ $(-\infty \le a < b \le \infty)$ is a *Hilbert space* relative to the *integral inner product*

$$f, g \mapsto (f, g) := \int_a^b f(t)\overline{g(t)}\, dt \in \mathbb{F}$$

generating the 2-*norm*

$$L_2(a, b) \ni f \mapsto \|f\|_2 := \left[\int_a^b |f(t)|^2\, dt \right]^{1/2}$$

(see Examples 4.1).

4.4 Certain Geometric Properties

The following are geometric properties inherent to inner product norm resonating with familiar ones from the two- and three-dimensional cases.

4.4.1 Polarization Identities

Proposition 4.2 (Polarization Identities).
(1) *In a real inner product space* $(X, (\cdot, \cdot), \| \cdot \|)$,

$$(x, y) = \frac{1}{4}\left[\|x + y\|^2 - \|x - y\|^2 \right], \quad x, y \in X.$$

(2) *In a complex inner product space* $(X, (\cdot, \cdot), \| \cdot \|)$,

$$(x, y) = \frac{1}{4}\left[\|x + y\|^2 - \|x - y\|^2 \right] + \frac{i}{4}\left[\|x + iy\|^2 - \|x - iy\|^2 \right], \quad x, y \in X.$$

Exercise 4.7. Prove.

Hint. Apply identity (4.2).

4.4.2 Parallelogram Law

The following geometric property is characteristic for an inner product norm.

Theorem 4.3 (Parallelogram Law). *In an inner product space* $(X, (\cdot, \cdot), \| \cdot \|)$, *the following parallelogram law holds:*

$$\|x + y\|^2 + \|x - y\|^2 = 2\left[\|x\|^2 + \|y\|^2 \right], \quad x, y \in X.$$

Conversely, if $(X, \| \cdot \|)$ is a normed vector space, whose norm satisfies the parallelogram law, then X is an inner product space, i. e., the norm is generated by an inner product $(\cdot, \cdot)$ on X:

$$\|x\| := (x, x)^{1/2}, \ x \in X.$$

Proof. The *parallelogram law* is derived directly from identity (4.2).

Exercise 4.8. Derive.

Suppose now that $(X, \| \cdot \|)$ is a normed vector space, whose norm satisfies the parallelogram law.

Let us first consider the *real case* and show that

$$(x, y) := \frac{1}{4} \left[\|x + y\|^2 - \|x - y\|^2 \right], \ x, y \in X,$$

is an *inner product* on X generating the norm $\| \cdot \|$.

The latter follows from the fact that

$$(x, x) := \|x\|^2, \ x \in X,$$

which also immediately implies *positive definiteness* for $(\cdot, \cdot)$.

The *symmetry* property also directly following from the definition, it remains to prove the *linearity* in the first argument.

To prove the *additivity* in the first argument, consider the function

$$F(x, y, z) := 4 \left[(x + y, z) - (x, z) - (y, z) \right], \ x, y, z \in X.$$

By the definition of $(\cdot, \cdot)$,

$$F(x, y, z) = \|x + y + z\|^2 - \|x + y - z\|^2 - \|x + z\|^2 + \|x - z\|^2 - \|y + z\|^2 + \|y - z\|^2. \quad (4.3)$$

Whence, since, by the *parallelogram law*,

$$\|x + y \pm z\|^2 = 2\|x \pm z\|^2 + 2\|y\|^2 - \|x - y \pm z\|^2,$$

we arrive at

$$F(x, y, z) = -\|x - y + z\|^2 + \|x - y - z\|^2 + \|x + z\|^2 - \|x - z\|^2 - \|y + z\|^2 + \|y - z\|^2. \quad (4.4)$$

Hence, using *absolute scalability* of norm, we have:

$$F(x, y, z) = \frac{1}{2}(4.3) + \frac{1}{2}(4.4)$$
$$= \frac{1}{2} \left[\|y + z + x\|^2 + \|y + z - x\|^2 \right] - \frac{1}{2} \left[\|y - z + x\|^2 + \|y - z - x\|^2 \right]$$

$$- \|y + z\|^2 + \|y - z\|^2 \qquad\qquad \text{by the } \textit{parallelogram law;}$$
$$= \|y + z\|^2 + \|x\|^2 - \|y - z\|^2 - \|x\|^2 - \|y + z\|^2 + \|y - z\|^2 = 0, \ x, y, z \in X,$$

which proves that

$$(x + y, z) = (x, z) + (y, z), \ x, y, z \in X,$$

i. e., the *additivity* in the first argument.

To prove the *homogeneity* in the first argument, consider the function

$$f(\lambda, x, y) := (\lambda x, y) - \lambda(x, y), \ \lambda \in \mathbb{R}, \ x, y \in X.$$

By the definition, and *absolute scalability* of norm,

$$f(0, x, y) = (0, y) = \frac{1}{4} \left[\|y\|^2 - \| - y\|^2 \right] = 0, \ x, y \in X,$$

i. e.,

$$(0, y) = 0(x, y), \ x, y \in X,$$

and

$$f(-1, x, y) = (-x, y) + (x, y) = \frac{1}{4} \left[\| - x + y\|^2 - \| - x - y\|^2 \right]$$
$$+ \frac{1}{4} \left[\|x + y\|^2 - \|x - y\|^2 \right] = 0, \ x, y \in X,$$

i. e.,

$$(-x, y) = -(x, y), \ x, y \in X.$$

Considering this and the demonstrated *additivity in the first argument*, for any $m \in \mathbb{Z}$, we have:

$$f(m, x, y) = (\operatorname{sgn} m(\underbrace{x + \cdots + x}_{|m| \text{ times}}), y) - m(x, y) = \operatorname{sgn} m \big[\underbrace{(x, y) + \cdots + (x, y)}_{|m| \text{ times}} \big] - m(x, y)$$
$$= \operatorname{sgn} m \cdot |m|(x, y) - m(x, y) = m(x, y) - m(x, y) = 0, \ x, y \in X,$$

i. e.,

$$(mx, y) = m(x, y), \ m \in \mathbb{Z}, \ x, y \in X.$$

For any $m \in \mathbb{Z}$ and $n \in \mathbb{N}$,

$$f(m/n, x, y) = ((m/n)x, y) - (m/n)(x, y) = m((1/n)x, y) - (m/n)(x, y)$$
$$= (m/n)n((1/n)x, y) - (m/n)(x, y) = (m/n)(x, y) - (m/n)(x, y) = 0,$$

i. e.,

$$f(\lambda, x, y) = 0, \; \lambda \in \mathbb{Q}, \; x, y \in X.$$

Considering the *continuity* of

$$f(\lambda x, y) = (\lambda x, y) - \lambda(x, y)$$
$$= \frac{1}{4} \left[\|\lambda x + y\|^2 - \|\lambda x - y\|^2 \right] - \frac{\lambda}{4} \left[\|x + y\|^2 - \|x - y\|^2 \right]$$

in λ for each fixed $x, y \in X$, we conclude that

$$f(\lambda, x, y) = 0, \; \lambda \in \mathbb{R}, \; x, y \in X,$$

i. e.,

$$(\lambda x, y) = \lambda(x, y), \; \lambda \in \mathbb{R}, \; x, y \in X,$$

which completes the proof for the *real case*.

The *complex case* is proved by defining an *inner product* on X as follows

$$(x, y) := \frac{1}{4} \left[\|x + y\|^2 - \|x - y\|^2 \right] + \frac{i}{4} \left[\|x + iy\|^2 - \|x - iy\|^2 \right], \; x, y \in X,$$

separating the real and imaginary parts.

The *additivity* and *homogeneity* in the first argument for *real scalars* follow from the *real case*.

Since, by *absolute scalability* of norm,

$$(ix, y) = \frac{1}{4} \left[\|ix + y\|^2 - \|ix - y\|^2 \right] + \frac{i}{4} \left[\|ix + iy\|^2 - \|ix - iy\|^2 \right]$$
$$= \frac{1}{4} \left[\|x - iy\|^2 - \|x + iy\|^2 \right] + \frac{i}{4} \left[\|x + y\|^2 - \|x - y\|^2 \right] = i(x, y), \; x, y \in X,$$

the *homogeneity* in the first argument holds for *complex scalars* as well, which completes the proof for the *complex case*. □

Remark 4.5. Observe that, in both real and complex cases, the only way to define the potential inner product is necessarily via the *Polarization Identities* (Proposition 4.2).

Corollary 4.1 (When Banach Space is Hilbert). *A Banach space* $(X, \| \cdot \|)$ *is a Hilbert space iff the parallelogram law holds.*

Exercise 4.9. Show that
(a) the Banach spaces $l_p^{(n)}$ and l_p $(1 \le p \le \infty, n \in \mathbb{N})$ are Hilbert spaces *iff* $p = 2$;
(b) the Banach space $(C[a, b], \| \cdot \|_\infty)$ is not a Hilbert space.

An important implication of the *Parallelogram Law* (Theorem 4.3) is as follows.

Proposition 4.3 (Strict Convexity of Inner Product Spaces). *An inner product space* $(X, (\cdot, \cdot), \|\cdot\|)$ *is strictly convex.*

Exercise 4.10. Prove.

Hint. Apply the *Parallelogram Law* (Theorem 4.3) and the *Midpoint Characterization of Strict Convexity* (Proposition 3.21) (see Section 3.6, Problem 19).

4.4.3 Orthogonality

The importance of the *orthogonality* concept, inherent to inner product spaces, for their theory and applications cannot be overestimated.

Definition 4.4 (Orthogonality). Two vectors x and y in an inner product space $(X, (\cdot, \cdot))$ are said to be *orthogonal* if

$$(x, y) = 0.$$

Notation. $x \perp y$.

Remark 4.6. Due to *conjugate symmetry* of inner product, the relationship of orthogonality of vectors is *symmetric*, i. e., $x \perp y \iff y \perp x$.

Examples 4.4.
1. In the (real or complex) space $l_2^{(n)}$ ($n \in \mathbb{N}$), the elements of the *standard unit basis*

$$e_1 := (1, 0, 0, \ldots, 0), e_2 := (0, 1, 0, \ldots, 0), \ldots, e_n := (0, 0, 0, \ldots, 1)$$

are a *pairwise orthogonal*, i. e., any two distinct vectors in it are orthogonal:

$$e_i \perp e_j, \ i, j = 1, \ldots, n, \ i \neq j.$$

2. In the (real or complex) space l_2, the elements of the *standard Schauder basis* $\{e_n := \{\delta_{nk}\}_{k=1}^{\infty}\}_{n \in \mathbb{N}}$, where δ_{nk} is the Kronecker delta, are *pairwise orthogonal*.
3. In the complex space $C[0, 2\pi]$ with the integral inner product

$$C[0, 2\pi] \ni f, g \mapsto (f, g) := \int_0^{2\pi} f(t)\overline{g(t)} \, dt,$$

the functions $x_k(t) := e^{ikt}$, $k \in \mathbb{Z}$, $t \in [0, 2\pi]$, where i is the *imaginary unit*, are *pairwise orthogonal*, as well as their real parts $\cos nt$, $n \in \mathbb{Z}_+$, and imaginary parts $\sin nt$, $n \in \mathbb{N}$, in the real space $(C[0, 2\pi], (\cdot, \cdot))$.
4. The functions of the prior example represent pairwise orthogonal equivalence classes in the space $L_2(0, 2\pi)$, complex and real, respectively.

Exercise 4.11. Verify.

Proposition 4.4 (Orthogonal Characterization of Zero Vector). *In an inner product space* $(X, (\cdot, \cdot))$,

$$x = 0 \Leftrightarrow \forall y \in X : x \perp y.$$

Exercise 4.12. Prove (cf. Remarks 4.1).

The following is a natural generalization of the classical *Pythagorean Theorem*.

Theorem 4.4 (Pythagorean Theorem). *For any pair of orthogonal vectors* x, y, *in an inner product space* $(X, (\cdot, \cdot), \| \cdot \|)$,

$$\|x \pm y\|^2 = \|x\|^2 + \|y\|^2.$$

Exercise 4.13.
(a) Prove.
(b) Show that, in a *real* inner product space, the converse is also true, i. e.,

$$x \perp y \Leftrightarrow \|x + y\|^2 = \|x\|^2 + \|y\|^2.$$

(c) Is converse true in a *complex* inner product space?

Hint. For (b) and (c), using identity (4.2), show that

$$\|x + y\|^2 = \|x\|^2 + \|y\|^2 \Leftrightarrow \mathrm{Re}(x, y) = 0,$$

then infer and find a *counterexample*, respectively.

4.5 Nearest Point Property

The following property of Hilbert spaces, important by itself, has far-reaching implications and is instrumental for proving the subsequent *Projection Theorem* (Theorem 4.6), which, in its turn, underlies the proof of the *Riesz Representation Theorem* (Theorem 7.1), and hence, is fundamental for the duality theory of such spaces.

Theorem 4.5 (Nearest Point Property). *Let* C *be a closed convex set in a Hilbert space* $(X, (\cdot, \cdot), \| \cdot \|)$. *Then, for each* $x \in X$, *there is a unique nearest point to* x *in* C, *i. e.,*

$$\forall x \in X \, \exists! \, y \in C : \|x - y\| = \rho(x, C) := \inf_{u \in C} \|x - u\|.$$

Proof. If $x \in C$ the statement is trivially true.

Exercise 4.14. Explain.

Suppose $x \notin C$. Then, by the *closedness* of C, $\rho(x, C) > 0$.
Choosing a sequence $\{y_n\}_{n=1}^{\infty}$ in C such that

$$\|x - y_n\| \to \rho(x, C), \ n \to \infty, \tag{4.5}$$

by the *Parallelogram Law* (Theorem 4.3), we have:

$$\|y_n - y_m\|^2 = \|y_n - x + x - y_m\|^2 = 2\|y_n - x\|^2 + 2\|x - y_m\|^2 - \|y_n + y_m - 2x\|^2$$

$$= 2\|y_n - x\|^2 + 2\|x - y_m\|^2 - 4\left\|\frac{y_n + y_m}{2} - x\right\|^2, \ m, n \in \mathbb{N}.$$

Since, by the *convexity* of C, $\frac{y_n + y_m}{2} \in C$, $m, n \in \mathbb{N}$, and hence

$$\left\|\frac{y_n + y_m}{2} - x\right\| \geq \rho(x, C), \ m, n \in \mathbb{N},$$

in view of (4.5), we infer

$$\|y_n - y_m\|^2 \leq 2\|y_n - x\|^2 + 2\|x - y_m\|^2 - 4\rho^2(x, C) \to 0, \ m, n \to \infty,$$

which implies that the sequence $\{y_n\}_{n=1}^{\infty}$ is *fundamental*.

Therefore, by the *completeness* of $(X, (\cdot, \cdot), \|\cdot\|)$ and the *closedness* of C,

$$\exists y \in C : y_n \to y, \ n \to \infty,$$

which in view of (4.5), by *continuity* of norm, implies that

$$\|x - y\| = \lim_{n \to \infty} \|x - y_n\| = \rho(x, C),$$

i. e., y is a *nearest point* to x in C, which completes the proof of the *existence*.

To prove the *uniqueness*, suppose

$$\|x - z\| = \rho(x, C)$$

for some $z \in C$. Then, by the *Parallelogram Law* (Theorem 4.3),

$$\|y - z\|^2 = \|y - x + x - z\|^2 = 2\|y - x\|^2 + 2\|x - z\|^2 - \|y + z - 2x\|^2$$

$$= 2\|y - x\|^2 + 2\|x - z\|^2 - 4\left\|\frac{y + z}{2} - x\right\|^2 = 4\rho^2(x, C) - 4\left\|\frac{y + z}{2} - x\right\|^2.$$

Since, by the *convexity* of C, $\frac{y+z}{2} \in C$, and hence,

$$\left\|\frac{y + z}{2} - x\right\| \geq \rho(x, C),$$

we infer

$$\|y - z\|^2 \leq 4\rho^2(x, C) - 4\rho^2(x, C) = 0,$$

which, by the *separation* norm axiom, implies that $z = y$ and completes the proof of the *uniqueness* and of the theorem. $\qquad\qquad\square$

Since every subspace is a convex set (see Examples 3.19), we immediately obtain the following corollary.

Corollary 4.2 (Nearest Point Property Relative to Closed Subspaces). *Let Y be a closed subspace in a Hilbert space $(X, (\cdot, \cdot), \|\cdot\|)$. Then, for each $x \in X$, there is a unique nearest point to x in Y, i.e.,*

$$\forall x \in X \ \exists! \, y \in Y : \ \|x - y\| = \rho(x, Y) := \inf_{u \in Y} \|x - u\|.$$

Cf. the *Nearest Point Property Relative to Finite-Dimensional Subspaces* (Proposition 3.14) and Remark 3.27.

Remarks 4.7.
- The prior theorem is also known as the *Closest Point Property* (see, e. g., [54]).
- The *closedness* condition on the convex set is essential for the existence of the closest points and cannot be dropped.

  Exercise 4.15. Give a corresponding example.

- The *completeness* condition on the space is also essential for the existence of the closest points and cannot be dropped.

  Example 4.5. In $(c_{00}, (\cdot, \cdot), \|\cdot\|_2)$, which is an incomplete inner product space, when treated as a subspace of the Hilbert space l_2 (see Examples 4.3), consider the nonempty set

  $$Y := \left\{ x := (x_n)_{n \in \mathbb{N}} \in c_{00} \ \middle| \ \sum_{k=1}^{\infty} \frac{1}{k} x_k = 0 \right\} = \{x \in c_{00} \mid x \perp y = 0 \text{ in } l_2\}$$

  where $y := (1/n)_{n \in \mathbb{N}}$.
  As is easily seen, Y is a *proper subspace* in c_{00}, and hence, in particular is a *convex set*.

  Exercise 4.16. Verify.

  Furthermore, Y is *closed* by the *Sequential Characterization of a Closed Set* (Theorem 2.18), since, if

  $$Y \ni x_n \to x \in c_{00}, \ n \to \infty, \text{ in } l_2,$$

  in view of $(x_n, y) = 0$, $n \in \mathbb{N}$, by *continuity* of inner product (Proposition 4.1),

  $$(x, y) = \lim_{n \to \infty} (x_n, y) = 0,$$

  which implies that $x \in Y$.
  However, as is shown in Example 4.7 of the following section, in the space $(c_{00}, (\cdot, \cdot), \|\cdot\|_2)$, to no point $x \in Y^c$, is there a nearest point in Y.

– Since the proof of the *uniqueness part* of the prior theorem is entirely based on the *Parallelogram Law* (Theorem 4.3) and uses the *convexity* only, we conclude that, in any inner product space, the *nearest point* in a convex set to a point of the space, if it exists, is *unique*. This is consistent with the *strict convexity* of inner product spaces (Proposition 4.3) and the *Nearest Point Property for Strictly Convex Spaces* (Theorem 3.18).

Exercise 4.17.
(a) Give an example of a Banach space, which is not a Hilbert space but where the analogue of the *Nearest Point Property* relative to closed convex sets holds.
(b) Give an example of a Banach space, in which the analogue of the *Nearest Point Property* relative to closed convex sets does not hold.

4.6 Projection Theorem

4.6.1 Orthogonal Complements

Definition 4.5 (Orthogonal Complement). In an inner product space $(X, (\cdot, \cdot), \| \cdot \|)$, the *orthogonal complement* $M^\perp$ of a nonempty set M is the set of all elements orthogonal to M, i. e.

$$M^\perp := \{z \in X \mid z \perp M, \text{ i. e., } z \perp y \text{ for all } y \in M\}.$$

Examples 4.6.
1. In an inner product space $(X, (\cdot, \cdot), \| \cdot \|)$, $\{0\}^\perp = X$ and $X^\perp = \{0\}$.
2. In the Euclidean 2-space $l_2^{(2)}(\mathbb{R})$,

$$\{(x, 0) \mid x \in \mathbb{R}\}^\perp = \{(0, y) \mid y \in \mathbb{R}\}.$$

3. For a fixed $n \in \mathbb{N}$, in l_2,

$$\{(x_k)_{k \in \mathbb{N}} \in l_2 \mid x_k = 0, \ k \geq n + 1\}^\perp = \{(x_k)_{k \in \mathbb{N}} \in l_2 \mid x_k = 0, \ k = 1, \ldots, n\}.$$

Exercise 4.18. Verify.

Proposition 4.5 (Orthogonal Complement is a Closed Subspace). *In an inner product space* $(X, (\cdot, \cdot), \| \cdot \|)$*, the orthogonal complement* $M^\perp$ *of a nonempty set* M *is a closed subspace and*

$$M \cap M^\perp \subseteq \{0\}.$$

Exercise 4.19. Prove.

Proposition 4.6 (Characterization of Orthogonal Complement of Subspace). *Let Y be a subspace in an inner product space $(X, (\cdot, \cdot), \|\cdot\|)$. Then, for any $x \in X$,*

$$x \in Y^{\perp} \iff \forall y \in Y : \|x - y\| \geq \|x\|,$$

i. e., $x \in Y^{\perp}$ iff 0 is the nearest point to x in Y.

Proof. "Only if" part. If $x \in Y^{\perp}$, then

$$\forall y \in Y : x \perp y$$

and, by the *Pythagorean Theorem* (Theorem 4.4),

$$\|x - y\|^2 = \|x\|^2 + \|y\|^2 \geq \|x\|^2.$$

"*If*" part. Let us prove this part *by contradiction* assuming the existence of an $x \in X$ such that

$$\forall y \in Y : \|x - y\| \geq \|x\|, \text{ but } x \not\perp Y.$$

Then, for each $y \in Y$ and any $\lambda \in \mathbb{F}$, since Y is a *subspace*, $\lambda y \in Y$ and, by identity (4.2),

$$\|x\|^2 - 2 \operatorname{Re} \bar{\lambda}(x, y) + |\lambda|^2 \|y\|^2 = \|x - \lambda y\|^2 \geq \|x\|^2.$$

Whence,

$$-2 \operatorname{Re} \bar{\lambda}(x, y) + |\lambda|^2 \|y\|^2 \geq 0, \ \lambda \in \mathbb{F}, \ y \in Y.$$

Since, by the assumption,

$$\exists y \in Y : (x, y) \neq 0,$$

setting $\lambda := t \frac{(x,y)}{|(x,y)|}$ with $t > 0$, we have:

$$\forall t > 0 : -2t |(x, y)| + t^2 \|y\|^2 \geq 0.$$

Whence,

$$\forall t > 0 : |(x, y)| \leq \frac{t}{2} \|y\|^2,$$

and letting $t \to 0+$, we infer that $(x, y) = 0$, which is a *contradiction* proving the "*if*" part. $\qquad \square$

4.6.2 Projection Theorem

Let us now prove the following key theorem.

Theorem 4.6 (Projection Theorem). *If Y is a closed subspace in a Hilbert space $(X, (\cdot, \cdot), \|\cdot\|)$, then every $x \in X$ has a unique decomposition*

$$x = y + z, \ y \in Y, \ z \in Y^{\perp},$$

where the element y, called the orthogonal projection of x on Y, is the nearest point to x in Y, and hence, the space X is decomposed into the orthogonal sum of the subspaces Y and $Y^{\perp}$:

$$X = Y \oplus Y^{\perp}.$$

Proof. Let $x \in X$ be arbitrary. By the *Nearest Point Property Relative to Closed Subspaces* (Corollary 4.2), we can choose $y \in Y$ to be the *unique nearest point* to x in Y. Then

$$x = y + z$$

with $z = x - y$.

Since Y is a subspace, for any $y' \in Y$, $y + y' \in Y$, and we have:

$$\|z - y'\| = \|x - (y + y')\| \geq \rho(x, Y) = \|x - y\| = \|z\|,$$

which, by the *Characterization of Orthogonal Complement of Subspace* (Proposition 4.6), implies that $z \in Y^{\perp}$.

The uniqueness of the decomposition immediately follows from the fact that the subspaces Y and $Y^{\perp}$ are *disjoint*, i. e.,

$$Y \cap Y^{\perp} = \{0\}$$

(see Proposition 4.5).

Hence, the subspaces Y and $Y^{\perp}$ are *complementary* and

$$X = Y \oplus Y^{\perp},$$

the direct sum decomposition naturally called *orthogonal*. □

Remark 4.8. As follows from the argument based on the *Characterization of Orthogonal Complement of Subspace* (Proposition 4.6) used in the foregoing proof, if, in an inner product space $(X, (\cdot, \cdot), \|\cdot\|)$, a y is the *nearest point* to an $x \in X$ in a subspace Y, which need not be closed, then $x - y \in Y^{\perp}$.

Proposition 4.7 (Twice Orthogonal Complement of Closed Subspace). *In a Hilbert space* $(X, (\cdot,\cdot), \|\cdot\|)$, *for every closed subspace* Y,

$$\left(Y^{\perp}\right)^{\perp} = Y.$$

Proof. It is obvious that $Y \subseteq (Y^{\perp})^{\perp}$.

To prove the inverse inclusion, consider an arbitrary $x \in (Y^{\perp})^{\perp}$. By the *Projection Theorem* (Theorem 4.6),

$$x = y + z$$

with some $y \in Y$ and $z \in Y^{\perp}$.

Since $x \in (Y^{\perp})^{\perp}$,

$$0 = (x, z) = (y + z, z) = (y, z) + \|z\|^2 = 0 + \|z\|^2 = \|z\|^2,$$

which implies that $z = 0$, and hence, $x = y \in Y$, i.e., $(Y^{\perp})^{\perp} \subseteq Y$.

Thus, we conclude that $(Y^{\perp})^{\perp} = Y$. $\square$

Remark 4.9. As the following example shows, in the *Projection Theorem* (Theorem 4.6), the requirement of the *completeness* of the space is essential and cannot be dropped.

Example 4.7. As is shown in Example 4.5,

$$Y := \{x \in c_{00} \mid x \perp y \text{ in } l_2\},$$

where $y := (1/n)_{n \in \mathbb{N}} \in l_2$, is a *proper closed subspace* in the incomplete inner product space $(c_{00}, (\cdot,\cdot), \|\cdot\|_2)$, the latter being treated as a subspace in l_2. Observe that the closedness of Y also immediately follows from the fact that

$$Y = c_{00} \cap \{y\}^{\perp},$$

where $\{y\}^{\perp}$ is the orthogonal complement of $\{y\}$ in l_2.

Then, for the orthogonal complement $Y^{\perp}$ of Y in $(c_{00}, (\cdot,\cdot), \|\cdot\|_2)$, we have:

$$Y^{\perp} = c_{00} \cap \left(\{y\}^{\perp}\right)^{\perp},$$

where $(\{y\}^{\perp})^{\perp}$ is the twice orthogonal complement of $\{y\}$ in l_2. By Proposition 4.17 (see Section 4.13, Problem 8), in l_2,

$$\left(\{y\}^{\perp}\right)^{\perp} = \text{span}(\{y\}) = \{\lambda y \mid \lambda \in \mathbb{F}\},$$

and hence, in $(c_{00}, (\cdot,\cdot), \|\cdot\|_2)$,

$$Y^{\perp} = c_{00} \cap \text{span}(\{y\}) = \{0\}.$$

Exercise 4.20. Explain.

Therefore,

$$c_{00} \neq Y \oplus Y^\perp = Y.$$

Furthermore, for any $x \in c_{00} \setminus Y$, assume that $y \in Y$ is the nearest point to x in Y. Then $x - y \in Y^\perp$ (see Remark 4.8), which, since $Y^\perp = \{0\}$, implies that $x = y \in Y$, which is a *contradiction* showing that, in $(c_{00}, (\cdot, \cdot), \|\cdot\|_2)$, to no point $x \in Y^c$ is there a nearest point in Y.

Remark 4.10. Example 4.7 also shows that the requirement of the *completeness* of the space is essential and cannot be dropped in Propositions 4.7 and 4.17. Indeed, in $(c_{00}, (\cdot, \cdot), \|\cdot\|_2)$,

$$(Y^\perp)^\perp = \{0\}^\perp = c_{00} \neq Y.$$

Proposition 4.8 (Characterization of Denseness of Subspace). *For a subspace Y in a Hilbert space $(X, (\cdot, \cdot), \|\cdot\|)$,*

$$\overline{Y} = X \iff Y^\perp = \{0\}$$

Proof. Follows immediately from the fact that

$$Y^\perp = \overline{Y}^\perp$$

(see Exercise 4.6) and the *Projection Theorem* (Theorem 4.6).

Exercise 4.21. Fill in the details. □

Remark 4.11. Example 4.7 also shows that, in the prior proposition, the requirement of the *completeness* of the space is essential and cannot be dropped. Indeed, in it, $\overline{Y} = Y \neq X$, but $Y^\perp = \{0\}$.

4.7 Completion

A completion procedure similar to that for normed vector spaces holds for inner product spaces.

Theorem 4.7 (Completion Theorem for Inner Product Spaces). *An arbitrary inner product space $(X, (\cdot, \cdot)_X, \|\cdot\|_X)$ over $\mathbb{F}$ can be isometrically embedded, as a dense subspace, in a Hilbert space $(\tilde{X}, (\cdot, \cdot)_{\tilde{X}}, \|\cdot\|_{\tilde{X}})$ over $\mathbb{F}$ called a completion of $(X, (\cdot, \cdot)_X, \|\cdot\|_X)$.*
Any two completions of $(X, (\cdot, \cdot)_X, \|\cdot\|_X)$ are isometrically isomorphic.

Proof. The statement follows from the *Completion Theorem for Normed Vector Spaces* (Theorem 3.9), considering that, in the constructed completion Banach space $(\tilde{X}, \|\cdot\|_{\tilde{X}})$

the equivalence classes of the *asymptotically equivalent* fundamental sequences of elements of X, the norm

$$\tilde{X} \ni [x] \mapsto \|[x]\|_{\tilde{X}} := \lim_{n \to \infty} \|x_n\|_X,$$

where the fundamental sequence $x := \{x_n\}_{n=1}^{\infty}$ is an arbitrary representative of the equivalence class $[x]$, is generated by the *inner product*

$$\tilde{X} \ni [x], [y] \mapsto ([x], [y])|_{\tilde{X}} := \lim_{n \to \infty} (x_n, y_n)_X, \quad x, y \in X, \tag{4.6}$$

where $x := \{x_n\}_{n=1}^{\infty}$ and $y := \{y_n\}_{n=1}^{\infty}$ are *arbitrary* representatives of the classes $[x]$ and $[y]$, respectively.

Exercise 4.22. Verify that the inner product $(\cdot, \cdot)_{\tilde{X}}$ is *well-defined* and generates the *norm* $\| \cdot \|_{\tilde{X}}$ on $\tilde{X}$.

Hint. The existence of the limit in (4.6) and its independence of the choice of the representative sequence follow from the *Cauchy–Schwarz inequality* (Theorem 4.2). □

Remark 4.12. It follows immediately that, if $(X_0, (\cdot, \cdot), \| \cdot \|)$ is a *dense subspace* of a Hilbert space $(X, (\cdot, \cdot), \| \cdot \|)$, the latter is a *completion* of the *former*.

Examples 4.8.
1. l_2 is a *completion* of $(c_{00}, (\cdot, \cdot), \| \cdot \|_2)$ (see Examples 4.3).
2. $L_2(a, b)$ $(-\infty < a < b < \infty)$ is a *completion* of $(C[a, b], (\cdot, \cdot), \| \cdot \|_2)$ or $(P, (\cdot, \cdot), \| \cdot \|_2)$, the latter interpreted as the subspaces of the equivalence classes represented by continuous on $[a, b]$ functions and polynomials, respectively (see, e. g., [16, 33, 43, 44, 47]).

4.8 Gram Determinant

In an inner product space, the notion of the *Gram*[3] *determinant*, familiar from linear algebra enables one to determine the nearest point to an element in a finite-dimensional subspace, the distance to it, and to characterize linear (in-)dependence for finite sets.

Definition 4.6 (Gram Matrix and Determinant). For a finite set of vectors $\{x_1, \ldots, x_n\}$ $(n \in \mathbb{N})$ in an inner product space $(X, (\cdot, \cdot), \| \cdot \|)$, the *Gram matrix* is defined as follows:

$$[(x_i, x_j)]_{i,j=1}^{n} = \begin{bmatrix} (x_1, x_1) & (x_1, x_2) & \cdots & (x_1, x_n) \\ (x_2, x_1) & (x_2, x_2) & \cdots & (x_2, x_n) \\ \vdots & \vdots & & \vdots \\ (x_n, x_1) & (x_n, x_2) & \cdots & (x_n, x_n) \end{bmatrix}.$$

---

3 Jørgen Pedersen Gram (1850–1916).

Its determinant

$$G(x_1,\ldots,x_n) := \det\left[(x_i,x_j)\right]_{i,j=1}^n = \begin{vmatrix} (x_1,x_1) & (x_1,x_2) & \cdots & (x_1,x_n) \\ (x_2,x_1) & (x_2,x_2) & \cdots & (x_2,x_n) \\ \vdots & \vdots & & \vdots \\ (x_n,x_1) & (x_n,x_2) & \cdots & (x_n,x_n) \end{vmatrix}$$

is called the *Gram determinant* of the set.

Remark 4.13. As immediately follows from *conjugate symmetry* of inner product, the Gram matrix is *self-adjoint* (or *Hermitian*[4]), i. e. it coincides with its *conjugate transpose*.

Exercise 4.23.
(a) Verify.
(b) Describe the Gram matrix and determinant for $n = 1$ and $n = 2$.

By the *Nearest Point Property Relative to Finite-Dimensional Subspaces* (Proposition 3.14), in a normed vector space $(X, \|\cdot\|)$, for each $x \in X$ and an arbitrary finite-dimensional subspace Y, there is a nearest point to x in Y. If, in addition, the space X is an inner product space, which is *strictly convex* (Proposition 4.3), by the *Nearest Point Property for Strictly Convex Spaces Relative to Finite-Dimensional Subspaces* (Corollary 3.8), such a point is also *unique*. The following statement describes how, in an inner product space $(X, (\cdot,\cdot), \|\cdot\|)$, to determine the nearest point to an element x in a finite-dimensional subspace Y and the distance to it via a *Gram determinant* approach.

Theorem 4.8 (Gram Determinant and the Nearest Point). *Let $\{x_1,\ldots,x_n\}$ $(n \in \mathbb{N})$ be a finite linearly independent set in an inner product space $(X, (\cdot,\cdot), \|\cdot\|)$. Then, for each $x \in X$, the nearest point to x in the n-dimensional subspace $Y := \mathrm{span}(\{x_1,\ldots,x_n\})$ is*

$$y = \sum_{k=1}^n \lambda_k x_k$$

with

$$\lambda_k = \frac{G_k}{G(x_1,\ldots,x_n)}, \quad k = 1,\ldots,n,$$

where G_k is the determinant obtained from $G(x_1,\ldots,x_n)$ by replacing the entries of its kth row by $(x,x_j), j = 1,\ldots,n$, and

$$\rho^2(x,Y) = \|x-y\|^2 = \frac{G(x_1,\ldots,x_n,x)}{G(x_1,\ldots,x_n)}.$$

---

4 Charles Hermite (1822–1901).

Proof. Since the subspace $Y = \text{span}(\{x_1, \ldots, x_n\})$ is *finite-dimensional* and an inner product space is *strictly convex* (Proposition 4.3), by the *Nearest Point Property for Strictly Convex Spaces Relative to Finite-Dimensional Subspaces* (Corollary 3.8), for each $x \in X$, there is a *unique nearest point* to x in Y, i. e.,

$$\forall x \in X \; \exists! \, y \in Y : \; \|x - y\| = \rho(x, Y) := \inf_{u \in Y} \|x - u\|.$$

Further,

$$x - y \in Y^{\perp}$$

(see Remark 4.8), and hence,

$$(x - y, x_k) = 0, \; k = 1, \ldots, n. \tag{4.7}$$

Since $y \in Y$, it is *uniquely* represented as a linear combination of $x_1, \ldots, x_n$:

$$y = \sum_{k=1}^{n} \lambda_k x_k \tag{4.8}$$

with some $\lambda_k \in \mathbb{F}$, $k = 1, \ldots, n$ (see Theorem 3.3).

The latter combined with (4.7), yields the following linear system relative to $\lambda_1, \ldots, \lambda_n$:

$$
\begin{array}{ccccccccc}
\lambda_1(x_1, x_1) & + & \lambda_2(x_2, x_1) & + & \cdots & + & \lambda_n(x_n, x_1) & = & (x, x_1) \\
\lambda_1(x_1, x_2) & + & \lambda_2(x_2, x_2) & + & \cdots & + & \lambda_n(x_n, x_2) & = & (x, x_2) \\
\vdots & & & & & & \vdots & & \\
\lambda_1(x_1, x_n) & + & \lambda_2(x_2, x_n) & + & \cdots & + & \lambda_n(x_n, x_n) & = & (x, x_n)
\end{array}
\tag{4.9}
$$

whose coefficient matrix, being *transpose* to the *Gram matrix* of $\{x_1, \ldots, x_n\}$, has the determinant equal to $G(x_1, \ldots, x_n)$ (see, e. g., [40]).

Since, by the uniqueness of the nearest point y and representation (4.8) for it, the system has a unique solution for any $x \in X$, in particular for $x = 0$, when it is *homogeneous*, we infer that

$$G(x_1, \ldots, x_n) \neq 0.$$

Hence, by *Cramer's*[5] *Rule* (see, e. g., [40]), in view of *transposition*,

$$\lambda_k = \frac{G_k}{G(x_1, \ldots, x_n)}, \; k = 1, \ldots, n.$$

---

5 Gabriel Cramer (1704–1752).

Further, in view of $x - y \in Y^{\perp}$, we have:

$$\rho^2(x, Y) = \|x - y\|^2 = (x - y, x - y) = (x - y, x) - (x - y, y) \qquad \text{since } (x - y, y) = 0;$$
$$= (x - y, x) = (x, x) - (y, x),$$

and hence, in view of (4.8),

$$\lambda_1(x_1, x) + \lambda_2(x_2, x) + \cdots + \lambda_n(x_n, x) = (x, x) - \rho^2(x, Y).$$

Appending this equation to system (4.9), we conclude that

$$\begin{vmatrix} (x_1, x_1) & (x_2, x_1) & \cdots & (x_n, x_1) & (x, x_1) \\ (x_1, x_2) & (x_2, x_2) & \cdots & (x_n, x_2) & (x, x_2) \\ \vdots & \vdots & & \vdots & \vdots \\ (x_1, x_n) & (x_2, x_n) & \cdots & (x_n, x_n) & (x, x_n) \\ (x_1, x) & (x_2, x) & \cdots & (x_n, x) & (x, x) - \rho^2(x, Y) \end{vmatrix} = 0.$$

Exercise 4.24. Explain why.

Whence, we obtain

$$\begin{vmatrix} (x_1, x_1) & (x_2, x_1) & \cdots & (x_n, x_1) & (x, x_1) \\ (x_1, x_2) & (x_2, x_2) & \cdots & (x_n, x_2) & (x, x_2) \\ \vdots & \vdots & & \vdots & \vdots \\ (x_1, x_n) & (x_2, x_n) & \cdots & (x_n, x_n) & (x, x_n) \\ (x_1, x) & (x_2, x) & \cdots & (x_n, x) & (x, x) \end{vmatrix} = \begin{vmatrix} (x_1, x_1) & (x_2, x_1) & \cdots & (x_n, x_1) & 0 \\ (x_1, x_2) & (x_2, x_2) & \cdots & (x_n, x_2) & 0 \\ \vdots & \vdots & & \vdots & \vdots \\ (x_1, x_n) & (x_2, x_n) & \cdots & (x_n, x_n) & 0 \\ (x_1, x) & (x_2, x) & \cdots & (x_n, x) & \rho^2(x, Y) \end{vmatrix}$$

and, using the expansion of the right-hand side determinant along the last column, in view of *transposition*, we arrive at

$$\rho^2(x, Y) = \|x - y\|^2 = \frac{G(x_1, \ldots, x_n, x)}{G(x_1, \ldots, x_n)}. \tag{4.10}$$

Exercise 4.25. Verify. ☐

Using Gram determinant one can characterize linear (in-)dependence for finite sets in inner product spaces.

Proposition 4.9 (Gram Determinant Characterization of Linear Independence). *In an inner product space $(X, (\cdot, \cdot), \| \cdot \|)$, a finite set $\{x_1, \ldots, x_n\}$ ($n \in \mathbb{N}$) is linearly independent iff*

$$G(x_1, \ldots, x_n) > 0.$$

Exercise 4.26. Prove.

Hint. Observe that $G(x_1) = (x_1, x_1) = \|x_1\|^2 > 0$ and use *induction* on formula (4.10).

Proposition 4.10 (Gram Determinant Characterization of Linear Dependence). *In an inner product space $(X, (\cdot, \cdot), \|\cdot\|)$, a finite set $\{x_1, \dots, x_n\}$ ($n \in \mathbb{N}$) is linearly dependent iff*

$$G(x_1, \dots, x_n) = 0.$$

Exercise 4.27. Prove.

Remark 4.14. Hence, in an inner product space $(X, (\cdot, \cdot), \|\cdot\|)$, for any finite set $\{x_1, \dots, x_n\}$ ($n \in \mathbb{N}$),

$$G(x_1, \dots, x_n) \geq 0,$$

with $G(x_1, \dots, x_n) = 0$ *iff* the set is *linearly dependent*.
   This fact can be considered as a generalization of the *Cauchy–Schwarz inequality*, which is obtained as a particular case when $n = 2$.

Exercise 4.28. Verify.

Example 4.9. In the (real or complex) pre-Hilbert space $(c_{00}, (\cdot, \cdot), \|\cdot\|_2)$, for the linearly independent set

$$\{x_1 := (1, 0, 0, \dots), x_2 := (-1, -1, 0, \dots)\},$$

and $x := (1, 1, 1, 0, \dots)$, by the *Gram determinant approach* described in the prior theorem, the nearest point to x in the subspace $Y := \mathrm{span}(\{x_1, x_2\})$ is

$$y = \frac{G_1}{G(x_1, x_2)} x_1 + \frac{G_2}{G(x_1, x_2)} x_2$$

and

$$\rho(x, Y) = \|x - y\|_2 = \sqrt{\frac{G(x_1, x_2, x)}{G(x_1, x_2)}}.$$

Whence with

$$G(x_1, x_2) = \begin{vmatrix} (x_1, x_1) & (x_1, x_2) \\ (x_2, x_1) & (x_2, x_2) \end{vmatrix} = \begin{vmatrix} 1 & -1 \\ -1 & 2 \end{vmatrix} = 1,$$

$$G_1 = \begin{vmatrix} (x, x_1) & (x, x_2) \\ (x_2, x_1) & (x_2, x_2) \end{vmatrix} = \begin{vmatrix} 1 & -2 \\ -1 & 2 \end{vmatrix} = 0,$$

$$G_2 = \begin{vmatrix} (x_1, x_1) & (x_1, x_2) \\ (x, x_1) & (x, x_2) \end{vmatrix} = \begin{vmatrix} 1 & -1 \\ 1 & -2 \end{vmatrix} = -1, \quad \text{and}$$

$$G(x_1, x_2, x) = \begin{vmatrix} (x_1, x_1) & (x_1, x_2) & (x_1, x) \\ (x_2, x_1) & (x_2, x_2) & (x_2, x) \\ (x, x_1) & (x, x_2) & (x, x) \end{vmatrix} = \begin{vmatrix} 1 & -1 & 1 \\ -1 & 2 & -2 \\ 1 & -2 & 3 \end{vmatrix} = \begin{vmatrix} 1 & -1 & 1 \\ 0 & 1 & -1 \\ 0 & 0 & 1 \end{vmatrix} = 1,$$

we obtain

$$y = -x_2 = (1, 1, 0, \ldots) \quad \text{and} \quad \rho(x, Y) = \|x - y\|_2 = 1.$$

Applying the prior theorem to the set $\{1, t, \ldots, t^n\}$ $(n \in \mathbb{Z}_+)$ in the inner product space $(C[a, b], (\cdot, \cdot), \|\cdot\|_2)$ $(-\infty < a < b < \infty)$, we obtain

Corollary 4.3 (Best Approximation Polynomial in $(C[a, b], (\cdot, \cdot), \|\cdot\|_2)$). *For each $f \in C[a, b]$ $(-\infty < a < b < \infty)$ and any $n \in \mathbb{Z}_+$, in $(C[a, b], (\cdot, \cdot), \|\cdot\|_2)$, there is a unique best approximation polynomial $p_n \in P_n$:*

$$\|f - p_n\|_2 := \left[\int_a^b |f(t) - p_n(t)|^2 \, dt \right]^{1/2} = \rho(x, P_n) := \inf_{u \in P_n} \|f - u\|_2,$$

i. e., a polynomial nearest to f in the $(n + 1)$-dimensional subspace P_n of all polynomial of degree at most n in $(C[a, b], (\cdot, \cdot), \|\cdot\|_2)$, which can be found via the Gram determinant approach of Theorem 4.8.

Remark 4.15. Unlike in the Banach space $(C[a, b], \|\cdot\|_\infty)$ (see Corollary 3.5), according to the prior statement, in the pre-Hilbert space $(C[a, b], (\cdot, \cdot), \|\cdot\|_2)$, for each $f \in C[a, b]$, not only does a best approximation polynomial exist, but it is also *unique*. This is consistent with the fact that the former space is not strictly convex, whereas the latter one is.

4.9 Orthogonal and Orthonormal Sets

Definition 4.7 (Orthogonal and Orthonormal Sets). A set $\{x_i\}_{i \in I}$ in an inner product space $(X, (\cdot, \cdot), \|\cdot\|)$ is called *orthogonal* if its elements are pairwise orthogonal, i. e.,

$$x_i \perp x_j, \; i, j \in I, \; i \neq j.$$

If further $\|x_i\| = 1$, for each $i \in I$, the orthogonal set is called *orthonormal*.

Remarks 4.16.
- Often, the terms *"orthogonal/orthonormal system"* and *"orthogonal/orthonormal sequence"*, when the set countably infinite, are used instead.
- For an orthonormal set $\{e_i\}_{i \in I}$,

$$(e_i, e_j) = \delta_{ij}, \; i, j \in I,$$

where δ_{ij} is the *Kronecker delta*.
- Any orthogonal set $\{x_i\}_{i \in I}$ with nonzero elements can be transformed into an orthonormal set $\{e_i\}_{i \in I}$ with the *same span* via the following simple *normalization procedure*:

$$e_i := \frac{x_i}{\|x_i\|}, \; i \in I.$$

Examples 4.10.

1. In the (real or complex) space $l_2^{(n)}$ ($n \in \mathbb{N}$), the *standard unit basis*

$$\{e_1 := (1,0,0,\ldots,0), e_2 := (0,1,0,\ldots,0), \ldots, e_n := (0,0,0,\ldots,1)\}$$

is an *orthonormal set* (see Examples 4.4).

2. In the (real or complex) space l_2, the *standard Schauder basis*

$$\{e_n := \{\delta_{nk}\}_{k=1}^{\infty}\}_{n\in\mathbb{N}},$$

where δ_{nk} is the Kronecker delta, is an *orthonormal set* (see Examples 4.4).

3. In the complex space $(C[0,2\pi], (\cdot,\cdot)_2)$, the set

$$\left\{x_k(t) := e^{ikt} \mid k \in \mathbb{Z}, t \in [0,2\pi]\right\}$$

is *orthogonal* (see Examples 4.4) and, in view of

$$\|x_k\|_2 = \sqrt{2\pi}, \ k \in \mathbb{Z},$$

it can be normalized into the following *orthonormal* one:

$$\left\{e_k(t) := \frac{e^{ikt}}{\sqrt{2\pi}} \mid k \in \mathbb{Z}, t \in [0,2\pi]\right\}$$

In the real $(C[0,2\pi], (\cdot,\cdot)_2)$ the set

$$\{1, \cos nt, \ \sin nt \mid n \in \mathbb{N}, t \in [0,2\pi]\}$$

is *orthogonal* (see Examples 4.4) and, in view of

$$\|1\|_2 = \sqrt{2\pi}, \quad \|\cos\cdot\|_2 = \|\sin n \cdot\|_2 = \sqrt{\pi}, \ n \in \mathbb{N},$$

it can be normalized into the following *orthonormal* one:

$$\left\{\frac{1}{\sqrt{2\pi}}, \frac{\cos nt}{\sqrt{\pi}}, \frac{\sin nt}{\sqrt{\pi}} \mid n \in \mathbb{N}, t \in [0,2\pi]\right\}.$$

4. The sets of the prior example are also orthogonal and orthonormal in the space $L_2(0,2\pi)$ (complex or real, respectively), their elements being interpreted as the equivalence classes represented by the corresponding functions.

Exercise 4.29. Verify.

Proposition 4.11 (Linear Independence of Orthogonal Sets). *In an inner product space* $(X, (\cdot,\cdot), \|\cdot\|)$, *an orthogonal set* $\{x_i\}_{i\in I}$ *with nonzero elements, in particular an orthonormal set, is linearly independent.*

Exercise 4.30. Prove.

Definition 4.8 (Complete Orthonormal Set). In an inner product space $(X, (\cdot, \cdot), \|\cdot\|)$, an orthonormal set $\{e_i\}_{i \in I}$, which is not a proper subset of any orthonormal set, i. e., a maximal orthonormal set relative to the set-theoretic inclusion $\subseteq$, is called *complete*.

The following statement resembles the *Basis Theorem* (Theorem 3.2). The same is true for the proofs.

Theorem 4.9 (Existence of a Complete Orthonormal Set). *In a nontrivial inner product space* $(X, (\cdot, \cdot), \|\cdot\|)$, *each orthonormal set S can be extended to a complete orthonormal set S'.*

Proof. Let S be an arbitrary orthonormal set in $(X, (\cdot, \cdot), \|\cdot\|)$.

In the collection $(\mathcal{O}, \subseteq)$ of all orthonormal subsets in X partially ordered by the set-theoretic inclusion $\subseteq$, consider an arbitrary *chain* $\mathcal{C}$ in $(\mathcal{O}, \subseteq)$.

The set

$$U := \bigcup_{C \in \mathcal{C}} C,$$

is also *orthonormal* and is an *upper bound* of $\mathcal{C}$ in $(\mathcal{O}, \subseteq)$.

Exercise 4.31. Verify.

By *Zorn's Lemma (Precise Version)* (Theorem A.6), there is a *maximal element S'* in $(\mathcal{O}, \subseteq)$, i. e., a *complete orthonormal set* in X, such that $S \subseteq S'$. □

Proposition 4.12 (Characterization of Complete Orthonormal Sets). *In an inner product space* $(X, (\cdot, \cdot), \|\cdot\|)$, *an orthonormal set $\{e_i\}_{i \in I}$ is complete iff*

$$\{e_i\}_{i \in I}^{\perp} = \{0\}.$$

Exercise 4.32. Prove.

Remarks 4.17.

– Theorem 4.9, in particular, establishes the *existence* of a complete orthonormal set in a nontrivial inner product space $(X, (\cdot, \cdot), \|\cdot\|)$.

Indeed, one can take a singleton $S := \{\frac{x}{\|x\|}\}$, where $x \in X \setminus \{0\}$ is an arbitrary nonzero element, and extend it to a complete orthonormal set S' in $(X, (\cdot, \cdot), \|\cdot\|)$.

– A complete orthonormal set in an inner product space, although *existent* by Theorem 4.9, *need not be unique.*

Thus, in the (real or complex) space $l_2^{(2)}$, the *standard unit basis*

$$\{e_1 := (1, 0), e_2 := (0, 1)\}$$

is a *complete orthonormal set*, as well as its clockwise rotation by $\pi/4$ (radians):

$$\left\{ e_1' := \frac{1}{\sqrt{2}}(1, -1), e_2' := \frac{1}{\sqrt{2}}(1, 1) \right\}$$

– In fact, if $S := \{e_i\}_{i \in I}$ is a complete orthonormal set in an inner product space $(X, (\cdot, \cdot), \|\cdot\|)$, for any $\theta \in (0, 2\pi)$, its clockwise rotation by θ $S' := \{e^{i\theta} e_j\}_{j \in I}$ (i in the exponent is the *imaginary unit*) is a complete orthonormal set in an inner product space $(X, (\cdot, \cdot), \|\cdot\|)$ as well.

Exercise 4.33. Verify.

Hint. Use the prior characterization.

Examples 4.11.

1. In the (real or complex) space $l_2^{(n)}$ ($n \in \mathbb{N}$), the *standard unit basis*

$$\{e_1 := (1, 0, 0, \ldots, 0), e_2 := (0, 1, 0, \ldots, 0), \ldots, e_n := (0, 0, 0, \ldots, 1)\}$$

   is a *complete orthonormal set*.

2. In the (real or complex) space l_2, the *standard Schauder basis*

$$\{e_n := \{\delta_{nk}\}_{k=1}^{\infty}\}_{n \in \mathbb{N}},$$

   where δ_{nk} is the Kronecker delta, is a *complete orthonormal sequence*.

3. In the *complex* space $(C[0, 2\pi], (\cdot, \cdot)_2)$, the set

$$\left\{\frac{e^{ikt}}{\sqrt{2\pi}} \,\middle|\, k \in \mathbb{Z}, t \in [0, 2\pi]\right\}$$

   is a *complete orthonormal sequence*.
   In the *real* space $(C[0, 2\pi], (\cdot, \cdot)_2)$,

$$\left\{\frac{1}{\sqrt{2\pi}}, \frac{\cos kt}{\sqrt{\pi}}, \frac{\sin kt}{\sqrt{\pi}} \,\middle|\, k \in \mathbb{Z}, t \in [0, 2\pi]\right\}$$

   is a *complete orthonormal sequence*.

4. The sets of the prior example are also *complete orthonormal sequences* in the Hilbert space $L_2(0, 2\pi)$ (complex or real, respectively), their elements being interpreted as the equivalence classes represented by the corresponding functions.

5. The set

$$S := \left\{e^{i\lambda t} \,\middle|\, -\infty < t < \infty\right\}_{\lambda \in \mathbb{R}}$$

   is an *uncountable complete orthonormal set* in a complex inner product space $(X, (\cdot, \cdot)_X)$ constructed as follows:

   – the vector space $X := \operatorname{span}(S)$, i.e., the set of all *"exponential polynomials"* of the form

$$x(t) := \sum_{k=1}^{n} a_k e^{i\lambda_k t},$$

   where $n \in \mathbb{N}$, $a_1, \ldots, a_n \in \mathbb{C}$, and $\lambda_1, \ldots, \lambda_n \in \mathbb{R}$;

– for arbitrary

$$x(t) := \sum_{k=1}^{n} a_k e^{i\lambda_k t}, \quad y(t) := \sum_{l=1}^{m} b_l e^{i\mu_l t} \in X,$$

the inner product (x, y) is defined by

$$(x, y) = \lim_{T \to \infty} \frac{1}{2T} \int_{-T}^{T} x(t)\overline{y(t)} \, dt = \lim_{T \to \infty} \frac{1}{2T} \sum_{k=1}^{n} \sum_{l=1}^{m} a_k \overline{b_l} \int_{-T}^{T} e^{i\lambda_k t} e^{-i\mu_l t} \, dt$$

$$= \sum_{k=1}^{n} \sum_{l=1}^{m} a_k \overline{b_l} \lim_{T \to \infty} \frac{1}{2T} \int_{-T}^{T} e^{i(\lambda_k - \mu_l)t} \, dt = \sum_{k=1}^{n} \sum_{l=1}^{m} a_k \overline{b_l} \delta(\lambda_k, \mu_l),$$

where $\delta(\lambda, \mu)$ is the *Kronecker delta*.

The *completion* $(\tilde{X}, (\cdot, \cdot)_{\tilde{X}})$ of $(X, (\cdot, \cdot)_X)$ is a Hilbert space, in which the set of the equivalence classes $\tilde{S} := \{[x] \mid x \in X\}$ is an *uncountable complete orthonormal set*.

Exercise 4.34. Verify 1 and 2.

Here is another characterization for complete orthonormal sets in a Hilbert space setting.

Proposition 4.13 (Characterization of Complete Orthonormal Sets). *In a Hilbert space* $(X, (\cdot, \cdot), \| \cdot \|)$, *an orthonormal set* $\{e_i\}_{i \in I}$ *is complete iff it is fundamental, i. e.,*

$$\overline{\text{span}\left(\{e_i\}_{i \in I}\right)} = X,$$

(cf. Definition 6.1), i. e., iff $\{e_i\}_{i \in I}$ *is a topological basis of* X.

Proof. Since, in an inner product space, by the *Coincidence of Orthogonal Complements Proposition* (Proposition 4.16) (Section 4.13, Problem 5),

$$\overline{\text{span}\left(\{e_i\}_{i \in I}\right)}^{\perp} = \{e_i\}_{i \in I}^{\perp},$$

by the *Characterization of Complete Orthonormal Sets* (Proposition 4.12), the *completeness* of $\{e_i\}_{i \in I}$ *is equivalent to*

$$\overline{\text{span}\left(\{e_i\}_{i \in I}\right)}^{\perp} = \{e_i\}_{i \in I}^{\perp} = \{0\}.$$

The latter, in a Hilbert space, by the *Projection Theorem* (Theorem 4.6), is *equivalent to*

$$\overline{\text{span}\left(\{e_i\}_{i \in I}\right)} = X,$$

i. e., to the *fundamentality* of $\{e_i\}_{i \in I}$. □

4.10 Gram–Schmidt Process

The *Gram–Schmidt*[6] *process* transforms a *linearly independent countable set* in an inner product space into an *orthogonal* or *orthonormal set* with the *same span*.

Given a linearly independent countable set $\{x_n\}_{n\in I}$, where $I = \{1, \dots, N\}$ with some $N \in \mathbb{N}$ or $I = \mathbb{N}$, in an inner product space $(X, (\cdot, \cdot), \|\cdot\|)$, one can inductively build new sets $\{y_n\}_{n\in I}$ and $\{e_n\}_{n\in I}$ as follows:

$$y_1 := x_1, \quad e_1 := \frac{y_1}{\|y_1\|},$$

and

$$y_n := x_n - \sum_{k=1}^{n-1} (x_n, e_k) e_k, \quad e_n := \frac{y_n}{\|y_n\|}, \quad n \in I, \ n \geq 2.$$

The output sets $\{y_n\}_{n\in I}$ and $\{e_n\}_{n\in I}$ are *orthogonal* and *orthonormal*, respectively, and

$$\forall n \in I: \ \mathrm{span}\,(\{e_1, \dots, e_n\}) = \mathrm{span}\,(\{y_1, \dots, y_n\}) = \mathrm{span}\,(\{x_1, \dots, x_n\}).$$

Exercise 4.35. Verify.

The latter implies that

$$\mathrm{span}\,(\{e_n\}_{n\in I}) = \mathrm{span}\,(\{y_n\}_{n\in I}) = \mathrm{span}\,(\{x_n\}_{n\in I}).$$

The process of building the orthogonal set $\{y_n\}_{n\in I}$ and the orthonormal set $\{e_n\}_{n\in I}$ is called the *Gram–Schmidt orthogonalization* and the *Gram–Schmidt orthonormalization*, respectively.

Example 4.12. Applying the *Gram–Schmidt orthonormalization* in the (real or complex) Hilbert space l_2 to the linearly independent set

$$\{x_1 := (1, 0, 0, \dots), x_2 := (-1, -1, 0, \dots)\},$$

we obtain $e_1 = \frac{x_1}{\|x_1\|_2} = (1, 0, 0, \dots)$ and

$$e_2 = \frac{x_2 - (x_2, e_1)e_1}{\|x_2 - (x_2, e_1)e_1\|_2} = \frac{(-1, -1, 0, \dots) + (1, 0, 0, \dots)}{\|(-1, -1, 0, \dots) + (1, 0, 0, \dots)\|_2} = (0, -1, 0, \dots),$$

i. e., the orthonormal set $\{e_1 = (1, 0, 0, \dots), e_2 = (0, -1, 0, \dots)\}$.

---

6 Erhard Schmidt (1876–1959).

Examples 4.13.

1. In $L_2(-1, 1)$, the *Gram–Schmidt orthonormalization* applied to the set

$$\{t^n\}_{n \in \mathbb{Z}_+}$$

yields the *complete orthonormal sequence* of the well-known *Legendre*[7] *polynomials*

$$\left\{ P_n(t) = \frac{\sqrt{n + 1/2}}{2^n n!} \frac{d^n}{dt^n} \left[(t^2 - 1)^n \right] \right\}_{n=0}^{\infty}.$$

2. In $L_2((-\infty, \infty), e^{-t^2} dt)$, the *Gram–Schmidt orthonormalization* applied to the set

$$\{t^n\}_{n \in \mathbb{Z}_+}$$

yields the *complete orthonormal sequence* of *Hermite polynomials*

$$\left\{ H_n(t) = \frac{(-1)^n}{\sqrt{\sqrt{\pi} 2^n n!}} e^{t^2} \frac{d^n}{dt^n} e^{-t^2} \right\}_{n=0}^{\infty}.$$

3. In $L_2((0, \infty), e^{-t} dt)$, the *Gram–Schmidt orthonormalization* applied to the set

$$\{t^n\}_{n \in \mathbb{Z}_+}$$

yields the *complete orthonormal sequence* of *Laguerre*[8] *polynomials*

$$\left\{ L_n(t) = (-1)^n e^t \frac{d^n}{dt^n} \left[t^n e^{-t} \right] \right\}_{n=0}^{\infty}.$$

See, e. g., [9].

4.11 Generalized Fourier Series

We begin our treatment of *generalized Fourier*[9] *series* in a Hilbert space with studying the case of a finite orthonormal set.

---

7 Adrien-Marie Legendre (1752–1833).
8 Edmond Laguerre (1834–1886).
9 Jean-Baptiste Joseph Fourier (1768–1830).

4.11.1 Finite Orthonormal Set

Theorem 4.10 (Finite Orthonormal Set). *Let* $\{e_1, \ldots, e_n\}$ $(n \in \mathbb{N})$ *be a finite orthonormal set in an inner product space* $(X, (\cdot, \cdot), \|\cdot\|)$. *Then, for each* $x \in X$, *the nearest point to* x *in the n-dimensional subspace* $Y := \mathrm{span}(\{e_1, \ldots, e_n\})$ *is*

$$y = \sum_{i=1}^{n} (x, e_i) e_i$$

with

$$x - y \in Y^{\perp} \quad and \quad \|y\|^2 = \sum_{i=1}^{n} |(x, e_i)|^2$$

and

$$\rho^2(x, Y) = \|x - y\|^2 = \|x\|^2 - \|y\|^2 = \|x\|^2 - \sum_{i=1}^{n} |(x, e_i)|^2,$$

which implies that

$$\sum_{i=1}^{n} |(x, e_i)|^2 \le \|x\|^2.$$

Further,

$$x \in Y \iff x = y \iff \|x - y\| = 0 \iff \sum_{i=1}^{n} |(x, e_i)|^2 = \|x\|^2.$$

Proof. Since the set $\{e_1, \ldots, e_n\}$ is *linearly independent*, being orthonormal (Proposition 4.11) and

$$G(e_1, \ldots, e_n) = 1, \quad G_i = (x, e_i), \ i = 1, \ldots, n,$$

(see Theorem 4.8 and Section 4.13, Problem 11).

Exercise 4.36. Verify.

Hence, by the *Gram Determinant and the Nearest Point Theorem* (Theorem 4.8), the nearest point to x in the n-dimensional subspace $Y := \mathrm{span}(\{e_1, \ldots, e_n\})$ is

$$y = \sum_{i=1}^{n} (x, e_i) e_i$$

with $x - y \in Y^{\perp}$ (see Remark 4.8).

By the *Generalized Pythagorean Theorem* (Theorem 4.17, Section 4.13, Problem 13), in view of *absolute scalability* of norm and considering that $\|e_i\| = 1$, $i = 1, \ldots, n$,

$$\|y\|^2 = \sum_{i=1}^{n} \|(x, e_i)e_i\|^2 = \sum_{i=1}^{n} |(x, e_i)|^2.$$

Further, since $x - y \in Y^{\perp}$, by the *Pythagorean Theorem* (Theorem 4.4),

$$\rho^2(x, Y) = \|x - y\|^2 = \|x\|^2 - \|y\|^2 = \|x\|^2 - \sum_{i=1}^{n} |(x, e_i)|^2,$$

which implies that

$$\sum_{i=1}^{n} |(x, e_i)|^2 \le \|x\|^2$$

and

$$x \in Y \Leftrightarrow x = y \Leftrightarrow \|x - y\| = 0 \Leftrightarrow \sum_{i=1}^{n} |(x, e_i)|^2 = \|x\|^2. \qquad \square$$

From the prior theorem and the *Projection Theorem* (Theorem 4.6), we immediately obtain the following

Corollary 4.4 (Finite Orthonormal Set). *Let* $\{e_1, \ldots, e_n\}$ ($n \in \mathbb{N}$) *be a finite orthonormal set in a Hilbert space* $(X, (\cdot, \cdot), \|\cdot\|)$. *Then, for each* $x \in X$, *the orthogonal projection of* x *on the n-dimensional subspace* $Y := \text{span}(\{e_1, \ldots, e_n\})$ *is*

$$y = \sum_{i=1}^{n} (x, e_i)e_i$$

with

$$x - y \in Y^{\perp} \quad and \quad \|y\|^2 = \sum_{i=1}^{n} |(x, e_i)|^2$$

and, the point y being the nearest to x in Y,

$$\rho^2(x, Y) = \|x - y\|^2 = \|x\|^2 - \|y\|^2 = \|x\|^2 - \sum_{i=1}^{n} |(x, e_i)|^2,$$

which implies that

$$\sum_{i=1}^{n} |(x, e_i)|^2 \le \|x\|^2.$$

Further,

$$x \in Y \Leftrightarrow x = y \Leftrightarrow \|x - y\| = 0 \Leftrightarrow \sum_{i=1}^{n} |(x, e_i)|^2 = \|x\|^2.$$

Examples 4.14.

1. In the (real or complex) space $l_2^{(n)}$ ($n \in \mathbb{N}$), the *orthogonal projection* of an arbitrary n-tuple $x := (x_1, \ldots, x_n)$ on the k-dimensional subspace Y_k spanned by the first k vectors of the *standard unit basis* $\{e_1, \ldots, e_n\}$, $k = 1, \ldots, n$, is

$$y = \sum_{i=1}^{k} (x, e_i) e_i = \sum_{i=1}^{k} x_i e_i = (x_1, \ldots, x_k, 0, \ldots, 0).$$

In particular, for $k = n$, $Y_n = l_2^{(n)}$ and

$$y = \sum_{i=1}^{n} (x, e_i) e_i = \sum_{i=1}^{n} x_i e_i = (x_1, \ldots, x_n) = x.$$

2. In the (real or complex) space l_2, for any fixed $n \in \mathbb{N}$, the *orthogonal projection* of an arbitrary sequence $x := (x_k)_{k \in \mathbb{N}}$ on the n-dimensional subspace Y_n spanned by the first n vectors of the standard Schauder basis $\{e_k\}_{k \in \mathbb{N}}$, $n \in \mathbb{N}$, is

$$y = \sum_{i=1}^{n} (x, e_i) e_i = \sum_{i=1}^{n} x_i e_i = (x_1, \ldots, x_n, 0, \ldots).$$

4.11.2 Arbitrary Orthonormal Set

Let us now proceed to the case of an arbitrary orthonormal set.

Theorem 4.11 (Arbitrary Orthonormal Set). *Let $S := \{e_i\}_{i \in I}$ be an orthonormal set in a Hilbert space $(X, (\cdot, \cdot), \|\cdot\|)$. Then, for each $x \in X$, the set of all indices $i \in I$ for which the generalized Fourier coefficients (x, e_i), $i \in I$, of x relative to S do not vanish, i. e.,*

$$N(x) := \{i \in I \mid (x, e_i) \neq 0\},$$

is countable, and hence, the summation

$$\sum_{i \in I} (x, e_i) e_i$$

is either a finite sum or an infinite series, called the generalized Fourier series of x relative to S.

If $\{i(n)\}_{n=1,\ldots,N}$ ($N \in \mathbb{N}$ or $N = \infty$) is an arbitrary countable arrangement of the set $N(x)$, then

$$\sum_{k=1}^{N} |(x, e_{i(k)})|^2 \leq \|x\|^2 \quad \text{(Bessel's Inequality)}$$

and

$$y = \sum_{k=1}^{N} (x, e_{i(k)}) e_{i(k)},$$

is the orthogonal projection of x on the closed subspace $Y = \overline{\text{span}(S)}$ *with*

$$x - y \in Y^{\perp} \quad and \quad \|y\|^2 = \sum_{k=1}^{N} |(x, e_{i(k)})|^2.$$

The point y is the nearest to x in Y with

$$\rho^2(x, Y) = \|x - y\|^2 = \|x\|^2 - \|y\|^2 = \|x\|^2 - \sum_{k=1}^{N} |(x, e_{i(k)})|^2.$$

Further,

$$x \in Y \Leftrightarrow x = y \Leftrightarrow \|x - y\| = 0 \Leftrightarrow \sum_{k=1}^{N} |(x, e_{i(k)})|^2 = \|x\|^2 \quad (Parseval's\ Identity).$$

Proof. Since the subspace $Y := \overline{\text{span}(S)}$ is closed, by the *Projection Theorem* (Theorem 4.6),

$$X = Y \oplus Y^{\perp},$$

i. e., each $x \in X$ can be uniquely represented as

$$x = y + z$$

with some $y \in Y$ and $z = x - y \in Y^{\perp}$, where y is the *orthogonal projection* of x on Y. For each $x \in X$,

$$N(x) := \{i \in I \mid (x, e_i) \neq 0\} = \bigcup_{n=1}^{\infty} N_n(x),$$

where

$$N_n(x) := \{i \in N(x) \mid |(x, e_i)| > 1/n\}, \ n \in \mathbb{N}.$$

By the *Finite Orthonormal Set Corollary* (Corollary 4.4), for each $n \in \mathbb{N}$, the set $N_n(x)$ is *finite*.

Exercise 4.37. Prove *by contradiction*.

Hence, by the properties of countable sets (Proposition 1.3), the set $N(x)$ is *countable*.

Let $\{i(n)\}_{n=1,\dots,N}$ ($N \in \mathbb{N}$ or $N = \infty$) be an arbitrary countable arrangement of the set $N(x)$.

The case of a *finite* $N(x)$, i. e., $N \in \mathbb{N}$, immediately follows from the *Finite Orthonormal Set Corollary* (Corollary 4.4).

Suppose that the set $N(x)$ is *countably infinite*, i. e., $N = \infty$. Then, by the *Finite Orthonormal Set Corollary* (Corollary 4.4),

$$\forall n \in \mathbb{N} : \sum_{k=1}^{n} |(x, e_{i(k)})|^2 \leq \|x\|^2,$$

whence, letting $n \to \infty$, we obtain *Bessel's*[10] *inequality*:

$$\sum_{k=1}^{\infty} |(x, e_{i(k)})|^2 \leq \|x\|^2,$$

which, since, for any $n, p \in \mathbb{N}$, by the *inner product axioms*, implies that

$$\left\| \sum_{k=n+1}^{n+p} (x, e_{i(k)}) e_{i(k)} \right\|^2 = \left(\sum_{k=n+1}^{n+p} (x, e_{i(k)}) e_{i(k)}, \sum_{l=n+1}^{n+p} (x, e_{i(l)}) e_{i(l)} \right)$$

$$= \sum_{k=n+1}^{n+p} \sum_{l=n+1}^{n+p} (x, e_{i(k)}) \overline{(x, e_{i(l)})} (e_{i(k)}, e_{i(l)}) \qquad \text{since } (e_{i(k)}, e_{i(l)}) = \delta_{i(k)i(l)};$$

$$= \sum_{k=n+1}^{n+p} |(x, e_{i(k)})|^2 \to 0, \ n \to \infty.$$

Whence, in view of the *completeness* of $(X, (\cdot, \cdot), \|\cdot\|)$, by *Cauchy's Convergence Test for Series* (Theorem 3.7), we infer that the series

$$\sum_{k=1}^{\infty} (x, e_{i(k)}) e_{i(k)},$$

converges in $(X, (\cdot, \cdot), \|\cdot\|)$.

The fact that

$$y := \sum_{k=1}^{\infty} (x, e_{i(k)}) e_{i(k)},$$

is the orthogonal projection of x on the subspace $Y = \overline{\operatorname{span}(S)}$, by the *Coincidence of Orthogonal Complements Proposition* (Proposition 4.16), follows from the fact

$$\forall n \in \mathbb{N} : x - y \perp e_{i(n)}.$$

Exercise 4.38. Verify.

---

10 Friedrich Wilhelm Bessel (1784–1846).

By the *Generalized Pythagorean Theorem* (Theorem 4.17, 4.13, Problem 13) and *continuity* of norm, in view of $\|e_i\| = 1, i = 1, \ldots, n$,

$$\|y\|^2 = \lim_{n \to \infty} \left\| \sum_{k=1}^{n} (x, e_{i(k)}) e_{i(k)} \right\|^2 = \lim_{n \to \infty} \sum_{k=1}^{n} \|(x, e_{i(k)}) e_{i(k)}\|^2$$

$$= \lim_{n \to \infty} \sum_{k=1}^{n} |(x, e_{i(k)})|^2 = \sum_{k=1}^{\infty} |(x, e_{i(k)})|^2.$$

By the *Projection Theorem* (Theorem 4.6), y is the *nearest point* to x in Y, which, by the *Pythagorean Theorem* (Theorem 4.4), implies that

$$\rho^2(x, Y) = \|x - y\|^2 = \|x\|^2 - \|y\|^2 = \|x\|^2 - \sum_{k=1}^{\infty} |(x, e_{i(k)})|^2,$$

and hence,

$$x \in Y \Leftrightarrow x = y \Leftrightarrow \|x - y\| = 0 \Leftrightarrow \sum_{k=1}^{\infty} |(x, e_{i(k)})|^2 = \|x\|^2,$$

the latter being called *Parseval's*[11] *identity*. □

Remarks 4.18.

- Thus, for an orthonormal set $S := \{e_i\}_{i \in I}$ in a Hilbert space $(X, (\cdot, \cdot), \| \cdot \|)$, without specifying a countable arrangement of I, we can write

$$\sum_{i \in I} (x, e_i) e_i,$$

$$\sum_{i \in I} |(x, e_i)|^2 \le \|x\|^2 \quad (Bessel's \ Inequality),$$

$$y = \sum_{i \in I} (x, e_i) e_i \text{ with } x - y \in Y^\perp \text{ and } \|y\|^2 = \sum_{i \in I} |(x, e_i)|^2,$$

$$\rho^2(x, Y) = \|x - y\|^2 = \|x\|^2 - \|y\|^2 = \|x\|^2 - \sum_{i \in I} |(x, e_i)|^2,$$

and

$$x \in Y \Leftrightarrow x = y \Leftrightarrow \|x - y\| = 0 \Leftrightarrow \sum_{i \in I} |(x, e_i)|^2 = \|x\|^2 \quad (Parseval's \ Identity).$$

- *Parseval's identity*, which can be regarded as a *Pythagorean Theorem*, has the following equivalent inner-product form:

$$(x, y) = \sum_{i \in I} (x, e_i)\overline{(y, e_i)}, \ x, y \in Y = \overline{\text{span}\,(S)}. \tag{4.11}$$

---

11 Marc-Antoine Parseval (1755–1836).

Exercise 4.39. Verify.

–   As follows from the proof of the prior theorem, for an orthonormal set $S := \{e_i\}_{i\in I}$ in a Hilbert space $(X, (\cdot,\cdot), \|\cdot\|)$ over $\mathbb{F}$ and an arbitrary numeric I-tuple $(c_i)_{i\in I} \in \mathbb{F}^I$ such that

$$c_i \neq 0 \text{ for countably many } i\text{'s and } \sum_{i\in I} |c_i|^2 < \infty,$$

the series

$$\sum_{i\in I} c_i e_i,$$

converges in $(X, (\cdot,\cdot), \|\cdot\|)$ and

$$\left\| \sum_{i\in I} c_i e_i \right\|^2 = \sum_{i\in I} |c_i|^2.$$

4.11.3 Orthonormal Sequence

As an important particular case, we obtain that of an orthonormal sequence ($I = \mathbb{N}$).

Corollary 4.5 (Orthonormal Sequence). *Let $\{e_n\}_{n\in\mathbb{N}}$ be a countably infinite orthonormal set (an orthonormal sequence) in a Hilbert space $(X, (\cdot,\cdot), \|\cdot\|)$. Then, for each $x \in X$,*

$$\sum_{n=1}^{\infty} |(x, e_n)|^2 \leq \|x\|^2 \quad \text{(Bessel's Inequality)}$$

and the generalized Fourier series $\sum_{n=1}^{\infty}(x, e_n)e_n$ of x relative to $\{e_n\}_{n\in\mathbb{N}}$ converges in X to the orthogonal projection

$$y = \sum_{n=1}^{\infty}(x, e_n)e_n$$

of x on the closed subspace $Y = \overline{\operatorname{span}(\{e_n\}_{n\in\mathbb{N}})}$ with

$$x - y \in Y^{\perp} \quad \text{and} \quad \|y\|^2 = \sum_{n=1}^{\infty} |(x, e_n)|^2,$$

which is the nearest point to x in Y with

$$\rho^2(x, Y) = \|x - y\|^2 = \|x\|^2 - \|y\|^2 = \|x\|^2 - \sum_{n=1}^{\infty} |(x, e_n)|^2.$$

Further,

$$x \in Y \Leftrightarrow x = y \Leftrightarrow \|x - y\| = 0 \Leftrightarrow \sum_{n=1}^{\infty} |(x, e_n)|^2 = \|x\|^2 \quad \text{(Parseval's Identity)}.$$

From Bessel's Inequality, we immediately obtain the following statement.

Corollary 4.6 (Generalized Fourier Coefficients Sequence). *Let $\{e_n\}_{n\in\mathbb{N}}$ be an orthonormal sequence in a Hilbert space $(X, (\cdot,\cdot), \|\cdot\|)$. Then*

$$(x, e_n) \to 0, \ n \to \infty.$$

Exercise 4.40. Prove.

4.12 Orthonormal Bases and Orthogonal Dimension

In addition to the *three* meanings that the notion of *basis* has in a Banach space, in a Hilbert space setting, due to the presense of *orthogonality*, it acquires an additional one discussed below.

Definition 4.9 (Orthonormal Basis). An orthonormal set $S := \{e_i\}_{i\in I}$ in a nontrivial Hilbert space $(X, (\cdot,\cdot), \|\cdot\|)$ is called an *orthonormal basis* of X if the generalized Fourier series representation

$$x = \sum_{i\in I}(x, e_i)e_i$$

holds for each $x \in X$.

Theorem 4.12 (Orthonormal Basis Characterizations). *An orthonormal set $S := \{e_i\}_{i\in I}$ in a nontrivial Hilbert space $(X, (\cdot,\cdot), \|\cdot\|)$ is an orthonormal basis of X iff any of the following equivalent conditions is satisfied.*
1. *S is complete, i. e., $S^\perp = \{0\}$.*
2. *S is a topological basis of X, i. e., $\overline{\text{span}(S)} = X$.*
3. *Parseval's identity*

$$\sum_{i\in I}|(x, e_i)|^2 = \|x\|^2$$

holds for each $x \in X$.

Proof. A rather effortless proof immediately follows from the definition, the *Arbitrary Orthonormal Set Theorem* (Theorem 4.11), and the *Characterization of Complete Orthonormal Sets* (Proposition 4.13).

Exercise 4.41. Fill in the details.   ☐

By the prior statement and the *Existence of a Complete Orthonormal Set Theorem* (Theorem 4.9), we obtain the following analogue of the *Basis Theorem* (Theorem 3.2).

Theorem 4.13 (Orthonormal Basis Theorem). *In a nontrivial Hilbert space $(X, (\cdot,\cdot), \|\cdot\|)$, each orthonormal set S can be extended to an orthonormal basis S' of X.*

Remark 4.19. An orthonormal basis in a Hilbert space, although *existent* by Theorem 4.13, *need not be unique* (see Remarks 4.17) (cf. Section 4.13, Problem 15).

Examples 4.15.
1. The complete orthonormal sets from Examples 4.11 and the complete orthonormal sequences of *Legendre, Hermite*, and *Laguerre polynomials* from Examples 4.13 are *orthonormal bases* in the corresponding Hilbert spaces.
2. The uncountable complete orthonormal set S from Examples 4.11 is an *orthonormal basis* in the *completion* $\tilde{X}$ of X.

Further, the following is an analogue of the *Dimension Theorem* (Theorem 3.4).

Theorem 4.14 (Dimension Theorem for Hilbert Spaces). *All orthonormal bases of a nontrivial Hilbert space have equally many elements.*

Proof. Let $S := \{e_i\}_{i \in I}$ and $S' := \{e'_j\}_{j \in J}$ be two orthonormal bases in a Hilbert space $(X, (\cdot, \cdot), \| \cdot \|)$, with and $|I|$ and $|J|$ being their cardinalities, respectively.

If S is *finite*, i.e., $|I| \in \mathbb{N}$, then S is a *Hamel basis* of X and the *algebraic dimension* of X is $|I|$. Since the orthonormal set S' is *linearly independent* (Proposition 4.11), it is also finite and

$$|J| \le |I|.$$

Hence, $|J| \in \mathbb{N}$ and symmetrically, we have:

$$|I| \le |J|,$$

which implies that

$$|I| = |J|.$$

Suppose that S is *infinite*, i. e. $|I| \ge \aleph_0$ (see Examples 1.1). This immediately implies that $|J| \ge \aleph_0$.

Exercise 4.42. Explain.

For each $i \in I$, the set

$$N'_i = \{j \in J \mid (e_i, e'_j) \neq 0\}$$

is *nonempty*, since otherwise the orthonormal set S' would be *incomplete*.

Exercise 4.43. Explain.

By the *Arbitrary Orthonormal Set Theorem* (Theorem 4.11), the nonempty set N'_i, $i \in I$, is *countable*.

Further,

$$\forall j \in J \ \exists i \in I : j \in N_i',$$

since otherwise the orthonormal set S would be *incomplete*.

Exercise 4.44. Explain.

Hence,

$$J := \bigcup_{i \in I} N_i',$$

which, by the arithmetic of cardinals (see, e. g., [21, 26, 38]), implies that

$$|J| \leq \aleph_0 |I| = |I|.$$

Symmetrically,

$$|I| \leq \aleph_0 |J| = |J|,$$

and hence,

$$|I| = |J|,$$

which completes the proof. $\square$

By the *Dimension Theorem for Hilbert Spaces*, the following notion is well-defined.

Definition 4.10 (Orthogonal Dimension of a Hilbert Space). The *orthogonal dimension* of a nontrivial Hilbert space $(X, (\cdot, \cdot), \| \cdot \|)$ is the *common cardinality* of all orthonormal bases of X.

The dimension of a trivial space is naturally defined to be 0.

Remark 4.20. The symbol $\dim X$ may be used contextually.

The case of a separable Hilbert space deserves special attention.

Theorem 4.15 (Orthogonal Dimension of a Separable Hilbert Space). *A Hilbert space* $(X, (\cdot, \cdot), \| \cdot \|)$ *is separable iff the orthogonal dimension of X does not exceed* $\aleph_0$, *i. e., every orthonormal basis of X is countable.*

Proof. The case of $X = \{0\}$ being trivial, let us regard that $X \neq \{0\}$.

"*If*" part. Suppose that X has a *countable* orthonormal basis. Then the finite linear combinations of the basis vectors with rational/complex rational coefficients form a *countable dense set* in X, which implies that the space X is *separable*.

Exercise 4.45. Explain.

"*Only if*" part. Suppose that X is *separable*. Then it has a *countably infinite dense subset* $M := \{x_n\}_{n \in \mathbb{N}}$, which is *fundamental*, since

$$\operatorname{span}(M) \supseteq \overline{M} = X.$$

Let us inductively construct a linearly independent subset M' of M with

$$\text{span}(M') = \text{span}(M).$$

Letting

$$n(1) := \min \{n \in \mathbb{N} \mid x_n \neq 0\},$$

we choose our first nonzero element $x_{n(1)} \in M$.
Letting

$$n(2) := \min \{n \in \mathbb{N} \mid x_{n(1)}, x_n \text{ are } \textit{linearly independent}\},$$

we choose our second element $x_{n(2)} \in M$ linearly independent of $x_{n(1)}$, if any.

Continuing inductively in this manner (using the *Axiom of Choice* (see Appendix A)), we obtain a *countable linearly independent subset* $M' := \{x_{n(i)}\}_{i\in I}$, where $I = \{1, \ldots, N\}$ with some $N \in \mathbb{N}$ or $I = \mathbb{N}$, of M such that

$$\text{span}(M') = \text{span}(M).$$

Exercise 4.46. Explain.

Hence,

$$\overline{\text{span}(M')} = \overline{\text{span}(M)} = X,$$

i. e., the set M' is *fundamental*.

Applying to $M' := \{x_{n(i)}\}_{i\in I}$ the *Gram–Schmidt orthonormalization* (see Section 4.10), we obtain a *countable orthonormal set* $S := \{e_i\}_{i\in I}$, which is *fundamental* as well as M' since

$$\text{span}(S) = \text{span}(M'),$$

and hence, by the *Orthonormal Basis Characterizations* (Theorem 4.12), is an *orthonormal basis* of X. □

Remarks 4.21.
- The *orthogonal dimension* of a separable Hilbert space is *equal* to its *algebraic dimension*, provided the space is finite-dimensional, and is *less* than its *algebraic dimension*, provided the space is infinite-dimensional.

Exercise 4.47. Explain.

- An orthonormal basis $\{e_n\}_{n\in\mathbb{N}}$ of an infinite-dimensional separable Hilbert space is also its *Schauder basis* (see Section 3.2.6), as is, in l_2, the *standard orthonormal basis*

$$\{e_n := \{\delta_{nk}\}_{k=1}^{\infty}\}_{n\in\mathbb{N}},$$

where δ_{nk} is the Kronecker delta (see Examples 4.11).

Finally, the following is an analogue of the *Isomorphism Theorem* (Theorem 3.5).

Theorem 4.16 (Isomorphism Theorem for Hilbert Spaces).
Two nontrivial Hilbert spaces $(X, (\cdot, \cdot)_X, \| \cdot \|_X)$ and $(Y, (\cdot, \cdot)_Y, \| \cdot \|_Y)$ are isometrically iso-morphic iff they have the same orthogonal dimension.

Proof. "Only if" part. Let $T : X \to Y$ be an *isometric isomorphism* between X and Y.

By the *Polarization Identities* (Proposition 4.2), along with the inner-product norm, T preserves the inner product, i. e.,

$$(x, y)_X = (Tx, Ty)_Y, \; x, y \in X,$$

and hence, T preserves *orthogonality*, i. e.

$$x \perp y \text{ in } X \; \Leftrightarrow \; Tx \perp Ty \text{ in } Y.$$

Exercise 4.48. Explain.

Therefore, a set S is an *orthonormal basis* in X iff $T(S)$ is an *orthonormal basis* in Y, which, since T is a *bijection*, implies that X and Y have the same orthogonal dimension.

"If" part. Suppose that X and Y have the same orthogonal dimension. Choosing orthonormal bases $S := \{e_i\}_{i \in I}$ in X and $S' := \{e_i'\}_{i \in I}$ in Y sharing the *indexing set I*, we can establish an *isometric isomorphism* T between X and Y by matching the vectors with the *identical Fourier series representations* relative to the bases S and S', respectively, as follows:

$$X \ni x = \sum_{i \in I} (x, e_i) e_i \mapsto Tx := \sum_{i \in I} (x, e_i) e_i' \in Y.$$

In particular, $Te_i = e_i', \; i \in I$.

The mapping $T : X \to Y$ is *well-defined* since, for each $x \in X$, by *Parseval's Identity*,

$$\sum_{i \in I} |(x, e_i)|^2 = \|x\|^2 < \infty,$$

which implies convergence for the series

$$\sum_{i \in I} (x, e_i) e_i'$$

in $(Y, (\cdot, \cdot)_Y, \| \cdot \|_Y)$ and the fact that

$$\|Tx\|^2 = \sum_{i \in I} |(x, e_i)|^2 = \|x\|^2$$

(see Remarks 4.18).

Hence, $T : X \to Y$ is *norm preserving*.

The mapping $T : X \to Y$ is, obviously, *linear* and also *onto* (i. e., *surjective*). Since, for each

$$y = \sum_{i \in I} (y, e_i') e_i' \in Y,$$

we can choose

$$x = \sum_{i \in I} (y, e_i') e_i \in X$$

so that $y = Tx$.

Thus, $T : X \to Y$ is an *isometric isomorphism* between X and Y. □

Remark 4.22. Therefore, two abstract Hilbert spaces differ from each other only in their *orthogonal dimension*.

From the *Orthogonal Dimension of a Separable Hilbert Space Theorem* (Theorem 4.15) and the *Isomorphism Theorem for Hilbert Spaces* (Theorem 4.16) we obtain the following direct corollary.

Corollary 4.7 (Isomorphism of Separable Hilbert Spaces). *A separable Hilbert space* $(X, (\cdot, \cdot), \| \cdot \|)$ *over* $\mathbb{F}$ *is isometrically isomorphic to either* $l_2^{(n)}(\mathbb{F})$ *with some* $n \in \mathbb{Z}_+$ *or to* $l_2(\mathbb{F})$.

Remarks 4.23.

– More generally, a Hilbert space $(X, (\cdot, \cdot), \| \cdot \|)$ over $\mathbb{F}$ with an orthonormal basis $S := \{e_i\}_{i \in I}$ is isometrically isomorphic to the Hilbert space defined as follows:

$$l_2(I, \mathbb{F}) := \left\{ x = (x_i)_{i \in I} \in \mathbb{F}^I \ \middle| \ x_i \neq 0 \text{ for } countably \ many \ i\text{'s and } \sum_{i \in I} |x_i|^2 < \infty \right\}$$

with the inner product

$$(x, y) = \sum_{i \in I} x_i \overline{y_i}.$$

Exercise 4.49. Describe an *orthonormal basis* in $l_2(I, \mathbb{F})$.

– For a set I of an arbitrary cardinality $|I|$, there is a Hilbert space $l_2(I, \mathbb{F})$ of orthogonal dimension $|I|$.

– The Hilbert space $\tilde{X}$ from Examples 4.11 is *not separable* and is isometrically isomorphic to $l_2(\mathbb{R}, \mathbb{C})$.

4.13 Problems

1. Prove

   Proposition 4.14 (Inner Product Separation Property). *For elements x and y in an inner product space* $(X, (\cdot, \cdot))$,

   $$x = y \iff \forall z \in X : (x, z) = (y, z).$$

2. (Cartesian Product of Inner Product Spaces). Let $(X_1, (\cdot, \cdot)_1, \|\cdot\|_1)$ and $(X_2, (\cdot, \cdot)_2, \|\cdot\|_2)$ be inner product spaces over $\mathbb{F}$.

    (a) Show that the Cartesian product $X = X_1 \times X_2$ is an inner product space relative to the inner product

    $$X_1 \times X_2 \ni x := (x_1, x_2), y := (y_1, y_2) \mapsto (x, y) := (x_1, y_1)_1 + (x_2, y_2)_2 \in \mathbb{F}$$

    generating the *product norm*

    $$X_1 \times X_2 \ni x := (x_1, x_2) \mapsto \|x\|_{X_1 \times X_2} = \sqrt{\|x_1\|_1^2 + \|x_2\|_2^2}$$

    (cf. Section 3.6, Problem 5).

    (b) Show that the product space $(X_1 \times X_2, (\cdot, \cdot), \|\cdot\|_{X_1 \times X_2})$ is Hilbert space *iff* each space $(X_i, (\cdot, \cdot)_i, \|\cdot\|_i)$, $i = 1, 2$, is a Hilbert space.

3. Prove

    Proposition 4.15 (Characterization of Convergence in Pre-Hilbert Spaces). *For a sequence $\{x_n\}_{n=1}^{\infty}$ in a pre-Hilbert space $(X, (\cdot, \cdot), \|\cdot\|)$,*

    $$x_n \to x \in X, \ n \to \infty, \ in \ (X, (\cdot, \cdot), \|\cdot\|)$$

    iff
    (1) $\forall y \in X : (x_n, y) \to (x, y), \ n \to \infty, and$
    (2) $\|x_n\| \to \|x\|, \ n \to \infty.$

4. Show that, in an inner product space $(X, (\cdot, \cdot), \|\cdot\|)$, for nonempty sets M and N,

    (a) $M \subseteq N \Rightarrow M^{\perp} \supseteq N^{\perp}$;
    (b) $M^{\perp} = \overline{M}^{\perp}$.

5. Prove

    Proposition 4.16 (Coincidence of Orthogonal Complements). *For a nonempty set M in an inner product space $(X, (\cdot, \cdot), \|\cdot\|)$,*

    $$M^{\perp} = \overline{\mathrm{span}(M)}^{\perp}.$$

6. In $l_2^{(3)}(\mathbb{R})$, for the subspace

    $$Y := \left\{(x, y, z) \in \mathbb{R}^3 \,\middle|\, x + y + z = 0\right\},$$

    (a) determine $Y^{\perp}$ and
    (b) for $x := (1, 2, 3)$, find the unique decomposition

    $$x = y + z, \ y \in Y, \ z \in Y^{\perp}.$$

7. In the (real or complex) space l_2, for the subspace,

    $$Y := \{(x_n)_{n \in \mathbb{N}} \in l_2 \mid x_{2n} = 0, \ n \in \mathbb{N}\},$$

(a) determine $Y^\perp$ and

(b) for $x := (1/n)_{n\in\mathbb{N}}$, find the unique decomposition

$$x = y + z, \; y \in Y, \; z \in Y^\perp.$$

8. Prove

   Proposition 4.17 (Twice Orthogonal Complement of a Set). *For a nonempty set M in a Hilbert space* $(X, (\cdot, \cdot), \|\cdot\|)$,

   $$(M^\perp)^\perp = \overline{\text{span}(M)}.$$

   Hint. Use Propositions 4.16 and 4.7.

9. In l_2, for the linearly independent set

   $$\{x_1 := (1, 0, 0, \dots), x_2 := (1, 1, 0, \dots)\}$$

   and $x := (1, 2, 3, 0, \dots)$, use the *Gram determinant approach* to find the nearest point to x in the subspace $Y := \text{span}(\{x_1, x_2\})$ and $\rho(x, Y)$.

10. In $(C[0, 1], (\cdot, \cdot), \|\cdot\|_2)$, use the *Gram determinant approach* to find the best approximation polynomial to $f(t) := e^t$ in $P_1 := \text{span}(\{1, t\})$ and $\rho(f, P_1)$.

11. Describe the Gram matrix and determinant for an *orthogonal/orthonormal set* $\{x_1, \dots, x_n\}$ ($n \in \mathbb{N}$).

12. In l_2, apply the *Gram–Schmidt process* to *orthonormalize* the set

    $$\{x_1 := (1, 1, 0, 0, \dots), x_2 := (1, 0, 1, 0, \dots), x_3 := (1, 1, 1, 0, \dots)\}.$$

13. Prove

    Theorem 4.17 (Generalized Pythagorean Theorem). *For a finite orthogonal set* $\{x_1, \dots, x_n\}$ ($n \in \mathbb{N}$) *in an inner product space* $(X, (\cdot, \cdot), \|\cdot\|)$,

    $$\left\| \sum_{i=1}^{n} x_i \right\|^2 = \sum_{i=1}^{n} \|x_i\|^2.$$

14. In l_2, for the orthonormal set $\{e_1, e_2, e_3\}$ obtained in Problem 12, find the *orthogonal projection* of $x := (1, 0, 0, \dots)$ on $Y := \text{span}(\{e_1, e_2\})$ and $\rho(x, Y)$.

15. * Prove that, if $\{e_i\}_{i\in\mathbb{N}}$ is an *orthonormal basis* for a separable Hilbert space $(X, (\cdot, \cdot), \|\cdot\|)$ and $\{e_j'\}_{j\in\mathbb{N}}$ is an *orthonormal set* such that

    $$\sum_{i=1}^{\infty} \|e_i - e_i'\|^2 < \infty,$$

    then $\{e_n'\}_{n=1}^{\infty}$ is also an *orthonormal basis* for $(X, (\cdot, \cdot), \|\cdot\|)$.

    Hint. First, show that $|(e_j - e_j', e_i)| = |(e_i - e_i', e_j')|$, $i, j \in \mathbb{N}$, then use the *Parseval's identity* approach.

5 Linear Operators and Functionals

In this chapter, we discuss an important class of mappings on vector and normed vector spaces, which are closely related to the linear structure of the spaces or both their linear and topological structures.

5.1 Linear Operators and Functionals

Linear operators on vector spaces are "married" to the linear structure of such spaces.

5.1.1 Definitions and Examples

Recall that $\mathbb{F}$ stands for the *scalar field* of $\mathbb{R}$ or $\mathbb{C}$.

Definition 5.1 (Linear Operator and Linear Functional). Let X and Y be vector spaces over $\mathbb{F}$.

A *linear operator* (also a *linear transformation* or a *linear mapping*) from X to Y is mapping

$$A : D(A) \to Y,$$

where $D(A) \subseteq X$ is a *subspace* of X, which preserves linear operations:

$$A(\lambda x + \mu y) = \lambda A x + \mu A y, \quad \lambda, \mu \in \mathbb{F}, \ x, y \in D(A),$$

i. e., is a *homomorphism* of $D(A)$ to Y.

The subspace $D(A)$ is called the *domain (of definition)* of A.

Notation. $(A, D(A))$.

If $Y = X$, i. e.,

$$A : X \supseteq D(A) \to X,$$

A is said to be a linear operator *in X*, or *on X* provided $D(A) = X$.

If $Y = \mathbb{F}$, i. e.,

$$A : D(A) \to \mathbb{F},$$

A is called a *linear functional*. These are customarily designated by the lower case letters and the notation $f(x)$ is used instead of Ax.

Examples 5.1.
1. On a vector space X over $\mathbb{F}$, multiplication by a number $\lambda \in \mathbb{F}$ is a linear operator (*endomorphism*) (see Examples 3.2).

https://doi.org/10.1515/9783110614039-005

In particular, we obtain the *zero operator* 0 for $\lambda = 0$ and the *identity operator I* for $\lambda = 1$.

Remark 5.1. Here and henceforth, 0 is used to designate zero operators or functionals, such a connotation being a rather common practice of symbol economy.

2. Recall that, by the *Representation Theorem* (Theorem 3.3), in a vector space X over $\mathbb{F}$ with a *basis* $B := \{x_i\}_{i \in I}$, each element $x \in X$ has a *unique representation* relative to B

$$x = \sum_{i \in I} c_i x_i,$$

in which all but a finite number of the coefficients $c_i \in \mathbb{F}$, $i \in I$, called the *coordinates* of x relative to B, are zero.

For each $j \in I$, the mapping

$$X \ni x = \sum_{i \in I} c_i x_i \mapsto c_j(x) := c_j \in \mathbb{F}$$

is a *linear functional* on X, called the *jth (Hamel) coordinate functional* relative to the basis B.

3. Multiplication by an $m \times n$ matrix $[a_{ij}]$ $(m, n \in \mathbb{N})$ with entries from $\mathbb{F}$

$$\mathbb{F}^n \ni x = (x_1, \ldots, x_n) \mapsto Ax := [a_{ij}] x \in \mathbb{F}^m$$

is a linear operator from $\mathbb{F}^n$ to $\mathbb{F}^m$ and, provided $m = n$, is a linear operator on $\mathbb{F}^n$ (see Examples 3.2).

In particular, for $m = 1$, we obtain a *linear functional* on $\mathbb{F}^n$

$$\mathbb{F}^n \ni x = (x_1, \ldots, x_n) \mapsto f(x) := \sum_{k=1}^{n} a_k x_k \in \mathbb{F}, \tag{5.1}$$

where $(a_1, \ldots, a_n) \in \mathbb{F}^n$.

Conversely, every linear functional on $\mathbb{F}^n$ ($n \in \mathbb{N}$) is of the form given by (5.1) with some $(a_1, \ldots, a_n) \in \mathbb{F}^n$.

4. On c_{00}, the operator of multiplication by a numeric sequence $a := (a_n)_{n \in \mathbb{N}} \in \mathbb{F}^{\mathbb{N}}$

$$c_{00} \ni x = (x_n)_{n \in \mathbb{N}} \mapsto Ax := (a_n x_n)_{n \in \mathbb{N}} \in c_{00}$$

is a *linear operator*.

5. Let $-\infty < a < b < \infty$.

(a) $C[a, b] \ni x \to [Ax](t) := \int_a^t x(s)\, ds$, $t \in [a, b]$, is a *linear operator* on $C[a, b]$.

(b) $C[a, b] \ni x \to f(x) := \int_a^b x(t)\, dt$ is a *linear functional* on $C[a, b]$.

(c) $C^1[a, b] \ni x \to [Ax](t) := \frac{d}{dt} x(t)$, $t \in [a, b]$, is a *linear operator* from $C^1[a, b]$ to $C[a, b]$ or in $C[a, b]$ with the *domain* $D(A) := C^1[a, b]$.

Exercise 5.1. Verify.

5.1.2 Kernel, Range, and Graph

Definition 5.2 (Kernel, Range, and Graph). Let X and Y be vector spaces over $\mathbb{F}$ and $(A, D(A))$ be a linear operator from X to Y.
–   The *kernel* (or *null space*) of A is

$$\ker A := \{x \in D(A) \mid Ax = 0\}\,.$$

–   The *range* of A is

$$R(A) := \{y \in Y \mid \exists x \in D(A) : y = Ax\}\,.$$

–   The *graph* of A is

$$G_A := \{(x, Ax) \in X \times Y \mid x \in D(A)\}\,.$$

Remarks 5.2. For a linear operator $(A, D(A))$ from X to Y,
– $A0 = 0$;
– $\ker A$ is a *subspace* of $D(A)$, and hence of X;
– $R(A)$ is a *subspace* of Y, and
– G_A is a *subspace* of $X \times Y$.

Exercise 5.2. Verify.

5.1.3 Rank-Nullity and Extension Theorems

5.1.3.1 Rank-Nullity Theorem
Definition 5.3 (Rank and Nullity). Let X and Y be vector spaces over $\mathbb{F}$ and $(A, D(A))$ be a linear operator from X to Y. The *rank* and the *nullity* of A are $\dim(R(A))$ and $\dim(\ker A)$, respectively.

The celebrated *Rank-Nullity Theorem* of linear algebra, which states that the *rank* and the *nullity of a matrix* add up to the number of its columns (see, e. g., [40]), allows the following generalization.

Theorem 5.1 (Rank-Nullity Theorem). *Let X and Y be vector spaces over $\mathbb{F}$ and $A : X \rightarrow Y$ be a linear operator. Then*

$$\dim(X/\ker A) = \dim(R(A))$$

and

$$\dim X = \dim(R(A)) + \dim(\ker A).$$

Proof. Consider the *quotient space* $X/\ker A$ of X modulo $\ker A$ and the mapping

$$X/\ker A \ni [x] = x + \ker A \mapsto \hat{A}[x] := Ax,$$

which is *well-defined*, i. e., for each $[x] \in X/\ker$, the value $\hat{A}[x]$ is independent of the choice of a representative of the *coset* $[x] := x + \ker A$, and is an *isomorphism* between $X/\ker A$ and $R(A)$.

Exercise 5.3. Verify.

Hint. Apply the *Characterization of Isomorphisms* (Proposition 3.12, Section 3.6, Problem 1).

Hence, by the *Isomorphism Theorem* (Theorem 3.5),

$$\dim(X/\ker A) = \dim(R(A)).$$

Since, by the *Codimension of a Subspace Corollary* (Corollary 3.4),

$$\dim X = \dim(\ker A) + \dim(X/\ker A),$$

we conclude that

$$\dim X = \dim(\ker A) + \dim(R(A)). \qquad \square$$

Proposition 5.1 (Null Space of a Linear Functional). *Let X be a vector space over $\mathbb{F}$ and $f : X \to \mathbb{F}$ be a linear functional on X. Then*

$$\dim(X/\ker f) = \begin{cases} 0, \\ 1, \end{cases}$$

i. e., either $\ker f = X$ for $f = 0$ or $\ker f$ is a hyperplane of X for $f \neq 0$.

Proof. By the *Rank-Nullity Theorem* (Theorem 5.1),

$$\dim(X/\ker f) = \dim(R(f)) = \begin{cases} 0, \\ 1, \end{cases}$$

since either $R(f) = \{0\}$ for $f = 0$ or $R(f) = \mathbb{F}$ for $f \neq 0$.

Exercise 5.4. Verify. $\qquad \square$

5.1.3.2 Extension Theorem
The following is a fundamental statement concerning the extendability of linear operators. Its proof ideologically foreshadows that of the celebrated *Hahn*[1]*–Banach Theorem* (the extension form) (see Section 6.1).

---
1 Hans Hahn (1879–1934).

Theorem 5.2 (Extension Theorem for Linear Operators). *Let X and Y be vector spaces over* $\mathbb{F}$ *and*

$$A : X \supseteq D(A) \rightarrow Y,$$

where D(A) is a subspace of X, be a linear operator from X to Y. Then there exists a linear operator $\tilde{A} : X \rightarrow Y$ *defined on the entire space X such that*

$$\tilde{A}x = Ax, \; x \in D(A),$$

i. e., there exists a linear extension $\tilde{A}$ *of A to the entire space X.*

Proof. If $D(A) = X$, then, trivially, $\tilde{A} = A$.

Suppose that $D(A) \neq X$ and let $\mathscr{E}$ be the set of all linear extensions of A (in particular, $A \in \mathscr{E}$) partially ordered by *extension*:

$$\forall B, C \in \mathscr{E} : \; B \leq C \; \Leftrightarrow \; C \text{ is an } extension \text{ of } B,$$

i. e.,

$$D(B) \subseteq D(C) \quad \text{and} \quad \forall x \in D(B) : \; Bx = Cx,$$

and let $\mathscr{C}$ be an *arbitrary chain* in $(\mathscr{E}, \leq)$.

Then

$$\tilde{D} := \bigcup_{B \in \mathscr{C}} D(B)$$

is a *subspace* of X and

$$\tilde{D} \ni x \mapsto Cx := Bx,$$

where $B \in \mathscr{C}$ is arbitrary with $x \in D(B)$, is a *well-defined linear operator* from X to Y, which is an *upper bound* of $\mathscr{C}$ in $(\mathscr{E}, \leq)$.

Exercise 5.5. Verify.

By *Zorn's Lemma* (Theorem A.5), $(\mathscr{E}, \leq)$ has a *maximal element* $(\tilde{A}, D(\tilde{A}))$, i. e., $(\tilde{A}, D(\tilde{A}))$ is a *maximal linear extension* of $(A, D(A))$.

This implies that $D(\tilde{A}) = X$. Indeed, otherwise, there exists an element $x \in D(\tilde{A})^c$ and, any

$$z \in \text{span}(D(\tilde{A}) \cup \{x\}) = D(\tilde{A}) \oplus \text{span}(\{x\})$$

can be *uniquely* represented as

$$z = u + \lambda x$$

with some $u \in D(\tilde{A})$ and $\lambda \in \mathbb{F}$.

By choosing an arbitrary element $y \in Y$, we can define a linear extension (an "*extension by one dimension*")

$$E : \mathrm{span}(D(\tilde{A}) \cup \{x\}) \to Y$$

of $\tilde{A}$, and hence, of A, as follows:

$$Ez := \tilde{A}u + \lambda y, \quad z = u + \lambda x \in \mathrm{span}(D(\tilde{A}) \cup \{x\})$$

(i. e., $Ex := y$).

Exercise 5.6. Verify.

Since, $E \in \mathscr{E}$ and $\tilde{A} < E$ (see Section A.2), we obtain a *contradiction* to the maximality of $(\tilde{A}, D(\tilde{A}))$ in $(\mathscr{E}, \leq)$.

Hence, $D(\tilde{A}) = X$ and $\tilde{A}$ is the desired extension of A. ☐

Remark 5.3. The procedure of "*extension by one dimension*" described in the proof of the prior theorem applies to any linear operator $(A, D(A))$ whose domain is not the entire space $(D(A) \neq X)$. Hence, as readily follows from this procedure, a linear extension of such a linear operator to the whole space, although *existent*, is *not unique* whenever the target space Y is *nontrivial* $(Y \neq \{0\})$, being dependent on the choice of the image $y \in Y$ for an $x \in D(\tilde{A})^c$.

5.2 Bounded Linear Operators and Functionals

Bounded linear operators on normed vector spaces are "married" to both their linear and topological structures.

5.2.1 Definitions, Properties, and Examples

The following is a remarkable inherent property of linear operators.

Theorem 5.3 (Continuity of a Linear Operator). *Let $(X, \| \cdot \|_X)$ and $(Y, \| \cdot \|_Y)$ be normed vector spaces over $\mathbb{F}$. If a linear operator $A : X \to Y$ is continuous at a point $x \in X$, then it is Lipschitz continuous on X.*

Proof. Due to the linearity of A, without loss of generality, we can regard that $x = 0$.

Exercise 5.7. Verify.

In view of the fact that $A0 = 0$ (see Remarks 5.2), the continuity of A at 0 implies that

$$\forall \varepsilon > 0 \; \exists \delta > 0 \; \forall x \in X \text{ with } \|x\|_X < \delta : \|Ax\|_Y < \varepsilon.$$

Then, by *absolute scalability* of norm, for any *distinct* $x', x'' \in X$,

$$\frac{\delta}{2\|x' - x''\|_X}(x' - x'') \in B_X(0, \delta),$$

and hence,

$$\left\| A\left[\frac{\delta}{2\|x' - x''\|_X}(x' - x'') \right] \right\|_Y < \varepsilon.$$

By the *linearity* of A and *absolute scalability* of norm, we infer that

$$\forall x', x'' \in X : \|Ax' - Ax''\|_Y = \|A(x' - x'')\|_Y \leq \frac{2\varepsilon}{\delta}\|x' - x''\|_X,$$

which completes the proof. □

Remark 5.4. Thus, a *linear operator* $A : (X, \|\cdot\|_X) \to (Y, \|\cdot\|_Y)$ is either Lipschitz continuous on X or is discontinuous at every point of X.

Definition 5.4 (Bounded Linear Operator and Operator Norm). Let $(X, \|\cdot\|_X)$ and $(Y, \|\cdot\|_Y)$ be normed vector spaces over $\mathbb{F}$. A *linear operator* $A : X \to Y$ is called *bounded* if

$$\exists M > 0 : \|Ax\|_Y \leq M\|x\|_X, \ x \in X,$$

in which case

$$\|A\| := \min\{M > 0 \mid \|Ax\|_Y \leq M\|x\|_X, \ x \in X\} = \sup_{x \in X\setminus\{0\}} \frac{\|Ax\|_Y}{\|x\|_X} = \sup_{\|x\|_X=1} \|Ax\|_Y$$

is a nonnegative number called the *norm* of the operator A.

Exercise 5.8.
(a) Check the consistency of the above definitions of *operator norm*.
(b) Show that *operator norm* can also be equivalently defined as follows:

$$\|A\| := \sup_{\|x\|_X \leq 1} \|Ax\|_Y.$$

(c) Give an example showing that, unless X is finite-dimensional, in the definitions of *operator norm*, sup cannot be replaced with max.

Hint. On $(c_{00}, \|\cdot\|_\infty)$, consider the operator of multiplication by the sequence $(1 - 1/n)_{n\in\mathbb{N}}$:

$$c_{00} \ni x = (x_n)_{n\in\mathbb{N}} \mapsto Ax := ((1 - 1/n)x_n)_{n\in\mathbb{N}} \in c_{00}$$

(see Examples 5.1).

Theorem 5.4 (Boundedness Characterizations). *Let $(X, \| \cdot \|_X)$ and $(Y, \| \cdot \|_Y)$ be normed vector spaces over $\mathbb{F}$. A linear operator $A : X \to Y$ is bounded iff any of the following equivalent conditions holds.*

1. *A maps the unit sphere/unit ball of $(X, \| \cdot \|_X)$ to a bounded set of $(Y, \| \cdot \|_Y)$.*
2. *A maps each bounded set of $(X, \| \cdot \|_X)$ to a bounded set of $(Y, \| \cdot \|_Y)$.*
3. *A is Lipschitz continuous on X, with $\|A\|$ being its best Lipschitz constant.*

Exercise 5.9. Prove.

Remark 5.5. Recall that, in particular, when $Y = \mathbb{F}$, linear operators are called *linear functionals* and the lower-case letters are used: $f : X \to \mathbb{F}$.

Thus, everything defined/proved for linear operators applies to linear functionals. In particular, the number

$$\|f\| := \sup_{\|x\|_X=1} |f(x)|$$

is the *norm* of a bounded linear functional $f : (X, \| \cdot \|) \to (\mathbb{F}, | \cdot |)$.

Examples 5.2.

1. On a normed vector space $(X, \| \cdot \|_X)$ over $\mathbb{F}$, multiplication by a number $\lambda \in \mathbb{F}$ is a bounded linear operator and $\|A\| = |\lambda|$.
    In particular, the *zero operator* 0 $(\lambda = 0)$ and the *identity operator* I $(\lambda = 1)$ are bounded liner operators on X with $\|0\| = 0$ and $\|I\| = 1$.
2. For normed vector spaces $(X, \| \cdot \|_X)$ and $(Y, \| \cdot \|_Y)$ over $\mathbb{F}$ and $n \in \mathbb{N}$, if $f_1, \ldots, f_n$ are *bounded linear functionals* on X and $y_1, \ldots, y_n$ are arbitrary elements in Y,

$$X \ni x \mapsto Ax := \sum_{k=1}^{n} f_k(x)y_k \in Y$$

    is a *bounded linear operator* and $\|A\| \le \sum_{k=1}^{n} \|f_k\| \|y_k\|_Y$.
3. For $m, n \in \mathbb{N}$ and $1 \le p, p' \le \infty$, multiplication by an $m \times n$ matrix $[a_{ij}]$ with entries from $\mathbb{F}$

$$l_p^{(n)} \ni x = (x_1, \ldots, x_n) \mapsto Ax := [a_{ij}] x \in l_{p'}^{(m)}$$

    is a bounded linear operator from $l_p^{(n)}$ to $l_{p'}^{(m)}$ and, provided $m = n$, a bounded linear operator on $l_p^{(n)}$.
    By *Hölder's Inequality for n-Tuples* (Theorem 2.2),

$$\|A\| \le \left\| (\|a_1\|_{p'}, \ldots, \|a_n\|_{p'}) \right\|_q,$$

    where a_j, $j = 1, \ldots, n$, are the *column vectors* of the matrix $[a_{ij}]$ and q is the *conjugate index* to p ($1/p + 1/q = 1$).

In particular, for $m = 1$, we obtain a *bounded linear functional* on $l_p^{(n)}$:

$$l_p^{(n)} \ni x = (x_1, \ldots, x_n) \mapsto f(x) := \sum_{k=1}^{n} a_k x_k \in \mathbb{F}$$

and, by *Hölder's Inequality for n-Tuples* (Theorem 2.2),

$$\|f\| \le \|a\|_q,$$

where $a := (a_1, \ldots, a_n)$ and q is the *conjugate index* to p ($1/p + 1/q = 1$).

4. On l_p ($1 \le p \le \infty$),

    (a) for a *bounded sequence* $a := (a_n)_{n \in \mathbb{N}} \in l_\infty$, the operator of multiplication

    $$l_p \ni x := (x_n)_{n \in \mathbb{N}} \mapsto Ax := (a_n x_n)_{n \in \mathbb{N}} \in l_p$$

    is a *bounded linear operator* with $\|A\| = \|a\|_\infty := \sup_{n \in \mathbb{N}} |a_n|$;

    (b) the *right-shift operator*

    $$l_p \ni x := (x_1, x_2, \ldots) \mapsto Rx := (0, x_1, x_2, \ldots) \in l_p$$

    and the *left-shift operator*

    $$l_p \ni x := (x_1, x_2, \ldots) \mapsto Lx := (x_2, x_3, x_4, \ldots) \in l_p$$

    are a *bounded linear operators* with $\|R\| = \|L\| = 1$, the right-shift operator being an *isometry*, i. e.,

    $$\forall x \in l_p : \|Rx\|_p = \|x\|_p;$$

    (c) for a sequence $a := (a_n)_{n \in \mathbb{N}} \in l_q$, where q is the *conjugate index* to p ($1/p + 1/q = 1$), by *Hölder's Inequality for Sequences* (Theorem 2.5),

    $$l_p \ni x := (x_n)_{n \in \mathbb{N}} \mapsto f(x) := \sum_{k=1}^{\infty} a_k x_k \in \mathbb{F}$$

    is a *bounded linear functional* with

    $$\|f\| \le \|a\|_q.$$

Remark 5.6. In fact, as is shown below (see Theorem 7.3), more precisely, $\|f\| = \|a\|_q$ and such functionals are the only bounded linear functionals on l_p ($1 \le p \le \infty$).

5. On $(c_{00}, \|\cdot\|_\infty)$, the operator of multiplication by an *unbounded numeric sequence* $a := (a_n)_{n \in \mathbb{N}} \in \mathbb{F}^{\mathbb{N}}$

    $$c_{00} \ni x := (x_n)_{n \in \mathbb{N}} \mapsto Ax := (a_n x_n)_{n \in \mathbb{N}} \in c_{00}$$

    is an *unbounded linear operator*.

6. On $(c, \|\cdot\|_\infty)$, the *limit functional*

$$c \ni x := (x_n)_{n\in\mathbb{N}} \mapsto l(x) := \lim_{n\to\infty} x_n \in \mathbb{F},$$

assigning to each convergent sequence its limit, is a *bounded linear functional* with $\|l\| = 1$ and $\ker l = c_0$.

7. On l_1, the *sum functional*

$$l_1 \ni x := (x_n)_{n\in\mathbb{N}} \mapsto s(x) := \sum_{n=1}^\infty x_n \in \mathbb{F},$$

assigning to each *absolutely summable sequence* the sum of the series composed of its terms, is a *bounded linear functional* with $\|s\| = 1$.

8. On $(C[a,b], \|\cdot\|_\infty)$ $(-\infty < a < b < \infty)$,
   (a) multiplication by a fixed function $m \in C[a,b]$

$$C[a,b] \ni x \to [Ax](t) := m(t)x(t) \in C[a,b]$$

   is a *bounded linear operator* with $\|A\| = \|m\|_\infty := \max_{a\le t\le b} |m(t)|$;
   (b) the *integration operator*

$$C[a,b] \ni x \to [Ax](t) := \int_a^t x(s)\, ds \in C^1[a,b]$$

   is a *bounded linear operator* with $\|A\| = b - a$;
   (c) the *integration functional*

$$C[a,b] \ni x \to f(x) := \int_a^b x(t)\, dt \in \mathbb{F}$$

   is a *bounded linear functional* with $\|f\| = b - a$;
   (d) for each $t \in [a,b]$, the *fixed-value functional*

$$C[a,b] \ni x \to f_t(x) := x(t) \in \mathbb{F}$$

   is a *bounded linear functional* with $\|f_t\| = 1$.

9. The *differentiation operator*

$$C^1[a,b] \ni x \to [Ax](t) := \frac{d}{dt}x(t) \in C[a,b]$$

   (a) is a *bounded linear operator* from $X := C^1[a,b]$ with the norm

$$\|x\| = \max\left[\max_{a\le t\le b}|x(t)|, \max_{a\le t\le b}|x'(t)|\right]$$

   to $Y := C[a,b]$ with the *maximum norm* $\|\cdot\|_\infty$, $\|A\| \le 1$ and
   (b) is an *unbounded linear operator* in $(C[a,b], \|\cdot\|_\infty)$ with the *domain* $D(A) := C^1[a,b]$.

Exercise 5.10. Verify.

Proposition 5.2 (Kernel of a Bounded Linear Operator). *Let $(X, \| \cdot \|_X)$ and $(Y, \| \cdot \|_Y)$ be normed vector spaces over $\mathbb{F}$. For a bounded linear operator $A : X \to Y$, its kernel, $\ker A$, is a closed subspace of $(X, \| \cdot \|_X)$.*

Exercise 5.11.
(a) Prove.
(b) Give an example showing that a linear operator $A : (X, \| \cdot \|_X) \to (Y, \| \cdot \|_Y)$ with a *closed kernel*, need not be bounded.

   Hint. On $(c_{00}, \| \cdot \|_\infty)$, consider the operator of multiplication by the sequence $(n)_{n \in \mathbb{N}}$:

$$c_{00} \ni x = (x_n)_{n \in \mathbb{N}} \mapsto Ax := (nx_n)_{n \in \mathbb{N}} \in c_{00}$$

   (see Examples 5.1).

Remarks 5.7.
–   However, for a *linear functional* $f : (X, \|\cdot\|_X) \to \mathbb{F}$ on a normed vector space $(X, \|\cdot\|_X)$, the closedness of the kernel is not only necessary, but sufficient for its boundedness (see Proposition 5.10, Section 5.4, Problem 5).
–   Thus, for a bounded linear operator $A : (X, \| \cdot \|_X) \to (Y, \| \cdot \|_Y)$, well-defined is the quotient space $(X/\ker A, \| \cdot \|_1)$ of X modulo $\ker A$, where $\| \cdot \|_1$ is the *quotient-space norm* (see Proposition 3.19, Section 3.6, Problem 10).

5.2.2 Space of Bounded Linear Operators, Dual Space

Here, we are to see that bounded linear operators from one normed vector space to another themselves form a normed vector space with various forms of convergence in it.

5.2.2.1 Space of Bounded Linear Operators
Theorem 5.5 (Space of Bounded Linear Operators). *Let $(X, \| \cdot \|_X)$ and $(Y, \| \cdot \|_Y)$ be normed vector spaces over $\mathbb{F}$.*
   The set $L(X, Y)$ of all bounded linear operators $A : X \to Y$ is a normed vector space over $\mathbb{F}$ relative to the pointwise linear operations

$$(A + B)x := Ax + Bx, \quad A, B \in L(X, Y), \ x \in X,$$
$$(\lambda A)x := \lambda Ax, \ \lambda \in \mathbb{F}, \quad A \in L(X, Y), \ x \in X, \tag{5.2}$$

and the operator norm

$$\|A\| := \sup_{\|x\|_X = 1} \|Ax\|_Y.$$

If the space $(Y, \| \cdot \|_Y)$ is Banach, then $L(X, Y)$ is also a Banach space.

Proof. The *vector space axioms* for $L(X, Y)$ are readily verified.

Exercise 5.12. Verify.

Let us verify the *norm axioms* for the operator norm.
Nonnegativity is obvious.
Separation holds as well, since, for an $A \in L(X, Y)$,

$$\|A\| = 0 \Leftrightarrow \|Ax\|_Y = 0 \quad \text{for all } x \in X \text{ with } \|x\|_X = 1,$$

the latter being equivalent to the fact that

$$\forall x \in X : Ax = 0, \text{ i. e., } A = 0.$$

Exercise 5.13. Verify.

Absolute scalability is easily verified, too.

Exercise 5.14. Verify.

Since, for any $A, B \in L(X, Y)$, by the *subadditivity* of $\| \cdot \|_Y$, for any $x \in X$,

$$\|(A + B)x\|_Y = \|Ax + Bx\|_Y \leq \|Ax\|_Y + \|Bx\|_Y \leq \|A\|\|x\|_X + \|B\|\|x\|_X$$
$$= [\|A\| + \|B\|] \|x\|_X.$$

Whence, we conclude that

$$\|A + B\| \leq \|A\| + \|B\|,$$

i. e., the operator norm is *subadditive*.
Thus, $(L(X, Y), \| \cdot \|)$ is an *normed vector space* over $\mathbb{F}$.
Suppose $(Y, \| \cdot \|_Y)$ is a Banach space and let $\{A_n\}_{n=1}^{\infty}$ be an arbitrary *Cauchy sequence* in $(L(X, Y), \| \cdot \|)$. Then, for each $x \in X$,

$$\|A_n x - A_m x\|_Y = \|(A_n - A_m)x\|_Y = \|A_n - A_m\|\|x\|_X \to 0, \ m, n \to \infty,$$

i. e., $\{A_n x\}_{n=1}^{\infty}$ is a *Cauchy sequence* in the space $(Y, \| \cdot \|_Y)$, which, due to the completeness of the latter, converges, and hence, we can define a *linear operator* from X to Y as follows:

$$X \ni x \mapsto Ax := \lim_{n \to \infty} A_n x \in Y$$

Exercise 5.15. Verify that the operator A is *linear*.

The operator A is *bounded*. Indeed, being *fundamental*, the sequence $\{A_n\}_{n=1}^{\infty}$ is *bounded* in $(L(X,Y), \|\cdot\|)$ (Theorem 2.21), i. e.,

$$\exists M > 0 \ \forall n \in \mathbb{N} : \ \|A_n\| \le M,$$

and hence, for each $x \in X$ and any $n \in \mathbb{N}$,

$$\|A_n x\|_Y \le \|A_n\| \|x\|_X \le M \|x\|_X.$$

Whence, in view of the *continuity* of norm (see Remarks 3.18), passing to the limit as $n \to \infty$ for each fixed $x \in X$, we obtain the estimate

$$\forall x \in X : \ \|Ax\|_Y \le M \|x\|_X,$$

which implies that $A \in L(X,Y)$.

Since $\{A_n\}_{n=1}^{\infty}$ is a *Cauchy sequence* in $(L(X,Y), \|\cdot\|)$,

$$\forall \varepsilon > 0 \ \exists N \in \mathbb{N} \ \forall m, n \ge N : \ \|A_n - A_m\| < \varepsilon,$$

and hence,

$$\forall x \in X, \ \forall m, n \ge N : \ \|A_n x - A_m x\|_Y \le \|A_n - A_m\| \|x\|_X \le \varepsilon \|x\|_X.$$

Fixing arbitrary $x \in X$ and $n \ge N$ and passing to the limit as $m \to \infty$, we obtain

$$\|A_n x - Ax\|_Y \le \varepsilon \|x\|_X, \ x \in X,$$

which implies that

$$\forall n \ge N : \ \|A_n - A\| \le \varepsilon,$$

i. e.,

$$A_n \to A, \ n \to \infty, \ \text{in} \ (L(X,Y), \|\cdot\|),$$

and thus, concludes proving the *completeness* of the operator space $(L(X,Y), \|\cdot\|)$ when the space $(Y, \|\cdot\|_Y)$ is complete. ☐

Remark 5.8. If $X = Y$, we use the notation $L(X)$, $(L(X), \|\cdot\|)$ being a *normed algebra* over $\mathbb{F}$ with *operator multiplication* defined as follows:

$$\forall A, B \in L(X) : \ (AB)x := A(Bx), \ x \in X,$$

associative and *bilinear* relative to operator addition and scalar multiplication, the operator norm being *submultiplicative*:

$$\forall A, B \in L(X) : \ \|AB\| \le \|A\| \|B\|.$$

If $(X, \|\cdot\|_X)$ is a Banach space, then $(L(X), \|\cdot\|)$ is a *Banach algebra* (see, e. g., [3]).

Exercise 5.16. Verify the aforementioned properties of the operator multiplication and *submultiplicativity* of the operator norm.

5.2.2.2 Dual Space

Let us introduce the concept of *dual space,* which is studied in greater detail in the subsequent chapters.

Definition 5.5 (Dual Space). For $Y = \mathbb{F}$ with the absolute-value norm, we call the space $L(X, Y)$ of bounded linear functionals on X the *dual space* of X and use the notation X^* for it.

Remark 5.9. As is shown below, the nontriviality of the dual space X^* for any nontrivial normed vector space $(X, \| \cdot \|_X)$ is guaranteed by the *Hahn–Banach Theorem* (the extension form) (see Section 6.1).

From the prior theorem, in view of the completeness of the target space $(\mathbb{F}, | \cdot |)$, we immediately obtain the following corollary.

Corollary 5.1 (Completeness of the Dual Space). *For each normed vector space, its dual is a Banach space.*

5.2.2.3 Uniform and Strong Convergence in $L(X, Y)$

Let us now discuss various forms of convergence in operator spaces.

Definition 5.6 (Uniform Convergence). Let $(X, \| \cdot \|_X)$ and $(Y, \| \cdot \|_Y)$ be normed vector spaces over $\mathbb{F}$.

The convergence of a sequence of operators $\{A_n\}_{n=1}^{\infty}$ to an operator A in the space $(L(X, Y), \| \cdot \|)$:

$$\|A_n - A\| := \sup_{\|x\|_X \leq 1} \|Ax - A_n x\|_Y \to 0, \ n \to \infty,$$

is called *uniform,* and the operator A is called the *uniform limit* of $\{A_n\}_{n=1}^{\infty}$.

Remark 5.10. The name is justified by the fact that such a convergence is equivalent to the uniform convergence of $\{A_n\}_{n=1}^{\infty}$ to A on the closed unit ball

$$B_X(0, 1) := \{x \in X \mid \|x\|_X \leq 1\},$$

of $(X, \| \cdot \|_X)$, or more generally, on any *bounded set* of $(X, \| \cdot \|_X)$.

Example 5.3. Let $(X, \| \cdot \|)$ be a normed vector space. Then the operator sequence $\{\frac{1}{n}I\}_{n=1}^{\infty}$, where I is the *identity operator* on X, *converges uniformly* to the *zero operator* since

$$\left\| \frac{1}{n} I \right\| = \frac{1}{n} \|I\| = \frac{1}{n} \to 0, \ n \to \infty.$$

Definition 5.7 (Strong Convergence). Let $(X, \| \cdot \|_X)$ and $(Y, \| \cdot \|_Y)$ be normed vector spaces over $\mathbb{F}$.

The pointwise convergence of a sequence of operators $\{A_n\}_{n=1}^{\infty}$ to an operator A in the space $(L(X, Y), \|\cdot\|)$:

$$\forall x \in X : A_n x \to Ax, \ n \to \infty, \ \text{in} \ (Y, \|\cdot\|_Y),$$

is called *strong*, and the operator A is called the *strong limit* of $\{A_n\}_{n=1}^{\infty}$.

Remarks 5.11.
- Uniform convergence implies strong convergence.

  Exercise 5.17. Verify.

- However, as the following example shows, the converse is not true, i. e., a strongly convergent operator sequence need not converge uniformly.

Example 5.4. The *left-shift operator* on l_p $(1 \le p < \infty)$

$$l_p \ni x = (x_1, x_2, \dots) \mapsto Lx := (x_2, x_3, x_4, \dots) \in l_p,$$

belongs to $L(l_p)$ since $\|Lx\|_p \le \|x\|_p$, $x \in l_p$ (see Examples 5.2).
For the operator sequence $\{L^n\}_{n=1}^{\infty} \subseteq L(l_p)$,

$$l_p \ni x = (x_1, x_2, \dots) \mapsto L^n x := (x_{n+1}, x_{n+2}, x_{n+3}, \dots) \in l_p,$$

and hence,

$$\forall x \in l_p : \|L^n x\|_p = \left[\sum_{k=n+1}^{\infty} |x_k|^p \right]^{1/p} \to 0, \ n \to \infty,$$

i. e., $\{L^n\}_{n=1}^{\infty}$ *strongly converges* to the *zero operator* in $L(l_p)$.
However,

$$\|L^n\| = 1, \ n \in \mathbb{N}.$$

Exercise 5.18. Verify.

Hint. Apply L^n $(n \in \mathbb{N})$ to the sequence $\{\delta_{(n+1)k}\}_{k=1}^{\infty}$, where δ_{nk} is the *Kronecker delta*,

Therefore, the operator sequence $\{L^n\}_{n=1}^{\infty}$ *does not converge uniformly* in $L(l_p)$.

Exercise 5.19. Explain.

5.3 Closed Linear Operators

The following defines a very important class of linear operators, which need not be bounded.

Definition 5.8 (Closed Linear Operator). Let $(X, \| \cdot \|_X)$ and $(Y, \| \cdot \|_Y)$ be normed vector spaces over $\mathbb{F}$. A linear operator $(A, D(A))$ from X to Y is called *closed* if its *graph*

$$G_A := \{(x, Ax) \mid x \in D(A)\}$$

is a *closed subspace* in the product space $X \times Y$ relative to the product norm

$$X \times Y \ni (x, y) \mapsto \|(x, y)\|_{X \times Y} := \sqrt{\|x\|_X^2 + \|y\|_Y^2}$$

(see Section 3.6, Problem 5).

Remark 5.12. The product norm $\| \cdot \|_{X \times Y}$ on $X \times Y$ can be replaced with the *equivalent* one

$$X \times Y \ni (x, y) \mapsto \|(x, y)\| := \|x\|_X + \|y\|_Y,$$

which may be a little easier to handle.

Exercise 5.20. Verify the *norm axioms* for the latter and the equivalence of $\| \cdot \|_{X \times Y}$ and $\| \cdot \|$ on $X \times Y$.

In view of the *componentwise* nature of convergence in a product space (see Proposition 2.17, Section 2.19, Problem 9), by the *Sequential Characterization of Closed Sets* (Theorem 2.18), we obtain the following statement.

Proposition 5.3 (Sequential Characterization of Closed Linear Operators). *Let* $(X, \| \cdot \|_X)$ *and* $(Y, \| \cdot \|_Y)$ *be normed vector spaces over* $\mathbb{F}$. *A linear operator* $(A, D(A))$ *from* X *to* Y *is closed iff, for any sequence* $\{x_n\}_{n=1}^{\infty} \subset D(A)$ *such that*

$$\lim_{n \to \infty} x_n = x \text{ in } (X, \| \cdot \|_X) \quad and \quad \lim_{n \to \infty} A x_n = y \text{ in } (Y, \| \cdot \|_Y),$$

the following is true:

$$x \in D(A) \quad and \quad y = Ax.$$

Exercise 5.21. Prove.

Remark 5.13. Provided $D(A) = X$, the condition $x \in D(A)$ holds automatically.

Using the prior sequential characterization, one can prove the following fact.

Proposition 5.4 (Characterization of Closedness for Bounded Linear Operators). *Let* $(X, \| \cdot \|_X)$ *and* $(Y, \| \cdot \|_Y)$ *be normed vector spaces. For a bounded linear operator* $A : X \supseteq D(A) \to Y$ *to be closed it is sufficient and, provided* $(Y, \| \cdot \|_Y)$ *is a Banach space, necessary that the domain* $D(A)$ *be a closed subspace in* $(X, \| \cdot \|_X)$.
In particular, each $A \in L(X, Y)$ *is a closed operator.*

Exercise 5.22. Prove.

Examples 5.5.
1. By the prior proposition, all bounded linear operators from Examples 5.2 are closed operators.
2. In the Banach space l_p $(1 \le p \le \infty)$, the linear operator A of multiplication by a numeric sequence $a := (a_n)_{n\in\mathbb{N}} \in \mathbb{F}^{\mathbb{N}}$:

$$(x_n)_{n\in\mathbb{N}} \mapsto Ax := (a_n x_n)_{n\in\mathbb{N}}$$

with the domain

$$D(A) := \{(x_n)_{n\in\mathbb{N}} \in l_p \mid (a_n x_n)_{n\in\mathbb{N}} \in l_p\}$$

is *closed* and is *bounded* (and on) iff $a = (a_n)_{n\in\mathbb{N}} \in l_\infty$ (see Examples 5.2).

Remark 5.14. Observe that $c_{00} \subseteq D(A)$, which, for $1 \le p < \infty$, immediately implies that $D(A)$ is *dense* in l_p, which makes A *densely defined*.

3. In the Banach space $(C[a,b], \|\cdot\|_\infty)$ $(-\infty < a < b < \infty)$, the *unbounded* linear differentiation operator

$$C^1[a,b] =: D(A) \ni x \to Ax := \frac{dx}{dt} \in C[a,b]$$

(see Examples 5.2) is *closed*.
4. In the Banach space $(c_0, \|\cdot\|_\infty)$, the *unbounded* linear operator of multiplication

$$c_{00} =: D(A) \ni x = (x_n)_{n\in\mathbb{N}} \mapsto Ax := (nx_n)_{n\in\mathbb{N}} \in c_{00}$$

(see Examples 5.2) is *not closed*.

Exercise 5.23. Verify.

Hints.
- Apply the *Sequential Characterization of Closed Linear Operators* (Proposition 5.3).
- For 3, use the *Total Change Formula*

$$x(t) = x(a) + \int_a^t x'(s)\, ds,$$

valid for every $x \in C^1[a,b]$.
- For 4, in the domain $D(A) = c_{00}$ consider the sequence

$$\{x_n := (1, 1/2^2, \dots, 1/n^2, 0, 0, \dots)\}_{n=1}^{\infty}.$$

Remark 5.15. Thus, a closed linear operator need not be bounded and a linear operator need not be closed.

Proposition 5.5 (Kernel of a Closed Linear Operator). *Let $(X, \|\cdot\|_X)$ and $(Y, \|\cdot\|_Y)$ be normed vector spaces over $\mathbb{F}$. For a closed linear operator $(A, D(A))$ from X to Y, its kernel, $\ker A$, is a closed subspace of $(X, \|\cdot\|_X)$.*

Exercise 5.24. Prove.

Remarks 5.16.

– The *Kernel of a Bounded Linear Operator Proposition* (Proposition 5.2) is now a direct corollary of Propositions 5.4 and 5.5.

– Thus, for a closed linear operator $A : (X, \|\cdot\|_X) \to (Y, \|\cdot\|_Y)$, well-defined is the quotient space $(X/\ker A, \|\cdot\|_1)$ of X modulo $\ker A$, where $\|\cdot\|_1$ is the *quotient-space norm* (see Section 3.6, Problem 10, cf. Remarks 5.7).

5.4 Problems

1. Prove

   Proposition 5.6 (Characterization of the Graph of a Linear Operator). *Let X and Y be vector spaces over $\mathbb{F}$. A subspace G of the product space $X \times Y$ is the graph of a linear operator $(A, D(A))$ from X to Y iff it does not contain points of the form $(0, y)$ with $y \neq 0$.*

2. Prove

   Proposition 5.7 (Hyperplane Characterization). *A subspace Y of a nontrivial vector space X over $\mathbb{F}$ is a hyperplane iff there exists a nontrivial linear functional $f : X \to \mathbb{F}$ such that $Y = \ker f$.*

3. Prove

   Proposition 5.8 (Boundedness of Linear Operators on Finite-Dimensional Spaces). *Let $(X, \|\cdot\|_X)$ and $(Y, \|\cdot\|_Y)$ be normed vector spaces over $\mathbb{F}$, the space X being finite-dimensional. Then each linear operator $A : X \to Y$ is bounded.*

   Remark 5.17. In particular, for $Y = \mathbb{F}$ with the absolute-value norm, we conclude that all linear functionals on a *finite-dimensional* normed vector space $(X, \|\cdot\|_X)$ are *bounded*.

4. Prove

   Proposition 5.9 (Existence of Unbounded Linear Operators). *Let $(X, \|\cdot\|_X)$ be an infinite-dimensional normed vector space and $(Y, \|\cdot\|_Y)$ be a nontrivial normed vector space $(Y \neq \{0\})$ over $\mathbb{F}$. Then there exists an unbounded linear operator $A : X \to Y$.*

   Hint. Use the *Extension Theorem for Linear Operators* (Theorem 5.2).

Remark 5.18. In particular, for $Y = \mathbb{F}$ with the absolute-value norm, we conclude that, on every *infinite-dimensional* normed vector space $(X, \|\cdot\|_X)$, there is an *unbounded linear functional*.

5. * Prove

Proposition 5.10 (Kernel of a Linear Functional). *Let $(X, \|\cdot\|)$ be a normed vector space over $\mathbb{F}$. A linear functional f on X is bounded iff $\ker f$ is closed.*

Hint. Prove the *"if part"* by contrapositive.

6. * Prove

Proposition 5.11 (Unboundedness of Hamel Coordinate Functionals). *Let $(X, \|\cdot\|)$ be an infinite dimensional Banach space over $\mathbb{F}$ with a basis $B := \{x_i\}_{i \in I}$. Then all but a finite number of the Hamel coordinate functionals*

$$X \ni x = \sum_{i \in I} c_i x_i \mapsto c_j(x) := c_j \in \mathbb{F},\ j \in I,$$

relative to B are unbounded.

Give an example showing that the *completeness* requirement for the space is essential and cannot be dropped.

7. Prove

Proposition 5.12 (Characterization of Boundedness of Linear Operators).   *Let $(X, \|\cdot\|_X)$ and $(Y, \|\cdot\|_Y)$ be normed vector spaces over $\mathbb{F}$. A linear operator $A : X \to Y$ is bounded iff the preimage $A^{-1}B_Y(0,1)$ under A of the open unit ball $B_Y(0,1) := \{y \in Y \mid \|y\|_Y < 1\}$ in $(Y, \|\cdot\|_Y)$ has a nonempty interior in $(X, \|\cdot\|_X)$.*

8. Prove

Proposition 5.13 (Completeness of the Range). *Let $(X, \|\cdot\|_X)$ be a Banach space, $(Y, \|\cdot\|_Y)$ be a normed vector space over $\mathbb{F}$, and $A \in L(X, Y)$ have a bounded inverse $A^{-1} : R(A) \to X$. Then $(R(A), \|\cdot\|_Y)$ is a Banach space.*

Give an example showing that the *completeness* requirement for $(X, \|\cdot\|_X)$ is essential and cannot be dropped.

9. Prove

Proposition 5.14 (Closedness of Inverse Operator). *Let $(X, \|\cdot\|_X)$ and $(Y, \|\cdot\|_Y)$ be normed vector spaces over $\mathbb{F}$, and $A : X \supseteq D(A) \to Y$ be a closed linear operator. Then the inverse operator $A^{-1} : Y \supseteq R(A) \to X$, if existent, is a closed linear operator.*

10. Let $(X, \|\cdot\|_X)$ and $(Y, \|\cdot\|_Y)$ be normed vector spaces over $\mathbb{F}$ and $A : X \to Y$ be a linear operator. Show that the *injective* linear operator

$$(X/\ker A, \|\cdot\|_1) \ni [x] := x + \ker A \mapsto \hat{A}[x] := Ax \in (Y, \|\cdot\|_Y),$$

where $\| \cdot \|_1$ is the *quotient-space norm* (see Section 3.6, Problem 10 and the proof of the *Rank-Nullity Theorem* (Theorem 5.1)),

(a) is *bounded* provided A is *bounded* and

(b) is *closed* provided A is *closed*.

6 Three Fundamental Principles of Linear Functional Analysis

The major statements of this chapter—the *Hahn–Banach Theorem* (the extension form), the *Uniform Boundedness Principle*, and the *Open Mapping Theorem* along with its equivalents: the *Inverse Mapping Theorem* and the *Closed Graph Theorem*—are the so-called *three fundamental principles of linear functional analysis*, because of their importance. The significance of these principles, along with their vast applicability, cannot be overestimated.

6.1 Hahn–Banach Theorem

In this section, we study the *extension* (or *analytic*) form of the celebrated *Hahn–Banach Theorem* and its various implications. The *separation* (or *geometric*) form discussion can be found in, e. g., [6, 7, 12].

6.1.1 Hahn–Banach Theorem for Real Vector Spaces

We start our discourse with the case of a real vector space, the proof of the following theorem being largely reminiscent of that of the *Extension Theorem for Linear Operators* (Theorem 5.2).

Theorem 6.1 (Hahn–Banach Theorem for Real Vector Spaces). *Let X be a real vector space and p be a real-valued function on X satisfying the following conditions:*
1. *$p(\lambda x) = \lambda p(x)$, $\lambda \geq 0$, $x \in X$.*   *Nonnegative Homogeneity/Scalability*
2. *$p(x + y) \leq p(x) + p(y)$, $x, y \in X$.*   *Subadditivity/Triangle Inequality*

If $f : X \supseteq D(f) \to \mathbb{R}$, where $D(f)$ is a subspace of X, is a real-valued linear functional such that

$$\forall x \in D(f) : f(x) \leq p(x), \qquad\qquad (6.1)$$

then f can be extended to a real-valued linear functional $F : X \to \mathbb{R}$ defined on the entire space X and such that

$$\forall x \in X : F(x) \leq p(x).$$

Proof. If $D(f) = X$, then, trivially, $F = f$.

Suppose that $D(f) \neq X$ and let $\mathscr{E}$ be the collection of all real-valued linear extensions g of f such that

$$g(x) \leq p(x), \quad x \in D(g),$$

https://doi.org/10.1515/9783110614039-006

where the *domain D(g)* of g is a *subspace* of X with $D(f) \subseteq D(g)$ (in particular, $f \in \mathscr{E}$). It is partially ordered by *extension*:

$$\forall g, h \in \mathscr{E} : g \leq h \Leftrightarrow h \text{ is an } extension \text{ of } g,$$

i. e.,

$$D(g) \subseteq D(h) \quad \text{and} \quad \forall x \in D(g) : g(x) = h(x),$$

and let $\mathscr{C}$ be an *arbitrary chain* in $(\mathscr{E}, \leq)$.

Then

$$Z := \bigcup_{g \in \mathscr{C}} D(g)$$

is a *subspace* of X and

$$Z \ni x \mapsto u(x) := g(x),$$

where $g \in \mathscr{C}$ is arbitrary with $x \in D(g)$, is a *well-defined linear functional*, which is a real-valued linear extension of any $g \in \mathscr{C}$ (cf. the proof of the *Extension Theorem for Linear Operators* (Theorem 5.2)).

Exercise 6.1. Verify.

Furthermore, since

$$\forall x \in Z \, \exists g \in \mathscr{C} : x \in D(g),$$

we infer that

$$u(x) := g(x) \leq p(x), \ x \in Z.$$

Hence, the real-valued functional u with the domain $D(u) := Z$ is an *upper bound* of $\mathscr{C}$ in $(\mathscr{E}, \leq)$.

By *Zorn's Lemma* (Theorem A.5), $(\mathscr{E}, \leq)$ has a *maximal element* $(F, D(F))$, i. e., $(F, D(F))$ is a *maximal real-valued linear extension* of $(f, D(f))$ such that

$$\forall x \in D(F) : F(x) \leq p(x).$$

This necessarily implies that $D(F) = X$. Otherwise, there exists an element $x \in D(F)^c$ and any

$$z \in \text{span}(D(F) \cup \{x\}) = D(F) \oplus \text{span}(\{x\})$$

can be *uniquely* represented as

$$z = y + \lambda x$$

with some $y \in D(F)$ and $\lambda \in \mathbb{R}$.

Fixing an arbitrary number $\mu \in \mathbb{R}$, we can define a real-valued linear extension (an "*extension by one dimension*")

$$E : \text{span}(D(F) \cup \{x\}) \to \mathbb{R}$$

of F, and hence, of f, as follows:

$$E(z) := F(y) + \lambda\mu, \ z = y + \lambda\mu \in \text{span}(D(F) \cup \{x\})$$

(i. e., $E(x) := y$).

Exercise 6.2. Verify.

It remains to be shown that the number $\mu \in \mathbb{R}$ can be chosen so that

$$\forall x \in D(E): \ E(x) \le p(x).$$

Indeed, since, by the *subadditivity* of p, for arbitrary $y_1, y_2 \in D(F)$ and each $x \in X$,

$$F(y_2) - F(y_1) = F(y_2 - y_1) = F((y_2 + x) + (-x - y_1)) \le p(y_2 + x) + p(-x - y_1),$$

we have:

$$-p(-x - y_1) - F(y_1) \le p(y_2 + x) - F(y_2), \ y_1, y_2 \in D(F),$$

and hence,

$$\sup_{y_1 \in D(F)} [-p(-x - y_1) - F(y_1)] \le \inf_{y_2 \in D(F)} [p(y_2 + x) - F(y_2)].$$

Thus, by choosing a $\mu \in \mathbb{R}$ satisfying

$$\sup_{y_1 \in D(F)} [-p(-x - y_1) - F(y_1)] \le \mu \le \inf_{y_2 \in D(F)} [p(y_2 + x) - F(y_2)], \tag{6.2}$$

we have:

$$\forall x \in X, \ \forall y \in D(F): \ -p(-x - y) \le F(y) + \mu \le p(y + x), \tag{6.3}$$

which implies that, for any $z = y + \lambda x \in D(E)$,

$$E(z) = F(y) + \lambda\mu \le p(y + \lambda x) = p(z). \tag{6.4}$$

Indeed, considering the *nonnegative scalability* of p,
- for $\lambda = 0$, (6.4) holds by the premise;
- for $\lambda > 0$, (6.4) follows from (6.3) by substituting y/λ for y and multiplying through by λ;
- for $\lambda < 0$, (6.4) also follows from (6.3) by substituting y/λ for y and multiplying through by λ.

Exercise 6.3. Explain.

Since, $E \in \mathscr{E}$ and $F < E$ (see Section A.2), we obtain a *contradiction* to the maximality of $(F, D(F))$ in $(\mathscr{E}, \le)$.

Hence, $D(F) = X$ and F is the desired extension of f. ☐

Remark 6.1. The procedure of *"extension by one dimension"* in the proof of the prior theorem, being a particular case of that from the proof of the *Extension Theorem for Linear Operators* (Theorem 5.2), applies to any linear functional $(f, D(f))$ satisfying (6.1) whose domain is not the entire space $(D(f) \neq X)$. Hence, as readily follows from this procedure, a linear extension to the whole space also meeting condition (6.1), al- though *existent, need not be unique*, being dependent on the choice of the image $\mu \in \mathbb{R}$ satisfying (6.2) for an $x \in D(F)^c$ (cf. Remark 5.3 and Section 6.7, Problem 1).

6.1.2 Hahn–Banach Theorem for Normed Vector Spaces

We continue with the case of a (real or complex) normed vector space.

Theorem 6.2 (Hahn–Banach Theorem for Normed Vector Spaces). *Let $(X, \| \cdot \|)$ be a normed vector space over $\mathbb{F}$ and $f : X \supseteq D(f) \to \mathbb{F}$, where $D(f)$ is a subspace of X, be a bounded linear functional.*

Then f can be extended to a bounded linear functional $F : X \to \mathbb{R}$ defined on the entire space X (i. e., $F \in X^*$) and such that

$$\|F\| := \sup_{\|x\|=1} |F(x)| = \sup_{x \in D(f), \|x\|=1} |f(x)| =: \|f\|.$$

Proof. If $D(f) = X$, then, trivially, $F = f$. Henceforth, suppose that $D(f) \neq X$.

Real Case: Let us first consider the case of a *real* X (i. e., $\mathbb{F} = \mathbb{R}$).

As is easily seen, the conditions of the *Hahn–Banach Theorem for Real Vector Spaces* (Theorem 6.1) are satisfied for the functional $(f, D(f))$ with

$$p(x) := \|f\| \|x\|, \ x \in X.$$

Hence, there exists a linear extension F of $(f, D(f))$ to the whole space X such that

$$F(x) \leq \|f\| \|x\|, \ x \in X.$$

Substituting $-x$ for x, we have:

$$-F(x) = F(-x) \leq \|f\| \| - x\| = \|f\| \|x\|, \ x \in X.$$

Hence,

$$|F(x)| \leq \|f\| \|x\|, \ x \in X,$$

which implies that the functional F is bounded and $\|F\| \leq \|f\|$.

On the other hand, since F is an extension of f, obviously, $\|f\| \leq \|F\|$, and hence, $\|F\| = \|f\|$.

Complex Case: Now, let us consider the case of a *complex X* (i. e., $\mathbb{F} = \mathbb{R}$).

$$f(x) = f_1(x) + if_2(x), \; x \in D(f),$$

where $f_1(x) := \mathrm{Re}[f(x)]$ and $f_2(x) := \mathrm{Im}[f(x)]$.

As can be easily verified, f_1 and f_2 are real-valued linear functionals on $D(f)$, with $(X, \|\cdot\|)$ considered as a *real* normed vector space.

Exercise 6.4. Verify.

They are related as follows:

$$f_2(x) = -f_1(ix), \; x \in D(f).$$

Indeed, by the *linearity* of f,

$$f_1(ix) + if_2(ix) = f(ix) = if(x) = if_1(x) - f_2(x), \; x \in D(f).$$

Furthermore,

$$|f_1(x)| \le |f(x)| \le \|f\|\|x\|, \; x \in D(f),$$

i. e., $\|f_1\| \le \|f\|$.

Considering $(X, \|\cdot\|)$ as a *real* normed vector space, by the proved real case, we infer that there is a real-valued bounded linear extension F_1 to X of the real-valued bounded linear functional $(f_1, D(f_1))$ with $\|F_1\| = \|f_1\| \le \|f\|$.

Then

$$F(x) := F_1(x) - iF_1(ix), \; x \in X,$$

is a complex-valued bounded linear extension to X of $(f, D(f))$ with $\|f\| \le \|F\|$.

Indeed, it is clear that F is *additive*, i. e.,

$$F(x + y) = F(x) + F(y), \; x, y \in X,$$

and that, for any $\lambda \in \mathbb{R}$,

$$F(\lambda x) = \lambda F(x), \; x \in X.$$

Exercise 6.5. Verify.

Further,

$$F(ix) := F_1(ix) - iF_1(-x) = F_1(ix) + iF_1(x) = iF(x), \; x \in X.$$

Thus, the functional F is *linear*.

Now, let us show that F is an *extension* of $(f, D(f))$.

Indeed,

$$\forall x \in D(f): \; F(x) := F_1(x) - iF_1(ix) = f_1(x) - if_1(ix) = f_1(x) + if_2(x) = f(x).$$

Finally, since

$$\forall x \in X \, \exists \theta = \theta(x) \in (-\pi, \pi] : \ F(x) = |F(x)|e^{i\theta},$$

where $\theta := \arg F(x) \in (-\pi, \pi]$ is the *principal value* of the argument of $F(x)$, ($\arg 0 := 0$), we have:

$$|F(x)| = e^{-i\theta}F(x) = F(e^{-i\theta}x) \qquad\qquad \text{since } F(e^{-i\theta}x) = |F(x)| \in \mathbb{R};$$
$$= \operatorname{Re} F(x) = F_1(e^{-i\theta}x) \leq \|f\|\|e^{-i\theta}x\| = \|f\|\|e^{-i\theta}\|\|x\| = \|f\|\|x\|, \ x \in X.$$

Therefore, the functional F is *bounded* and $\|F\| \leq \|f\|$.

On the other hand, since F is an extension of f, $\|f\| \leq \|F\|$.

Thus, $\|F\| = \|f\|$, which completes the proof. $\qquad\qquad\qquad\qquad\qquad\qquad\square$

Remarks 6.2.

– Observe that, in the prior proof and henceforth, we use the same symbol $\|\cdot\|$ to designate the norm of a linear functional, such an economy of symbols being a rather common practice.

– Unless f is *densely defined* (i. e., $\overline{D(f)} = X$), an extension $F \in X^*$ of f with the same norm ($\|F\| = \|f\|$), although *existent*, is *not unique* (Section 6.7, Problem 1).

Exercise 6.6. Explain why the extension $F \in X^*$ of f with the same norm ($\|F\| = \|f\|$) is *unique* when $\overline{D(f)} = X$.

6.2 Implications of the Hahn–Banach Theorem

The *Hahn–Banach Theorem* has a plethora of fascinating nontrivial implications, several of which we consider here.

6.2.1 Separation and Norm Realization

Proposition 6.1 (Separation of Element from Subspace). *If Y is a nondense subspace of a normed vector space $(X, \|\cdot\|)$ over $\mathbb{F}$ (i. e., $\overline{Y} \neq X$), then, for any element $x \in X$ with*

$$\rho(x, Y) := \inf_{y \in Y} \|x - y\| > 0,$$

there exists a functional $f \in X^*$ with $\|f\| = 1$ such that

$$f(x) = \rho(x, Y) \quad \text{and} \quad f(y) = 0, \ y \in Y \ (i.\,e., Y \subseteq \ker f).$$

Proof. Let an element $x \in X$ with $\rho(x, Y) > 0$ be arbitrary.

Exercise 6.7. Explain why such an element exists.

Hence $x \in Y^c$ and each

$$z \in \text{span}(Y \cup \{x\}) = Y \oplus \text{span}(\{x\})$$

can be uniquely represented as

$$z = y + \lambda x$$

with some $y \in Y$ and $\lambda \in \mathbb{F}$, and we can define on the subspace $\text{span}(Y \cup \{x\})$ a *linear functional* as follows:

$$\text{span}(Y \cup \{x\}) \ni z = y + \lambda x \mapsto g(z) := \rho(x, Y)\lambda. \tag{6.5}$$

Exercise 6.8. Verify the *linearity* of g.

As is easily seen

$$g(x) = \rho(x, Y) \text{ and } g(y) = 0, \ y \in Y. \tag{6.6}$$

For $z = y + \lambda x \in \text{span}(Y \cup \{x\})$ with $\lambda \in \mathbb{F} \setminus \{0\}$, in view of the fact that $-y/\lambda \in Y$ and (6.5), we infer that

$$\|z\| = \|y + \lambda x\| = |\lambda| \|y/\lambda + x\| = |\lambda| \|x - (-y/\lambda)\| \geq |\lambda| \rho(x, Y)$$
$$= \frac{|g(z)|}{\rho(x, Y)} \rho(x, Y) = |g(z)|.$$

Therefore,

$$|g(z)| \leq \|z\|, \ z \in \text{span}(Y \cup \{x\}),$$

which implies that linear functional g is *bounded* on $\text{span}(Y \cup \{x\})$ and

$$\|g\| \leq 1. \tag{6.7}$$

Let $\{y_n\}_{n=1}^{\infty}$ be a sequence of elements in Y such that

$$\|x - y_n\| \to \rho(x, Y), \ n \to \infty.$$

Then, in view of (6.6),

$$\forall n \in \mathbb{N} : \rho(x, Y) = g(x) - g(y_n) = g(x - y_n) \leq \|g\| \|x - y_n\|.$$

Passing to the limit as $n \to \infty$, we arrive at

$$\rho(x, Y) \leq \|g\| \rho(x, Y).$$

Dividing through by $\rho(x, Y) > 0$, we infer that

$$1 \leq \|g\|. \tag{6.8}$$

Thus, by (6.7) and (6.8),

$$\|g\| = 1.$$

By the *Hahn–Banach Theorem for Normed Vector Spaces* (Theorem 6.2), there is an extension $f \in X^*$ of g from $\text{span}(Y \cup \{x\})$ such that $\|f\| = \|g\| = 1$ and, by (6.6),

$$f(x) = g(x) = \rho(x, Y) \quad \text{and} \quad f(y) = g(y) = 0, \ y \in Y. \qquad \square$$

We immediately obtain the following two corollaries.

Corollary 6.1 (Separation of Element from Closed Subspace). *If Y is a closed proper subspace of a normed vector space $(X, \|\cdot\|)$, then, for any element $x \in Y^c$, there exists a functional $f \in X^*$ with $\|f\| = 1$ such that*

$$f(x) = \rho(x, Y) > 0 \quad \text{and} \quad f(y) = 0, \ y \in Y \ (i.\,e., Y \subseteq \ker f).$$

Exercise 6.9. Prove.

Corollary 6.2 (Norm Realizing Functional). *For any element x in a nontrivial normed vector space $(X, \|\cdot\|)$, there exists a functional $f \in X^*$ with $\|f\| = 1$ such that*

$$f(x) = \|x\|.$$

Exercise 6.10. Prove.

Hint. For $x = 0$ the statement is trivial. For $x \neq 0$, apply Corollary 6.1 relative to the trivial subspace $Y := \{0\}$.

Remark 6.3. In particular, the latter implies that the following profound conclusion: a *nontrivial* normed vector space $(X, \|\cdot\|)$ has a *nontrivial dual* X^*.

From the prior corollary, we also immediately obtain the following two.

Corollary 6.3 (Norm of an Element). *For any element x in a nontrivial normed vector space $(X, \|\cdot\|)$,*

$$\|x\| = \sup_{f \in X^*, \|f\|=1} |f(x)| = \max_{f \in X^*, \|f\|=1} |f(x)|.$$

Exercise 6.11. Prove.

Corollary 6.4 (Separation of Elements). *Two elements x and y of a nontrivial normed vector space $(X, \|\cdot\|)$ are distinct iff there exists a functional $f \in X^*$ such that*

$$f(x) \neq f(y).$$

Equivalently, two elements x and y of a nontrivial normed vector space $(X, \|\cdot\|)$ coincide iff

$$\forall f \in X^* : f(x) = f(y).$$

Exercise 6.12. Prove.

6.2.2 Characterization of Fundamentality

Definition 6.1 (Fundamental Set in a Normed Vector Space). A set S in a normed vector space $(X, \| \cdot \|)$ is said to be *fundamental* if its span is dense in X, i. e.,

$$\overline{\text{span}(S)} = X$$

(cf. Proposition 4.13).

Remarks 6.4.
-   Observe that the underlying set S need not be linearly independent. Recall, that a linearly independent fundamental set is called a *topological basis* of the space (see Section 3.2.6).
-   The concept of fundamentality of a set is used above in the *Characterization of a Complete Orthonormal Set* (Proposition 4.13) and the *Orthogonal Dimension of a Separable Hilbert Space Theorem* (Theorem 4.15).

Proposition 6.2 (Characterization of Fundamental Sets). *A set S in a normed vector space $(X, \| \cdot \|)$ is fundamental iff*

$$\forall f \in X^* : f(x) = 0, \, x \in S \, (i.\,e., S \subseteq \ker f) \Rightarrow f = 0 \, (i.\,e., \ker f = X).$$

Proof. The case of a trivial space X being vacuous, we regard X to be nontrivial, i. e., $X \neq \{0\}$.

"*Only if*" part. Suppose a set S is fundamental in $(X, \| \cdot \|)$, i. e.,

$$\overline{\text{span}(S)} = X. \tag{6.9}$$

Then, for any $f \in X^*$, such that

$$f(x) = 0, \, x \in S,$$

by the *linearity* of f,

$$f(x) = 0, \, x \in \text{span}(S).$$

Exercise 6.13. Explain.

Further, in view of (6.9), by the *continuity* of f on X,

$$f(x) = 0, \, x \in X.$$

Exercise 6.14. Explain.

"If" part. Let us prove this part *by contrapositive*, assuming that S is not fundamental in $(X, \| \cdot \|)$, i. e.,

$$\overline{\text{span}(S)} \neq X.$$

Then

$$\exists x \in \overline{\text{span}(S)}^c$$

and, by the *Separation of Element from Closed Subspace Corollary* (Corollary 6.1), there exists a functional $f \in X^*$, $f \neq 0$, such that

$$f(x) = \rho(x, \overline{\text{span}(S)}) > 0 \quad \text{and} \quad f(y) = 0, \ y \in \overline{\text{span}(S)}.$$

Hence, in view of the inclusion $S \subseteq \overline{\text{span}(S)}$,

$$f(y) = 0, \ y \in S,$$

which completes the proof. $\qquad\qquad\qquad\qquad\qquad\qquad\qquad\qquad\qquad\qquad\square$

As an immediate corollary, we obtain the following.

Corollary 6.5 (Characterization of Dense Subspaces). *A subspace Y in a normed vector space $(X, \| \cdot \|)$ is dense iff*

$$\forall f \in X^* : f(x) = 0, \ x \in Y \ (i.\,e., Y \subseteq \ker f) \Rightarrow f = 0 \ (i.\,e., \ker f = X).$$

Exercise 6.15. Prove.

6.2.3 Sufficiency for Separability

An interesting implication of the *Characterization of Fundamental Sets* (Proposition 6.2) is the following sufficient condition for separability in terms of the dual space.

Theorem 6.3 (Sufficiency for Separability in Terms of Dual Space). *If the dual space $(X^*, \| \cdot \|)$ of a normed vector space $(X, \| \cdot \|)$ is separable, then so is $(X, \| \cdot \|)$.*

Proof. Since any subspace of a separable metric space is separable (see Proposition 2.19, Section 2.19, Problem 17), there is a countable dense subset $\{f_n\}_{n\in\mathbb{N}}$ on the *unit sphere* of $(X^*, \| \cdot \|)$ ($\|f_n\| = 1$, $n \in \mathbb{N}$) and

$$\forall n \in \mathbb{N} \ \exists x_n \in X \text{ with } \|x_n\| = 1 : \ |f_n(x_n)| \geq 1/2.$$

Exercise 6.16. Explain.

Let us show that the set $\{x_n\}_{n\in\mathbb{N}}$ is *fundamental* in $(X, \|\cdot\|)$, i. e.,

$$\overline{\operatorname{span}(\{x_n\}_{n\in\mathbb{N}})} = X. \tag{6.10}$$

Indeed, assuming the *opposite*, by the *Characterization of Fundamental Sets* (Proposition 6.2), implies that

$$\exists f \in X^*, f \neq 0 : f(x_n) = 0, \ n \in N.$$

Without loss of generality, we can regard that $\|f\| = 1$.

Exercise 6.17. Explain.

Hence,

$$\frac{1}{2} \leq \|f_n(x_n)\| = \|f_n(x_n) - f(x_n)\| = \|(f_n - f)(x_n)\| \leq \|f_n - f\|\|x_n\|$$
$$= \|f_n - f\|, \ n \in \mathbb{N},$$

which *contradicts* the denseness of $\{f_n\}_{n\in\mathbb{N}}$ on the unit sphere in $(X^*, \|\cdot\|)$.

Thus, the countable set $\{x_n\}_{n\in\mathbb{N}}$ is *fundamental* in $(X, \|\cdot\|)$, which implies that X is *separable*, a countable dense subset of X being formed by the linear combinations of the elements of $\{x_n\}_{n\in\mathbb{N}}$ with *rational/complex rational* coefficients. □

Remark 6.5. The converse statement is not true. As is shown below (see Remarks 7.4), the dual l_1^* of the separable space l_1 is isometrically isomorphic to l_∞, and hence, is not separable (see Examples 2.11).

6.2.4 Isometric Embedding Theorems

The *Norm Realizing Functional Corollary* (Corollary 6.2) makes it <no text>possible to establish the following profound fact.

Theorem 6.4 (Isometric Embedding Theorem). *For a normed vector space $(X, \|\cdot\|)$ over $\mathbb{F}$, there is a nonempty set T such that $(X, \|\cdot\|)$ is isometrically isomorphic to a subspace of the Banach space $(M(T, \mathbb{F}), \|\cdot\|_\infty)$, i. e., $(X, \|\cdot\|)$ can be isometrically embedded in $(M(T, \mathbb{F}), \|\cdot\|_\infty)$.*

Proof. Let $\{x_t\}_{t\in T}$, where T is a nonempty indexing set, be a *dense* subset in X.

Exercise 6.18. Explain why such a set always exists.

Then, by the *Norm Realizing Functional Corollary* (Corollary 6.2),

$$\forall t \in T \ \exists f(\cdot, t) \in X^* \text{ with } \|f(\cdot, t)\| = 1 : f(x_t, t) = \|x_t\|. \tag{6.11}$$

Furthermore,

$$\forall t \in T \, \forall x \in X : \, \|f(x,t)\| \leq \|f(\cdot,t)\|\|x\| = \|x\|, \tag{6.12}$$

which implies that, for any fixed $x \in X$, the $\mathbb{F}$-valued function $f(x, \cdot)$ is *bounded* on T.

In view of the *linearity* of $f(\cdot, t)$ for each fixed $t \in T$ and (6.12), we infer that the mapping

$$X \ni x \mapsto Rx := f(x, \cdot) \in M(T, \mathbb{F})$$

is a *linear operator*. It is also *bounded* with $\|R\| \leq 1$. Indeed, by (6.12)

$$\forall x \in X : \, \|Rx\|_\infty = \|f(x, \cdot)\|_\infty := \sup_{t \in T} |f(x,t)| \leq \|x\|.$$

Furthermore, the restriction of R to the set $\{x_t\}_{t \in T}$ is an *isometry*. Indeed, by (6.12) and (6.11),

$$\forall t \in T : \, \|Rx_t\|_\infty = \|f(x_t, \cdot)\|_\infty := \sup_{s \in T} |f(x_t, s)| = |f(x_t, t)| = \|x_t\|,$$

which, by the denseness of $\{x_t\}_{t \in T}$ in $(X, \|\cdot\|)$ and the *continuity* of R on $(X, \|\cdot\|)$, implies that

$$\forall x \in X : \, \|Rx\|_\infty = \|x\|,$$

i. e., R is an *isometric isomorphism* from $(X, \|\cdot\|)$ to $(M(T, \mathbb{F}), \|\cdot\|_\infty)$.

Exercise 6.19. Explain.

This completes the proof. □

As a particular case, we obtain the following corollary.

Corollary 6.6 (Isometric Embedding Theorem for Separable Spaces). *A separable normed vector space $(X, \|\cdot\|)$ over $\mathbb{F}$ is isometrically isomorphic to a subspace of the Banach space $l_\infty(\mathbb{F})$, i. e., $(X, \|\cdot\|)$ can be isometrically embedded in $l_\infty(\mathbb{F})$.*

Exercise 6.20. Prove.

6.2.5 Second Dual Space and Canonical Isomorphism

The *Hahn–Banach Theorem* also enables us to introduce the concepts of *second dual space* and *canonical isomorphism* fundamental for the *duality theory* of normed vector spaces, the subject of our next chapter.

Definition 6.2 (Second Dual Space). For a normed vector space $(X, \|\cdot\|)$, the dual space $(X^*)^*$ of the dual space X^* is called the *second dual space* (or *bidual space*) of X.

Notation. X^{**}.

The following statement, derived from the *Norm of an Element Corollary* (Corollary 6.3), is basic for the duality theory of normed vector spaces.

Theorem 6.5 (Canonical Isomorphism Theorem). *For each fixed element x in a normed vector space* $(X, \| \cdot \|)$ *over* $\mathbb{F}$, *the mapping*

$$X \ni x \mapsto Jx := F_x \in X^{**} : F_x(f) := f(x), \, f \in X^*,$$

is an isometric isomorphism from $(X, \| \cdot \|)$ *to the second dual space* $(X^{**}, \| \cdot \|)$, *called the canonical isomorphism (or the natural embedding), which isometrically embeds* $(X, \| \cdot \|)$ *in* $(X^{**}, \| \cdot \|)$.

Proof. For each fixed $x \in X$, the mapping

$$X^* \ni f \mapsto F_x(f) := f(x) \in \mathbb{F}$$

is a *linear functional* on the *dual space* X^*.

Exercise 6.21. Verify.

As follows from the *Norm of an Element Corollary* (Corollary 6.3),

$$\|F_x\| = \sup_{f \in X^*, \, \|f\|=1} |F_x(f)| = \sup_{f \in X^*, \, \|f\|=1} |f(x)| = \|x\|,$$

which implies that

$$\forall x \in X : F_x \in X^{**} \text{ and } \|F_x\| = \|x\|.$$

Hence, the mapping

$$X \ni x \mapsto Jx := F_x \in X^{**} : F_x(f) := f(x), \, f \in X^*,$$

is an *isometric isomorphism* from $(X, \| \cdot \|)$ to the second dual space $(X^{**}, \| \cdot \|)$. ☐

6.2.6 Closed Complemented Subspaces

Recall that the *Existence of a Complementary Subspace Theorem* (Theorem 3.6) establishes the existence of a *complementary subspace* (see Definition 3.14) for an arbitrary subspace of a vector space. Adding the factor of closedness to this in a normed vector space makes the situation more exciting.

Definition 6.3 (Closed Complemented Subspace). A *closed subspace* of Y in a normed vector space $(X, \| \cdot \|)$ is called *closed complemented subspace* to X, if there is a *closed complementary subspace* Z in $(X, \| \cdot \|)$.

Remarks 6.6.

- By the *Existence of a Complementary Subspace Theorem* (Theorem 3.6), in a finite-dimensional normed vector space $(X, \| \cdot \|)$, where all subspaces are closed (Theorem 3.13), every subspace is *closed complemented*.

  Exercise 6.22. Verify.

- As follows from the *Projection Theorem* (Theorem 4.6), every closed subspace in a Hilbert space is *orthogonally complemented*, and hence, is *closed complemented*.
- Furthermore, as is shown in [30], a *real* Banach space $(X, \| \cdot \|)$, whose every closed subspace is *closed complemented*, is *bicontinuously isomorphic* to a Hilbert space, i. e., the norm $\| \cdot \|$ is equivalent to a norm generated by an inner product.
- Generally, a closed subspace in a normed vector space need not be complemented. A counterexample built in l_1 can be found in [47, Section 5.7, Problem 9].

Based on the *Biorthogonal Sets Proposition* (Proposition 6.12) (see Section 6.7, Problem 2), which follows from the *Separation of Element from Closed Subspace Corollary* (Corollary 6.1), we prove the following statement.

Theorem 6.6 (Complementedness of Finite-Dimensional Subspaces). *A finite-dimensional subspace Y of a normed vector space $(X, \| \cdot \|)$ over $\mathbb{F}$ is closed complemented.*

Proof. Let Y be a finite-dimensional subspace of a normed vector space $(X, \| \cdot \|)$ with $\dim X = n$ $(n \in \mathbb{N})$ and $B := \{x_1, \dots, x_n\}$ be its *basis*.

By the *Biorthogonal Sets Proposition* (Proposition 6.12) (Section 6.7, Problem 2), there is a *biorthogonal set* of bounded linear functionals $\{f_1, \dots, f_n\} \subset X^*$:

$$f_i(x_j) = \delta_{ij}, \ i, j = 1, \dots, n, \tag{6.13}$$

where δ_{ij} is the *Kronecker delta*.
    Each *null space*

$$Z_i := \ker f_i, \ i = 1, \dots, n,$$

being a *closed hyperplane* in $(X, \| \cdot \|)$ (Proposition 5.10, Section 5.4, Problem 5), by the *Properties of Closed Sets* (Proposition 2.19), their intersection

$$Z := \bigcap_{i=1}^{n} Z_i$$

is a *closed subspace* of $(X, \| \cdot \|)$.
    For each $x \in X$, we have the *decomposition*

$$x = y + z, \tag{6.14}$$

where

$$y := \sum_{j=1}^{n} f_j(x)x_j \in Y$$

and

$$z := x - y = x - \sum_{j=1}^{n} f_j(x)x_j \in Z.$$

Indeed, in view of the *biorthogonality*, for each $i = 1, \ldots, n$,

$$f_i(z) = f_i\left(x - \sum_{j=1}^{n} f_j(x)x_j\right) = f_i(x) - \sum_{j=1}^{n} f_j(x)f_i(x_j) = f_i(x) - f_i(x) = 0.$$

Let us show that such a decomposition is *unique*. Indeed, suppose that, for an $x \in X$, decomposition (6.14) holds with some $y \in Y$ and $z \in Z$.

Considering that, by the *Representation Theorem* (Theorem 3.3), y in (6.14) is uniquely represented as

$$y = \sum_{j=1}^{n} \lambda_j x_j$$

with some $\lambda_j \in \mathbb{F}, j = 1, \ldots, n$, and $f_i(z) = 0, i = 1, \ldots, n$, in view of (6.13), we have:

$$f_i(x) = f_i\left(\sum_{j=1}^{n} \lambda_j x_j + z\right) = \sum_{j=1}^{n} \lambda_j f_i(x_j) + f_i(z) = \lambda_i, \ i = 1, \ldots, n,$$

and hence,

$$y = \sum_{j=1}^{n} f_j(x)x_j,$$

which proves the uniqueness of decomposition (6.14).

Thus, the direct-sum decomposition

$$X = Y \oplus Z$$

holds and the proof is complete. □

6.3 Weak and Weak* Convergence

In the context, it appears pertinent and timely to introduce *"weaker"* types of convergence, as the *Hahn–Banach Theorem* makes it possible to establish several essential facts concerning them.

Definition 6.4 (Weak Convergence). A sequence of elements $\{x_n\}_{n=1}^{\infty}$ in a normed vector space $(X, \| \cdot \|)$ is said to *weakly converge* to an element $x \in X$ if

$$\forall f \in X^* : f(x_n) \to f(x),\ n \to \infty,$$

the element x being called the *weak limit* of $\{x_n\}_{n=1}^{\infty}$.

Notations.

$$x_n \xrightarrow{w} x,\ n \to \infty \quad \text{or} \quad \text{w-}\lim_{n\to\infty} x_n = x.$$

Proposition 6.3 (Uniqueness of Weak Limit). *The weak limit of a sequence in a normed vector space, if existent, is unique.*

Exercise 6.23. Prove.

Hint. Apply the *Separation of Elements Corollary* (Corollary 6.4).

Remark 6.7. The convergence of a sequence in a normed vector space $(X, \|\cdot\|)$ relative to the norm metric is referred to as *strong convergence*.

Proposition 6.4 (Strong Implies Weak). *For a sequence of elements $\{x_n\}_{n=1}^{\infty}$ in a normed vector space $(X, \|\cdot\|)$,*

$$x_n \to x \in X,\ n \to \infty \Rightarrow x_n \xrightarrow{w} x,\ n \to \infty.$$

Exercise 6.24. Prove.

Remark 6.8. The converse statement is not true, i. e., a weakly convergent sequence need not converge in the strong sense.

Indeed, as is shown below (see the *Linear Bounded Functionals on Certain Hilbert Spaces Corollary* (Corollary 7.1)),

$$\forall f \in l_2^* \ \exists! y = \{y_k\}_{k=1}^{\infty} \in l_2 \ \forall x = \{x_k\}_{k=1}^{\infty} \in l_2 : f(x) = \sum_{k=1}^{\infty} x_k y_k,$$

and hence, the standard orthonormal basis sequence $\{e_n := \{\delta_{nk}\}_{k=1}^{\infty}\}_{n=1}^{\infty}$, where δ_{nk} is the *Kronecker delta*, weakly converges to 0, but does not (strongly) converge in l_2.

Exercise 6.25. Explain (cf. a more general *Weak Convergence of an Orthonormal Sequence Proposition* (Proposition 7.6), Section 7.6, Problem 1).

Remark 6.9. In Section 6.4.2 we are to prove a general *Characterization of Weak Convergence* (Theorem 6.10).

Definition 6.5 (Weak* Convergence). A sequence of functionals $\{f_n\}_{n=1}^{\infty}$ in the dual space $(X^*, \|\cdot\|)$ of a normed vector space $(X, \|\cdot\|)$ is said to *weakly* converge* to a functional $f \in X^*$ if

$$\forall x \in X : f_n(x) \to f(x),\ n \to \infty,$$

the functional f being called the *weak* limit* of $\{f_n\}_{n=1}^{\infty}$.

Notations.

$$f_n \xrightarrow{\ w^* \ } f, \ n \to \infty \quad \text{or} \quad w^*\text{-}\lim_{n\to\infty} f_n = f.$$

Proposition 6.5 (Uniqueness of Weak* Limit). *The weak* limit of a sequence of functionals in the dual space $(X^*, \|\cdot\|)$ of a normed vector space $(X, \|\cdot\|)$, if existent, is unique.*

Exercise 6.26. Prove.

Proposition 6.6 (Weak Implies Weak*). *For a sequence of functionals $\{f_n\}_{n=1}^{\infty}$ in the dual space X^* of a normed vector space $(X, \|\cdot\|)$,*

$$f_n \xrightarrow{\ w \ } f \in X^*, \ n \to \infty \Rightarrow f_n \xrightarrow{\ w^* \ } f, \ n \to \infty.$$

Exercise 6.27. Prove.

Hint. Apply the *Canonical Isomorphism Theorem* (Theorem 6.5).

Remarks 6.10.
- As is shown in the following example, the converse is not true, i. e., a weak* convergent sequence in X^* need not weakly converge.
- Thus, for a sequence $\{f_n\}_{n=1}^{\infty}$ in the dual space $(X^*, \|\cdot\|)$ of a normed vector space $(X, \|\cdot\|)$, there are three kinds of convergence to a functional $f \in X^*$: *strong* (i. e., in the norm of X^*), *weak*, and *weak*. The *strong convergence* implies *weak convergence*, which, in its turn, implies *weak* convergence*, the converse statements not being true.
- In the normed vector space $(L(X, Y), \|\cdot\|)$ of bounded linear operators from a normed vector space $(X, \|\cdot\|_X)$ to a normed vector space $(Y, \|\cdot\|_Y)$, for an operator sequence $\{A_n\}_{n=1}^{\infty}$, there are three kinds of convergence to an operator $A \in L(X, Y)$:
  (1) *uniform*, i. e., the convergence in the norm of $(L(X, Y), \|\cdot\|)$:

  $$A_n \to A, \ n \to \infty, \ \text{in } L(X, Y) \ \Leftrightarrow \ \|A_n - A\| = \sup_{\|x\|_X = 1} \|A_n x - Ax\|_Y \to 0, \ n \to \infty;$$

  (2) *strong*, i. e., the *strong convergence* of the sequence $\{A_n x\}_{n=1}^{\infty}$ in $(Y, \|\cdot\|_Y)$ for each $x \in X$:

  $$A_n \xrightarrow{\ s \ } A, \ n \to \infty \ \Leftrightarrow \ \forall x \in X : A_n x \to Ax, \ n \to \infty, \ \text{in } Y;$$

  (3) *weak*, i. e., the *weak convergence* of the sequence $\{A_n x\}_{n=1}^{\infty}$ in $(Y, \|\cdot\|_Y)$ for each $x \in X$:

  $$A_n \xrightarrow{\ w \ } A, \ n \to \infty \ \Leftrightarrow \ \forall x \in X \ \forall f \in Y^* : f(A_n x) \to f(Ax), \ n \to \infty.$$

The *uniform convergence* implies *strong convergence*, which, in its turn, implies *weak convergence*, the converse statements not being true.

Examples 6.1.

1. As follows from the *Representation Theorem for c_0^** (Theorem 7.6) and the *Representation Theorem for l_p^** ($1 \le p < \infty$) (Theorem 7.3), $c_0^* = l_1$ and $l_1^* = l_\infty$, the equalities being understood in the sense of the spaces' being isometrically isomorphic under the mappings defined in the corresponding theorems.

    The sequence of the *Schauder coordinate functionals* $\{c_n\}_{n=1}^\infty$ in c_0^* (see Section 6.6.2.3), which corresponds to the sequence $\{(\delta_{nk})_{k\in\mathbb{N}}\}_{n=1}^\infty \in l_1$, is weak* convergent to the zero functional in c_0^* since

    $$\forall x := (x_k)_{k\in\mathbb{N}} \in c_0 : c_n(x) = x_n \to 0, \ n \to \infty,$$

    (see Theorem 7.6), but does not weakly converge in c_0^* since, for the $F \in l_1^*$ corresponding to the element $(1,1,1,\dots) \in l_\infty$,

    $$F(f_n) = \sum_{k=1}^\infty \delta_{nk} = 1 \not\to 0, \ n \to \infty,$$

    (see Theorem 7.3).

2. As is seen from Example 5.4, for the *left-shift operator* in l_2,

    $$l_2 \ni x = (x_1, x_2, \dots) \mapsto Lx := (x_2, x_3, x_4, \dots) \in l_2,$$

    the operator sequence $\{A_n := L^n\}_{n=1}^\infty$ *converges strongly* to the *zero operator* in $L(l_2)$, but, since $\|A_n\| = 1$, $n \in \mathbb{N}$, but does not converge uniformly.

3. The operator sequence $\{B_n\}_{n=1}^\infty \subseteq L(l_2)$

    $$l_2 \ni x := (x_k)_{k\in\mathbb{N}} \mapsto B_n x := x_1(\delta_{nk})_{k\in\mathbb{N}} = (\underbrace{0,\dots,0}_{n-1 \text{ terms}}, x_1, 0, \dots) \in l_2 \ n \in \mathbb{N},$$

    converges weakly, but *not strongly*, to the *zero operator* in $L(l_2)$.

Exercise 6.28. Verify 3.

(a) Show that $\|B_n\| = 1$, $n \in \mathbb{N}$.

(b) Use the fact stated in Remark 6.8 to show that

$$\forall x \in l_2 \ \forall f \in l_2^* : f(B_n x) \to 0, \ n \to \infty,$$

i.e., $B_n \xrightarrow{w} 0$, $n \to \infty$.

(c) Show that

$$\forall x := (x_k)_{k\in\mathbb{N}} \in l_2 \text{ with } x_1 \ne 0 \ B_n x \not\to 0, \ n \to \infty, \text{ in } l_2,$$

and thus, $\{B_n\}_{n=1}^\infty$ does not converge strongly in $L(l_2)$.

6.4 Uniform Boundedness Principle, the Banach–Steinhaus Theorem

6.4.1 Uniform Boundedness Principle

The following fundamental principle states that, for a set of bounded linear operators defined on a Banach space, boundedness in operator norm (*uniform boundedness*) is equivalent to *pointwise boundedness*. It also has a number of profound implications and far reaching applications.

Theorem 6.7 (Uniform Boundedness Principle). *Let* $(X, \| \cdot \|_X)$ *and* $(Y, \| \cdot \|_Y)$ *be normed vector spaces over* $\mathbb{F}$ *and* $\{A_i\}_{i \in I}$ *be a set of bounded linear operators in* $(L(X, Y), \| \cdot \|)$.[1] *Then for*

$$\sup_{i \in I} \|A_i\| < \infty \qquad \textit{(uniform boundedness)} \qquad (6.15)$$

it is necessary and, provided $(X, \| \cdot \|_X)$ *is a Banach space, sufficient that*

$$\forall x \in X : \sup_{i \in I} \|A_i x\|_Y < \infty \qquad \textit{(pointwise boundedness)}. \qquad (6.16)$$

Proof. Necessity. Proving that uniform boundedness (6.15) implies pointwise boundedness (6.16) is straightforward, the implication holding true without the additional assumption on the domain space $(X, \| \cdot \|_X)$ to be complete.

Exercise 6.29. Prove.

   Sufficiency. Suppose that the domain space $(X, \| \cdot \|_X)$ is Banach and that (6.16) holds.
   For, each $n \in \mathbb{N}$, consider the set

$$X_n := \bigcap_{i \in I} \{x \in X \mid \|A_i x\| \leq n\},$$

which, by the *continuity* of the operators A_i, $i \in I$, (see the *Boundedness Characterizations* (Theorem 5.4)) and in view of the *Characterization of Continuity* (Theorem 2.52) and the *Properties of Closed Sets* (Proposition 2.19), is *closed* in $(X, \| \cdot \|_X)$.

Exercise 6.30. Explain.

   Since, by the premise,

$$\forall x \in X \; \exists n(x) \in \mathbb{N} \; \forall i \in I : \; \|A_i x\|_Y \leq n(x),$$

---

1 Due to Stefan Banach (1892–1945) and Hugo Steinhaus (1887–1972).

we infer that

$$X = \bigcup_{n=1}^{\infty} X_n.$$

Whence, by the *Baire Category Theorem* (Theorem 2.31) (see also Corollary 2.5),

$$\exists N \in \mathbb{N} : \text{int}(X_N) \neq \emptyset,$$

i. e., the set X_N is *not nowhere dense* in $(X, \| \cdot \|_X)$, and hence,

$$\exists x_0 \in X, \ \exists r > 0 : \overline{B}(x_0, r) := \{x \in X \mid \|x - x_0\|_X \leq r\} \subseteq X_N.$$

Then

$$\forall i \in I, \ \forall x \in X \text{ with } \|x\|_X \leq r : \ \|A_i(x_0 + x)\|_Y \leq N,$$

and hence, by the *linearity of A_i, $i \in I$*, and *subadditivity* of norm,

$$\forall i \in I, \ \forall x \in X \text{ with } \|x\|_X \leq r : \ \|A_i x\|_Y = \|A_i x + A_i x_0 - A_i x_0\|_Y$$
$$Q \leq \|A_i x + A_i x_0\|_Y + \|A_i x_0\|_Y = \|A_i(x_0 + x)\|_Y + \|A_i x_0\|_Y \leq 2N.$$

Whence, by the *linearity of A_i, $i \in I$*, and *absolute scalability* of norm, we infer that

$$\forall i \in I, \ \forall x \in X \text{ with } \|x\|_X \leq 1 : \ \|A_i x\|_Y = \frac{1}{r} \|A_i(rx)\|_Y \qquad\qquad \text{since } \|rx\|_X \leq r;$$
$$\leq \frac{2N}{r},$$

which implies that

$$\forall i \in I : \ \|A_i\| := \sup_{\|x\|_X \leq 1} \|A_i x\|_Y \leq \frac{2N}{r}$$

completing the proof of the sufficiency and the entire statement. □

In particular, for $(Y, \| \cdot \|_Y) = (\mathbb{F}, |\cdot|)$, we obtain the following version of the *Uniform Boundedness Principle* for *bounded linear functionals*.

Corollary 6.7 (Uniform Boundedness Principle for Functionals). *Let $(X, \| \cdot \|_X)$ be a normed vector spaces over $\mathbb{F}$ and $\{f_i\}_{i \in I}$ be a set of bounded linear functionals in the dual space $(X^*, \| \cdot \|)$. Then for*

$$\sup_{i \in I} \|f_i\| < \infty$$

it is necessary and, provided $(X, \| \cdot \|_X)$ is a Banach space, sufficient that

$$\forall x \in X : \ \sup_{i \in I} |f_i(x)| < \infty.$$

We immediately obtain the following corollary.

Proposition 6.7 (Boundedness of a Weakly* Convergent Sequence). *Every weakly* convergent sequence $(f_n)_{n \in \mathbb{N}}$ in the dual space $(X^*, \| \cdot \|)$ of a Banach space $(X, \| \cdot \|)$ is bounded.*

Exercise 6.31. Prove.

Remark 6.11. As the following example shows, the condition of *completeness* of the domain space $(X, \| \cdot \|_X)$ in the *sufficiency* of the *Uniform Boundedness Principle* (Theorem 6.7) is essential and cannot be dropped.

Example 6.2. In the *incomplete* normed vector space $(c_{00}, \| \cdot \|_\infty)$ (see Examples 3.13), consider the countable set $\{f_n\}_{n \in \mathbb{N}}$ of linear functionals defined as follows:

$$c_{00} \ni x := (x_k)_{k \in \mathbb{N}} \mapsto f_n(x) := \sum_{k=1}^{n} x_k, \ n \in \mathbb{N}.$$

Exercise 6.32. Verify the *linearity* of f_n, $n \in \mathbb{N}$.

Since

$$\forall\, n \in \mathbb{N}, \ \forall\, x := (x_k)_{k \in \mathbb{N}} \in X : \ |f_n(x)| \le \sum_{k=1}^{n} |x_k| \le n \sup_{k \in \mathbb{N}} |x_k| = n\|x\|_\infty,$$

we infer that

$$\forall\, n \in \mathbb{N} : f_n \in X^* \text{ with } \|f_n\| \le n.$$

Further, for each $x := (x_1, \dots, x_m, 0, 0, \dots) \in c_{00}$ with some $m \in \mathbb{N}$,

$$|f_n(x)| \le \sum_{k=1}^{m} |x_k| \le m \sup_{k \in \mathbb{N}} |x_k| = m\|x\|_\infty, \ n \in \mathbb{N},$$

and hence,

$$\forall\, x \in c_{00} : \sup_{n \in \mathbb{N}} |f_n(x)| < \infty,$$

i. e., the set $\{f_n\}_{n \in \mathbb{N}}$ is *pointwise bounded*.
However, since for $x_n := (\underbrace{1, \dots, 1}_{n \text{ terms}}, 0, 0, \dots) \in c_{00}$, $n \in \mathbb{N}$,

$$\|x_n\|_\infty = 1 \quad \text{and} \quad |f_n(x_n)| = n,$$

we infer that $\|f_n\| = n$, $n \in \mathbb{N}$, and hence,

$$\sup_{n \in \mathbb{N}} \|f_n\| = \infty,$$

i. e., the set $\{f_n\}_{n \in \mathbb{N}}$ is *not uniformly bounded*.

Remark 6.12. The *Uniform Boundedness Principle* (Theorem 6.7), found by Lebesgue in 1908 when studying convergence of Fourier series (see [16]), and stated in its general form and published in 1927 by Stefan Banach and Hugo Steinhaus, is often referred to as the *Banach–Steinhaus Theorem*, which, when stated as follows, is not the same as the former, but a rather close statement characterizing the *strong convergence* of a sequence of bounded linear operators.

6.4.2 Banach–Steinhaus Theorem

Theorem 6.8 (Banach–Steinhaus Theorem). *Let $(X, \|\cdot\|_X)$ be a Banach space, $(Y, \|\cdot\|_Y)$ be a normed vector space, and $\{A_n\}_{n=1}^{\infty}$ be a sequence of bounded linear operators in $(L(X, Y), \|\cdot\|)$.*

For the limit $\lim_{n\to\infty} A_n x$ to exist in $(Y, \|\cdot\|_Y)$ for each $x \in X$ it is necessary and, provided $(Y, \|\cdot\|_Y)$ is a Banach space, sufficient that

(1) *there exists a fundamental set S in $(X, \|\cdot\|_X)$ such that*

$$\forall x \in S : \exists \lim_{n\to\infty} A_n x \text{ in } (Y, \|\cdot\|_Y);$$

(2) $\sup_{n\in\mathbb{N}} \|A_n\| < \infty,$

in which case

$$Ax := \lim_{n\to\infty} A_n x, \ x \in X,$$

is a bounded linear operator in $L(X, Y)$ with

$$\|A\| \le \varliminf_{n\to\infty} \|A_n\| := \lim_{n\to\infty} \inf_{k\ge n} \|A_k\| \le \sup_{n\in\mathbb{N}} \|A_n\|.$$

Proof. Necessity. Suppose that, for each $x \in X$, the sequence $\{A_n x\}_{n=1}^{\infty}$ converges in $(Y, \|\cdot\|_Y)$. This implies (1) with $S = X$ and, since a convergent sequence is *bounded* (Proposition 2.21), by the *Uniform Boundedness Principle* (Theorem 6.7), implies (2) as well.

Sufficiency. Suppose that conditions (1) and (2) hold, with $(Y, \|\cdot\|_Y)$ being a Banach space.

The existence of $\lim_{n\to\infty} A_n x$ in $(Y, \|\cdot\|_Y)$ on a *fundamental set* $S \subseteq X$ in $(X, \|\cdot\|_X)$, by the linearity of A_n, $n \in \mathbb{N}$, implies that

$$\forall x \in \text{span}(S) : \exists \lim_{n\to\infty} A_n x \text{ in } (Y, \|\cdot\|_Y).$$

Since $\overline{\text{span}(S)} = X$,

$$\forall x \in X, \ \forall \varepsilon > 0 \ \exists y = y(x, \varepsilon) \in \text{span}(S) :$$

$$\|x - y\| < \frac{\varepsilon}{3(\sup_{n\in\mathbb{N}} \|A_n\| + 1)}. \tag{6.17}$$

Being convergent, the sequence $\{A_n y\}_{n=1}^{\infty}$ is a *fundamental* in $(Y, \|\cdot\|_Y)$ (Proposition 2.21), i. e.,

$$\exists N \in \mathbb{N} \; \forall m, n \geq N : \|A_n y - A_m y\|_Y < \varepsilon/3. \tag{6.18}$$

Hence, for each $x \in X$ and any $m, n \geq N$,

$$\begin{aligned}
\|A_n x - A_m x\| &\leq \|A_n x - A_n y\| + \|A_n y - A_m y\| + \|A_m y - A_m x\| \\
&= \|A_n (x - y)\| + \|A_n y - A_m y\| + \|A_m (y - x)\| \\
&\leq \|A_n\|\|x - y\| + \|A_n y - A_m y\| + \|A_m\|\|y - x\| \quad \text{by (6.17) and (6.18);} \\
&< \varepsilon/3 + \varepsilon/3 + \varepsilon/3 = \varepsilon,
\end{aligned}$$

which implies that, for each $x \in X$, $\{A_n x\}_{n=1}^{\infty}$ is a *Cauchy sequence* in $(Y, \|\cdot\|_Y)$, and hence, since the latter is a Banach space, *converges*.

Thus, the operator

$$X \ni x \mapsto Ax := \lim_{n\to\infty} A_n x \in Y$$

is *well defined* and *linear*.

Exercise 6.33. Explain.

Since, by *continuity* of norm,

$$\forall x \in X : \|Ax\|_Y = \lim_{n\to\infty} \|A_n x\|_Y = \lim_{n\to\infty} \|A_n x\|_Y \leq \left[\lim_{n\to\infty} \|A_n\|\right] \|x\|_X,$$

we conclude that $A \in L(X, Y)$ and

$$\|A\| \leq \lim_{n\to\infty} \|A_n\| \leq \sup_{n\in\mathbb{N}} \|A_n\|. \qquad \square$$

In particular, for the Banach space $(Y, \|\cdot\|_Y) = (\mathbb{F}, |\cdot|)$, we obtain the following version of the *Banach–Steinhaus Theorem* characterizing *weak* convergence* in the dual space $(X^*, \|\cdot\|)$ of a Banach space $(X, \|\cdot\|)$.

Theorem 6.9 (Banach–Steinhaus Theorem for Functionals). *Let $(X, \|\cdot\|)$ be a Banach space over $\mathbb{F}$ and $\{f_n\}_{n=1}^{\infty}$ be a sequence of bounded linear functionals in the dual space $(X^*, \|\cdot\|)$.*

For each $x \in X$, there exists $\lim_{n\to\infty} f_n(x) \in \mathbb{F}$ iff
(1) there exists a fundamental set S in $(X, \|\cdot\|_X)$ such that

$$\forall x \in S : \exists \lim_{n\to\infty} f_n(x) \in \mathbb{F};$$

(2) $\sup_{n\in\mathbb{N}} \|f_n\| < \infty$,

in which case

$$f(x) := \lim_{n\to\infty} f_n(x), \; x \in X,$$

is a bounded linear functional in X^* with

$$\|f\| \leq \varliminf_{n\to\infty} \|f_n\| := \lim_{n\to\infty} \inf_{k\geq n} \|f_k\| \leq \sup_{n\in\mathbb{N}} \|f_n\|.$$

The *Banach–Steinhaus Theorem for Functionals* (Theorem 6.9) enables us to prove the following general characterization of weak convergence.

Theorem 6.10 (Characterization of Weak Convergence). *In a normed vector space $(X, \|\cdot\|)$ over $\mathbb{F}$, a sequence $\{x_n\}_{n=1}^\infty$ converges weakly to an element $x \in X$, i. e.,*

$$\forall f \in X^* : \lim_{n\to\infty} f(x_n) = f(x)$$

iff
(1) *there exists a fundamental set S in the dual space $(X^*, \|\cdot\|)$ such that*

$$\forall f \in S : \lim_{n\to\infty} f(x_n) = f(x)$$

and
(2) $\sup_{n\in\mathbb{N}} \|x_n\| < \infty$,

in which case

$$\|x\| \leq \varliminf_{n\to\infty} \|x_n\| := \lim_{n\to\infty} \inf_{k\geq n} \|x_k\| \leq \sup_{n\in\mathbb{N}} \|x_n\|.$$

Exercise 6.34. Prove.

Hint. Replace the sequence $\{x_n\}_{n=1}^\infty$ and the element x with the sequence $\{Jx_n\}_{n=1}^\infty$ and the functional Jx in the bidual space $(X^{**}, \|\cdot\|)$, where $J : X \to X^{**}$ is the *canonical isomorphism*, then apply the *Banach–Steinhaus Theorem for Functionals* (Theorem 6.9).

Remark 6.13. In the next chapter, we consider the *Characterization of Weak Convergence in Reflexive Spaces* (Theorem 7.18), which is a version of the prior characterization in a more restrictive setting, but without the *a priori* existence of a weak limit in the space.

From the prior characterization, we immediately obtain the following useful corollary.

Corollary 6.8 (Boundedness of a Weakly Convergent Sequence). *Every weakly convergent sequence in a normed vector space is bounded.*

From the *Banach–Steinhaus theorem* (Theorem 6.8), we also immediately obtain the following corollaries.

Corollary 6.9 (Strong Limit of Sequence of Bounded Linear Operators). *Let $(X, \| \cdot \|_X)$ be a Banach space, $(Y, \| \cdot \|_Y)$ be a normed vector space, and $\{A_n\}_{n=1}^{\infty}$ be a sequence of bounded linear operators in $(L(X, Y), \| \cdot \|)$ such that the limit $\lim_{n\to\infty} A_n x$ exists in $(Y, \| \cdot \|_Y)$ for each $x \in X$. Then*

$$Ax := \lim_{n\to\infty} A_n x, \ x \in X,$$

is a bounded linear operator in $L(X, Y)$ with

$$\|A\| \le \varliminf_{n\to\infty} \|A_n\| := \lim_{n\to\infty} \inf_{k\ge n} \|A_k\| \le \sup_{n\in\mathbb{N}} \|A_n\|.$$

Corollary 6.10 (Weak* Limit of Sequence of Bounded Linear Functionals). *Let $\{f_n\}_{n=1}^{\infty}$ be a sequence of functionals in the dual space X^* of a Banach space $(X, \| \cdot \|)$ such that $\lim_{n\to\infty} f_n(x) \in \mathbb{F}$ exists for each $x \in X$. Then*

$$f(x) := \lim_{n\to\infty} f_n(x), \ x \in X,$$

is a bounded linear functional in X^* with

$$\|f\| \le \varliminf_{n\to\infty} \|f_n\| := \lim_{n\to\infty} \inf_{k\ge n} \|f_k\| \le \sup_{n\in\mathbb{N}} \|f_n\|.$$

Remark 6.14. The condition of *completeness* of the domain space $(X, \| \cdot \|_X)$ here is essential and cannot be dropped.

Indeed, the sequence $(f_n)_{n\in\mathbb{N}}$ of bounded linear functionals on the *incomplete* normed vector space $(c_{00}, \| \cdot \|_\infty)$ from Example 6.2 converges for each $x \in X$, however the limit linear functional

$$c_{00} \ni x := (x_k)_{k\in\mathbb{N}} \mapsto f(x) := \sum_{k=1}^{\infty} x_k$$

(the sum is actually *finite!*) is *unbounded*.

Exercise 6.35. Verify.

6.5 Applications of the Uniform Boundedness Principle

Let us now consider some applications of the *Uniform Boundedness Principle* (Theorem 6.7) (for more, see, e. g., [16]).

6.5.1 Weak Boundedness

Definition 6.6 (Weakly Bounded Set). A set S in a normed vector space $(X, \|\cdot\|)$ is said to be *weakly bounded* if

$$\forall f \in X^* : \sup_{x \in S} |f(x)| < \infty.$$

The following statement, based on the *Canonical Isomorphism Theorem* (Theorem 6.5) and the *Uniform Boundedness Principle* (Theorem 6.7), shows that the seemingly new notion of weak boundedness is actually indistinguishable from boundedness in the regular sense.

Theorem 6.11 (Equivalence of Weak Boundedness and Boundedness). *A set S in a normed vector space $(X, \|\cdot\|)$ is weakly bounded iff it is bounded, i. e.,* $\sup_{x \in S} \|x\| < \infty$.

Proof. *"If" part.* Suppose that S is *bounded.* Then

$$\forall f \in X^* : \sup_{x \in S} |f(x)| \le \|f\| \sup_{x \in S} \|x\| < \infty.$$

"Only if" part. Suppose that S is *weakly bounded*, i. e.,

$$\forall f \in X^* : \sup_{x \in S} |f(x)| < \infty.$$

Since, by the *Canonical Isomorphism Theorem* (Theorem 6.5), for each fixed $x \in X$, the functional

$$X^* \ni f \mapsto F_x(f) = f(x) \in \mathbb{F}$$

belongs to the *second dual space* X^{**} with

$$\|F_x\| = \|x\|. \tag{6.19}$$

For the set of functionals $\{F_x\}_{x \in S}$ in $(X^{**}, \|\cdot\|)$,

$$\forall f \in X^* : \sup_{x \in S} |F_x(f)| = \sup_{x \in S} |f(x)| < \infty.$$

Whence, since the dual $(X^*, \|\cdot\|)$ is a *Banach space* (Corollary 5.1), by the *Uniform Boundedness Principle* (Theorem 6.7), we infer that the set of functionals $\{F_x\}_{x \in S}$ is bounded in $(X^{**}, \|\cdot\|)$, i. e.,

$$\sup_{x \in S} \|F_x\| < \infty,$$

which, in view of (6.19), implies that

$$\sup_{x \in S} \|x\| < \infty,$$

and hence, the set S is *bounded* in $(X, \|\cdot\|)$. $\qquad\square$

Remark 6.15. The prior theorem is consistent with the fact that every weakly convergent sequence in a normed vector space is bounded (see Corollary 6.8).

6.5.2 Matrix Methods of Convergence and Summability

6.5.2.1 Definitions and Examples
The following approach allows to turn divergent sequences and series into convergent ones.

Definition 6.7 (*T-Convergence and Summability*). Let

$$
T := [c_{ij}]_{i,j=1}^{\infty} =
\begin{bmatrix}
c_{11} & c_{12} & \cdots & c_{1n} & \cdots \\
c_{21} & c_{22} & \cdots & c_{2n} & \cdots \\
\vdots & \vdots & & \vdots & \\
c_{m1} & c_{m2} & \cdots & c_{mn} & \cdots \\
\vdots & \vdots & & \vdots &
\end{bmatrix}
$$

be a doubly infinite matrix with entries from the scalar field $\mathbb{F}$ ($\mathbb{F} = \mathbb{C}$ or $\mathbb{F} = \mathbb{R}$).

- An $\mathbb{F}$-termed sequence $\{x_n\}_{n=1}^{\infty}$ is said to be *T-convergent* and have a *T-limit* $l \in \mathbb{F}$ if, for each $n \in \mathbb{N}$, the series

$$
\sum_{k=1}^{\infty} c_{nk} x_k
$$

converges, the corresponding sequence of sums

$$
\left\{ t_n := \sum_{k=1}^{\infty} c_{nk} x_k \right\}_{n=1}^{\infty}
$$

converges in the regular sense, and

$$
\lim_{n \to \infty} t_n = l.
$$

Notation. $\lim_{n \to \infty} x_n = l(T)$.

- An $\mathbb{F}$-termed series $\sum_{k=1}^{\infty} x_k$ is said to be *T-summable* and have a *T-sum* $s \in \mathbb{F}$ if the sequence of its partial sums

$$
\left\{ s_n := \sum_{k=1}^{n} x_k \right\}_{n=1}^{\infty}
$$

is *T-convergent* and

$$
\lim_{n \to \infty} s_n = s(T).
$$

Notation. $\sum_{k=1}^{\infty} x_k = s(T)$.

Such a method of convergence and summability associated with the *matrix* $T = [c_{ij}]_{i,j=1}^{\infty}$ is called the *T-method*.

Examples 6.3.

1. For the T-method associated with the *zero matrix* T, i.e., $t_{ij} := 0$, $i,j \in \mathbb{N}$, any $\mathbb{F}$-termed sequence $\{x_n\}_{n=1}^{\infty}$ is transformed into the zero sequence, and hence, is T-convergent to 0.

2. For the T-method associated with the *identity matrix*

$$T := \begin{bmatrix} 1 & 0 & 0 & 0 & \dots \\ 0 & 1 & 0 & 0 & \dots \\ 0 & 0 & 1 & 0 & \dots \\ \vdots & \vdots & \vdots & \vdots & \end{bmatrix},$$

i.e., $t_{ij} := \delta_{ij}$, $i,j \in \mathbb{N}$, where δ_{ij} is the *Kronecker delta*, any $\mathbb{F}$-termed sequence $\{x_n\}_{n=1}^{\infty}$ remains unchanged, and hence is T-convergent *iff* it is convergent in the regulars sense, in which case

$$\lim_{n \to \infty} x_n(T) = \lim_{n \to \infty} x_n.$$

3. For the T-method associated with the matrix

$$T := \begin{bmatrix} 1 & 1 & 0 & 0 & \dots \\ 0 & 1 & 1 & 0 & \dots \\ 0 & 0 & 1 & 1 & \dots \\ \vdots & \vdots & \vdots & \vdots & \end{bmatrix},$$

the divergent sequence $\{(-1)^n\}_{n=1}^{\infty}$ is T-convergent with

$$\lim_{n \to \infty} x_n(T) = 0.$$

4. For the T-method associated with the matrix

$$T := \begin{bmatrix} 1 & -1 & 0 & 0 & \dots \\ 0 & 1 & 1 & 0 & \dots \\ 0 & 0 & 1 & -1 & \dots \\ \vdots & \vdots & \vdots & \vdots & \end{bmatrix},$$

the convergent sequence $(1, 1, 1, \dots)$ is T-divergent, the generated sequence being

$$(0, 2, 0, 2, 0, \dots).$$

5. The important particular case of a T-method is the *Cesàro*[2] *method* associated with the *Cesàro matrix* $T := [c_{ij}]_{i,j=1}^{\infty}$, where

$$c_{ij} := \begin{cases} \frac{1}{i} & \text{for } 1 \le j \le i, \\ 0 & \text{for } j > i, \ i,j \in \mathbb{N}, \end{cases}$$

---

2 Ernesto Cesàro (1859–1906).

i. e.,

$$T := \begin{bmatrix} 1 & 0 & 0 & 0 & \cdots \\ 1/2 & 1/2 & 0 & 0 & \cdots \\ 1/3 & 1/3 & 1/3 & 0 & \cdots \\ \vdots & \vdots & \vdots & \vdots \end{bmatrix}.$$

As we see below, under the *Cesàro method*, every convergent sequence is also T-convergent to the same limit. Moreover, the divergent sequence $\{(-1)^n\}_{n=1}^{\infty}$ becomes T-convergent (see Section 6.7, Problem 6).

Exercise 6.36. Verify 1–4.

6.5.2.2 Regularity

The natural question, of course, is:

When does a T-method transform each convergent sequence into a convergent one with the same limit?

Definition 6.8 (Regularity of a T-Method). A *T-method* is said to be *regular* if every numeric sequence $\{x_n\}_{n=1}^{\infty}$ convergent in the usual senses is also T-convergent and

$$\lim_{n \to \infty} x_n(T) = \lim_{n \to \infty} x_n.$$

The following theorem gives a characterization of regularity and its proof is largely based on the *Uniform Boundedness Principle* (Theorem 6.7).

Theorem 6.12 (Toeplitz Regularity Theorem). *The T-method associated with a doubly infinite matrix $T = [c_{ij}]_{i,j=1}^{\infty}$ is regular iff T satisfies the following conditions:*[3]
(i) $\sup_{i \in \mathbb{N}} \sum_{j=1}^{\infty} |c_{ij}| < \infty$;
(ii) $\sum_{j=1}^{\infty} c_{ij} \to 1, i \to \infty$;
(iii) $\forall j \in \mathbb{N} : c_{ij} \to 0, i \to \infty$.

Remark 6.16. In particular, condition (i) implies that the series

$$\sum_{j=1}^{\infty} c_{ij}$$

converges absolutely for each $i \in \mathbb{N}$.

Proof. "*Only if*" part. Suppose that the T-method is *regular*.

---

3 Otto Toeplitz (1881–1940).

Recall that the space $(c, \| \cdot \|_\infty)$ of convergent sequences with the *supremum norm* is *Banach* (see Examples 3.13).

For each fixed pair $i, n \in \mathbb{N}$, the mapping

$$c \ni x := \{x_k\}_{k=1}^\infty \mapsto f_{in}(x) = \sum_{j=1}^n c_{ij}x_j \in \mathbb{F}$$

is a *linear functional*, which is *bounded* since

$$\forall x \in c: \ |f_{in}(x)| = \left| \sum_{j=1}^n c_{ij}x_j \right| \leq \sum_{j=1}^n |c_{ij}||x_j| \leq \left[\sum_{j=1}^n |c_{ij}| \right] \sup_{k\in\mathbb{N}} |x_k| = \left[\sum_{j=1}^n |c_{ij}| \right] \|x\|_\infty.$$

Hence,

$$\forall i, n \in \mathbb{N}: \ f_{in} \in c^* \text{ with } \|f_{in}\| \leq \sum_{j=1}^n |c_{ij}|. \tag{6.20}$$

By the *regularity* of the T-method,

$$\forall x := \{x_k\}_{k=1}^\infty \in c, \ \forall i \in \mathbb{N}: \ \lim_{n\to\infty} f_{in}(x) = \lim_{n\to\infty} \sum_{j=1}^n c_{ij}x_j = \sum_{j=1}^\infty c_{ij}x_j =: t_i(x) \in \mathbb{F},$$

and hence, in view of the *completeness* of the domain space $(c, \| \cdot \|_\infty)$, by the *Strong Limit of Sequence of Bounded Linear Operators Corollary* (Corollary 6.9),

$$\forall i \in \mathbb{N}: \ t_i(x) := \lim_{n\to\infty} f_{in}(x), \ x \in c,$$

is a *bounded linear functional* on $(c, \| \cdot \|_\infty)$, i. e.,

$$\forall i \in \mathbb{N}: \ t_i \in c^*.$$

Furthermore, by the *regularity* of the T-method,

$$\forall x = \{x_k\}_{k=1}^\infty \in c: \ \lim_{i\to\infty} t_i(x) = \lim_{k\to\infty} x_k =: l(x),$$

i. e., the sequence of functionals $\{t_i\}_{i=1}^\infty$ *weakly* converges* in c^* to the *limit functional* $l \in c^*$ (see Examples 5.2):

$$w^*\text{-}\lim_{i\to\infty} t_i = l.$$

Hence, by the *Boundedness of a Weakly* Convergent Sequence Proposition* (Proposition 6.7),

$$M := \sup_{i\in\mathbb{N}} \|t_i\| < \infty.$$

For an arbitrary fixed $i \in \mathbb{N}$ and any $n \in \mathbb{N}$, choosing

$$x_{in} := (\underbrace{e^{-i\theta_{i1}}, \ldots, e^{-i\theta_{in}}}_{n \text{ terms}}, 0, 0, \ldots) \in c_{00} \subset c \text{ with } \|x_{in}\|_\infty = 1,$$

where $\theta_{ij} := \arg c_{ij} \in (-\pi, \pi]$ is the *principal value* of the argument of c_{ij}, $i, j \in \mathbb{N}$, $(\arg 0 := 0)$, and hence,

$$c_{ij} = |c_{ij}|e^{i\theta_{ik}}, \; j = 1, \ldots, n,$$

i designating the *imaginary unit*, except when in subscript, in view of (6.20), we have:

$$\forall i \in \mathbb{N}: \; t_i(x_{in}) = \sum_{j=1}^{n} |c_{ij}| = \|t_i\| \leq M.$$

Exercise 6.37. Explain.

Whence, passing to the limit as $n \to \infty$ for each $i \in \mathbb{N}$, we conclude that

$$\forall i \in \mathbb{N}: \; \sum_{j=1}^{\infty} |c_{ij}| \leq M,$$

which implies that condition (i) is met.

Further, choosing the sequence

$$e_0 := (1, 1, 1, \ldots) \in c$$

convergent to 1, by the *regularity* of the T-method, we have:

$$t_i(e_0) = \sum_{j=1}^{\infty} c_{ij} \to 1, \; i \to \infty.$$

which implies that condition (ii) is also satisfied.

Finally, for the set of *vanishing* sequences

$$\{e_j := \{\delta_{jk}\}_{k=1}^{\infty}\}_{j \in \mathbb{N}},$$

where δ_{jk} is the *Kronecker delta*, by the *regularity* of the T-method, we have:

$$\forall j \in \mathbb{N}: \; t_i(e_j) = c_{ij} \to 0, \; i \to \infty,$$

which implies that condition (iii) is met as well and completes the proof of the "*only if*" part.

"*If*" part. Suppose that T satisfies conditions (i)–(iii).

As is known from the proof of the "*only if*" part, for any pair $i, n \in \mathbb{N}$, the mapping

$$c \ni x := \{x_k\}_{k=1}^{\infty} \mapsto f_{in}(x) = \sum_{j=1}^{n} c_{ij} x_j \in \mathbb{F}$$

is a *bounded linear functional* on $(c, \|\cdot\|_\infty)$, i. e.,

$$\forall i, n \in \mathbb{N}: \; f_{in} \in c^*,$$

by condition (i),

$$\forall\, i, n \in \mathbb{N},\ \forall\, x \in c:\ |f_{in}(x)| = \left|\sum_{j=1}^{n} c_{ij} x_j\right| \le \sum_{j=1}^{n} |c_{ij}||x_j| \le \left[\sum_{j=1}^{\infty} |c_{ij}|\right] \sup_{k\in\mathbb{N}} |x_k| \le M\|x\|_\infty,$$

where

$$M := \sup_{k\in\mathbb{N}} \sum_{j=1}^{\infty} |c_{kj}| < \infty,$$

which implies that

$$\forall\, i, n \in \mathbb{N}:\ \|f_{in}\| \le M. \tag{6.21}$$

For the sequences

$$e_0 := (1,1,1,\dots) \in c \ \text{ and } \ e_j := (\delta_{jk})_{k\in\mathbb{N}} \in c_0 \subset c,\ j \in \mathbb{N},$$

where δ_{jk} is the *Kronecker delta*,

$$\forall\, i \in \mathbb{N}:\ \lim_{n\to\infty} f_{in}(e_j) = \begin{cases} \sum_{k=1}^{\infty} c_{ik} & \text{if } j = 0, \\ c_{ij} & \text{if } j \in \mathbb{N}. \end{cases} \tag{6.22}$$

Since the set $\{e_j\}_{j\in\mathbb{Z}_+}$, being a *Schauder basis* (see Section 3.16), is *fundamental* in $(c, \|\cdot\|_\infty)$, from (6.21) and (6.22), by the *Banach–Steinhaus Theorem for Functionals* (Theorem 6.9), we infer that

$$\forall\, i \in \mathbb{N},\ \forall\, x := \{x_k\}_{k=1}^{\infty} \in c:\ \exists\, \lim_{n\to\infty} f_{in}(x) = \sum_{j=1}^{\infty} c_{ij} x_j =: t_i(x) \in \mathbb{F}$$

and

$$\forall\, i \in \mathbb{N}:\ t_i \in c^* \ \text{with, in view of (6.21),}\ \|t_i\| \le \sup_{n\in\mathbb{N}} \|f_{in}\| \le M.$$

Since, by (6.22),

$$\forall\, i \in \mathbb{N}:\ t_i(e_j) = \begin{cases} \sum_{k=1}^{\infty} c_{ik} & \text{if } j = 0, \\ c_{ij} & \text{if } j \in \mathbb{N}, \end{cases}$$

by conditions (ii) and (iii),

$$\lim_{i\to\infty} t_i(e_j) = \begin{cases} 1 & \text{if } j = 0, \\ 0 & \text{if } j \in \mathbb{N}, \end{cases}$$

i. e.,

$$\forall\, j \in \mathbb{Z}_+:\ \lim_{i\to\infty} t_i(e_j) = l(e_j), \tag{6.23}$$

where $l \in c^*$ is the *limit functional* (see Examples 5.2).

Since the set $\{e_j\}_{j \in \mathbb{Z}_+}$ is *fundamental* in $(c, \|\cdot\|_\infty)$, by the *Banach–Steinhaus Theorem for Functionals* (Theorem 6.9), we infer that

$$\forall x = \{x_n\}_{n=1}^\infty : \lim_{n \to \infty} x_n(T) := \lim_{n \to \infty} t_n(x) = l(x) := \lim_{n \to \infty} x_n,$$

which proves the *regularity* of the T-method. □

Exercise 6.38. Apply the *Toeplitz Regularity Theorem* (Theorem 6.12) to determine which of the T-methods from Examples 6.3 are regular and which are not.

Remark 6.17. Under condition (i) of the *Toeplitz Regularity Theorem* (Theorem 6.12), the mapping

$$l_\infty \ni x := \{x_n\}_{n=1}^\infty \mapsto Tx := \left\{ t_n := \sum_{k=1}^\infty c_{nk} x_k \right\}_{n=1}^\infty \in l_\infty$$

is a well-defined *bounded linear operator* on l_∞, i. e., $T \in L(l_\infty)$.

If conditions (ii) and (iii) are also met, by the *Toeplitz Regularity Theorem* (Theorem 6.12),

$$T(c) \subseteq c \quad \text{and} \quad T(c_0) \subseteq c_0$$

and the mapping

$$l_\infty \supseteq Y \ni x := \{x_n\}_{n=1}^\infty \mapsto t(x) := l(Tx) \in \mathbb{F},$$

where

$$Y := \{x \in l_\infty \mid Tx \in c\} = T^{-1}(c)$$

is a subspace of l_∞ with $c \subseteq Y$, is a *linear bounded functional*.

Furthermore, by the regularity of the T-method,

$$\forall x \in c : t(x) := l(Tx) = l(x),$$

and hence,

$$c_0 \subseteq \ker t.$$

Exercise 6.39. Verify.

6.5.2.3 Cesàro Method

The important example of a *regular* T-method is the *Cesàro method* associated with the *Cesàro matrix* $T := [c_{ij}]_{i,j=1}^\infty$, where

$$c_{ij} := \begin{cases} \frac{1}{i} & \text{for } 1 \le j \le i, \\ 0 & \text{for } j > i, \ i, j \in \mathbb{N}, \end{cases}$$

(see Examples 6.3).

Exercise 6.40. Apply the *Toeplitz Regularity Theorem* (Theorem 6.12) to verify that the *Cesàro method* is *regular*.

Thus, we obtain the following

Corollary 6.11 (Regularity of the Cesàro Method). *For any convergent numeric se-quence* $\{x_n\}_{n=1}^{\infty}$,

$$\lim_{n\to\infty} \frac{1}{n} \sum_{k=1}^{n} x_k = \lim_{n\to\infty} x_n.$$

Remark 6.18. For a sequence $\{x_n\}_{n=1}^{\infty}$ *Cesàro convergent* to a number l, we write

$$\lim_{n\to\infty} x_n = l(C,1).$$

For a series $\sum_{n=1}^{\infty} x_n$ *Cesàro summable* to a number s, we write

$$\sum_{n=1}^{\infty} x_n = s(C,1).$$

See Section 6.7, Problems 6 and 7.

6.6 Open Mapping, Inverse Mapping, and Closed Graph Theorems

The following three equivalent statements: the *Open Mapping Theorem* (*OMT*), the *Inverse Mapping Theorem* (*IMT*), and the *Closed Graph Theorem* (*CGT*) form the next fundamental principle of linear functional analysis, the last one in our ordering.

6.6.1 Open Mapping Theorem

Theorem 6.13 (Open Mapping Theorem). *Let* $(X, \|\cdot\|_X)$ *and* $(Y, \|\cdot\|_Y)$ *be Banach spaces over* $\mathbb{F}$ *and* $A : X \to Y$ *be a surjective bounded linear operator from* X *onto* Y *(i.e., $A \in L(X,Y)$ and $R(A) = Y$). Then A is an open mapping, i.e., the image $A(G)$ under A of each open set G in $(X, \|\cdot\|_X)$ is an open set in $(Y, \|\cdot\|_Y)$.*

Proof. Since

$$X = \bigcup_{n=1}^{\infty} B_X(0,n),$$

by the fact that image preserves unions (Exercise 1.4) and the *surjectivity* and *linearity* of A,

$$Y = R(A) = A\left(\bigcup_{n=1}^{\infty} B_X(0,n)\right) = \bigcup_{n=1}^{\infty} A\left(B_X(0,n)\right) = \bigcup_{n=1}^{\infty} nA\left(B_X(0,1)\right).$$

Whence, in view of the *completeness* of $(Y, \|\cdot\|_Y)$, as follows from the *Baire Category Theorem* (Theorem 2.31) (see also Corollary 2.5),

$$\exists N \in \mathbb{N}: \text{int}(\overline{N \cdot A\,(B_X(0,1))}) \neq \emptyset,$$

i. e., the set $N \cdot AB_X(0,1)$ is *not nowhere dense* in $(Y, \|\cdot\|_Y)$. This, since the linear operator of multiplication by a *nonzero* number $\lambda \in \mathbb{F} \setminus \{0\}$

$$Y \ni y \mapsto \lambda y \in Y$$

is a *homeomorphism* of $(Y, \|\cdot\|_Y)$ (see Examples 5.2), in view of the *linearity* of A, implies that

$$\forall \delta > 0: \text{int}(\overline{A\,(B_X(0,\delta))}) \neq \emptyset.$$

Exercise 6.41. Explain.

Hint. $\forall \delta > 0: \text{int}(\overline{A(B_X(0,\delta))}) = \frac{\delta}{N} \text{int}(\overline{N \cdot A(B_X(0,1))})$

By the *joint continuity* of the difference mapping

$$(x,y) \mapsto x - y,$$

for $(X, \|\cdot\|_X)$,

$$\exists \delta > 0: B_X(0,1) \supseteq B_X(0,\delta) - B_X(0,\delta).$$

Hence, by the *linearity* of A and the *joint continuity* of the difference mapping for $(Y, \|\cdot\|_Y)$,

$$\overline{A\,(B_X(0,1))} \supseteq \overline{A\,(B_X(0,\delta)) - A\,(B_X(0,\delta))} \supseteq \overline{A\,(B_X(0,\delta))} - \overline{A\,(B_X(0,\delta))}$$
$$\supseteq \text{int}(\overline{A\,(B_X(0,\delta))}) - \text{int}(\overline{A\,(B_X(0,\delta))}).$$

Observe that

$$\text{int}(\overline{A\,(B_X(0,\delta))}) - \text{int}(\overline{A\,(B_X(0,\delta))}) = \bigcup_{x \in \text{int}(\overline{A(B_X(0,\delta))})} \left[\text{int}(\overline{A\,(B_X(0,\delta))}) - x\right].$$

is an *open set* in $(Y, \|\cdot\|_Y)$ containing 0.

Exercise 6.42. Explain.

Hence, for the set $\overline{A(B_X(0,1))}$, 0 is an *interior point*, i. e.,

$$\exists \delta > 0: B_Y(0,\delta) \subseteq \overline{A\,(B_X(0,1))}. \tag{6.24}$$

We are to show that 0 is an *interior point* for the set $AB_X(0,1)$ (without closure!) as well, i. e., we are to prove the following stronger version of inclusion (6.24):

$$\exists \delta > 0 : \; B_Y(0,\delta) \subseteq A\,(B_X(0,1)). \tag{6.25}$$

By (6.24), in view of the *linearity* of A,

$$\exists \delta > 0 : \; B_Y(0,\delta) \subseteq \overline{A\,(B_X(0,1/3))},$$

and hence,

$$\forall n \in \mathbb{Z}_+ : \; B_Y(0,\delta/3^n) \subseteq \overline{A\,(B_X(0,1/3^{n+1}))}. \tag{6.26}$$

By (6.26) with $n = 0$,

$$\forall y \in B_Y(0,\delta) \; \exists x_0 \in B_X(0,1/3) : \; \|y - Ax_0\|_Y < \delta/3.$$

Exercise 6.43. Explain.

Since, $y - Ax_0 \in B_Y(0,\delta/3)$, by (6.26) with $n = 1$,

$$\exists x_1 \in B_X(0,1/3^2) : \; \|y - Ax_0 - Ax_1\|_Y < \delta/3^2.$$

Continuing inductively, we obtain a sequence of elements $\{x_n\}_{n=0}^{\infty}$ in $(X, \|\cdot\|)$ such that

$$\forall n \in \mathbb{Z}_+ : \; \|x_n\|_X \le 1/3^{n+1} \tag{6.27}$$

and, by the *linearity* of A,

$$\forall n \in \mathbb{N} : \; \left\| y - A\left[\sum_{k=0}^{n-1} x_k\right] \right\|_Y = \left\| y - \sum_{k=0}^{n-1} Ax_k \right\|_Y < \delta/3^n. \tag{6.28}$$

In view of (6.27), by the *Comparison Test*, the series

$$\sum_{k=0}^{\infty} x_k$$

converges absolutely in $(X, \|\cdot\|_X)$, which, in view of the *completeness* of $(X, \|\cdot\|_X)$, by the *Series Characterization of a Banach Space* (Theorem 3.8), implies that it converges in $(X, \|\cdot\|_X)$, i. e.,

$$\exists x \in X : \; x = \sum_{k=0}^{\infty} x_k := \lim_{n \to \infty} \sum_{k=0}^{n-1} x_k.$$

Then (6.27), by *subadditivity* of norm, implies that

$$\|x\|_X \le \sum_{k=0}^{\infty} \|x_k\|_X \le \sum_{k=0}^{\infty} \frac{1}{3^{k+1}} = \frac{1/3}{1-1/3} = \frac{1}{2} < 1,$$

and hence,

$$x \in B_X(0,1).$$

Passing to the limit in (6.28) as $n \to \infty$, by the *boundedness*, and hence, the *continuity* of the linear operator A (Theorem 5.4), *continuity* of norm, and the *Squeeze Theorem*, we infer that

$$\|y - Ax\|_Y = 0,$$

which, by the norm *separation* axiom, implies that

$$y = Ax.$$

Thus, we have shown that

$$\exists \delta > 0 \, \forall y \in B_Y(0,\delta) \, \exists x \in B_X(0,1) : \, y = Ax,$$

and hence, inclusion (6.25) does hold as desired, i. e., 0 is an *interior point* of the set $A(B_X(0,1))$. This, by the linearity of A, implies that, for an arbitrary $x \in X$ and any $\varepsilon > 0$, Ax is an *interior point* of the set

$$A\left(B_X(x,\varepsilon)\right),$$

i. e.,

$$\forall x \in X, \, \forall \varepsilon > 0 \, \exists \delta > 0 : \, B_Y(Ax,\delta) \subseteq A\left(B_X(x,\varepsilon)\right). \tag{6.29}$$

Exercise 6.44. Verify.

Hint. Show that $\forall x \in X, \, \forall \varepsilon > 0 : \, A(B_X(x,\varepsilon)) = Ax + \varepsilon A(B_X(0,1))$.

Now, let G be an arbitrary nonempty *open* set in $(X, \|\cdot\|_X)$. Then,

$$\forall x \in G \, \exists \varepsilon = \varepsilon(x) > 0 : \, B_X(x,\varepsilon) \subseteq G.$$

Whence,

$$A\left(B_X(x,\varepsilon)\right) \subseteq A(G)$$

and,

$$\exists \delta = \delta(x,\varepsilon) > 0 : \, B_Y(Ax,\delta) \subseteq A\left(B_X(x,\varepsilon)\right).$$

Therefore, we have the inclusion

$$B_Y(Ax,\delta) \subseteq A\left(B_X(x,\varepsilon)\right) \subseteq A(G),$$

which proves the *openness* of the image $A(G)$ in $(Y, \|\cdot\|_Y)$, and thus, the fact that A is an *open mapping*, completing the proof. $\square$

244 — 6 Three Fundamental Principles of Linear Functional Analysis

Remarks 6.19. *A priori*, we regard the operator $A : X \rightarrow Y$ to be *linear*.

- The condition of the *completeness* of the domain space $(X, \|\cdot\|_X)$ in the *Open Mapping Theorem* (*OMT*), provided other conditions hold, is *not essential* since a surjective bounded linear operator $A \in L(X, Y)$, with the target space $(Y, \|\cdot\|_Y)$ being *complete*, can be uniquely extended by continuity to a surjective bounded linear operator $\tilde{A} \in L(\tilde{X}, Y)$, where $\tilde{X}$ is a *completion* of X (see Theorem 3.9).

Exercise 6.45. Explain.

- As the following examples show, the *three* other conditions of the *OMT*:
  (i) the *completeness* of the target space $(Y, \|\cdot\|_Y)$,
  (ii) the *boundedness* of the linear operator $A : X \rightarrow Y$ ($A \in L(X, Y)$), and
  (iii) the *surjectivity* of the linear operator $A : X \rightarrow Y$ ($R(A) = Y$),
  are essential and none of them can be dropped.

Examples 6.4.
1. The differentiation operator

$$C^1[a, b] \ni x \rightarrow [Ax](t) := \frac{d}{dt}x(t) \in C[a, b]$$

$(-\infty < a < b < \infty)$ is a *surjective bounded linear operator* from the Banach space $X = C^1[a, b]$ with the norm

$$\|x\| := \max\left[\max_{a \leq t \leq b} |x(t)|, \max_{a \leq t \leq b} |x'(t)|\right], \ x \in C^1[a, b],$$

onto the incomplete normed vector space $Y = C[a, b]$ with the *integral norm*

$$\|x\|_1 := \int_a^b |x(t)|\, dt, \ x \in C[a, b],$$

(see Examples 5.2) that is *not an open mapping* (see Section 6.7, Problem 9).

2. Let $(X, \|\cdot\|)$ be an *infinite dimensional* Banach space with a Hamel basis $\{e_i\}_{i \in I}$, where, without loss of generality, we can regard $\|e_i\| = 1$, $i \in I$. Observe that, since, by the *Basis of a Banach Space Theorem* (Theorem 3.15), $\{e_i\}_{i \in I}$ is *uncountable*, we can choose a *countably infinite* subset $J := \{i_n\}_{n \in \mathbb{N}}$ of I and consider the *bijective unbounded linear operator* on X defined on $\{e_i\}_{i \in I}$ as follows:

$$Ae_i := \lambda_i e_i, \ i \in I,$$

where $\{\lambda_i\}_{i \in I}$ is an *unbounded set of nonzero numbers* with

$$\lambda_{i_n} := 1/n, \ n \in \mathbb{N}$$

The operator $A : X \to X$ is *not an open mapping* since its inverse $A^{-1} : X \to X$ is *unbounded*. Indeed,

$$A^{-1}e_i := (1/\lambda_i)e_i, \ i \in I,$$

with

$$A^{-1}e_{i_n} := ne_{i_n}, \ n \in \mathbb{N}.$$

3. If $(X, \|\cdot\|_X)$ and $(Y, \|\cdot\|_Y)$ are nontrivial Banach spaces over $\mathbb{F}$ $(X, Y \neq \{0\})$, then the *zero operator* $0 : X \to Y$ is a *nonsurjective* bounded linear operator that is *not an open mapping*.

Exercise 6.46. Explain and verify.

6.6.2 Inverse Mapping Theorem and Applications

6.6.2.1 Inverse Mapping Theorem
As an immediate corollary of the *Open mapping Theorem* (Theorem 6.13), we obtain the following statement.

Theorem 6.14 (Inverse Mapping Theorem). *Let $(X, \|\cdot\|_X)$ and $(Y, \|\cdot\|_Y)$ be Banach spaces over $\mathbb{F}$ and $A : X \to Y$ be a bijective bounded linear operator from X onto Y $(A \in L(X, Y))$. Then the inverse $A^{-1} : Y \to X$ is a bounded linear operator from Y onto X $(A^{-1} \in L(Y, X))$.*

Exercise 6.47. Prove.

Hint. Apply the *Characterization of Continuity* (Theorem 2.52) (see Section 2.19, Problem 14) and the *Boundedness Characterizations* (Theorem 5.4).

Remarks 6.20.
-   As is shown below (see Section 6.6.4), the *Open Mapping Theorem* (Theorem 6.13) is equivalent to the *Inverse Mapping Theorem* (Theorem 6.14).
-   The *Inverse Mapping Theorem* is also called the *Bounded Inverse Theorem*.

Let us now consider two profound applications of the *OMT*.

6.6.2.2 Application: Equivalence of Banach Norms
Theorem 6.15 (Equivalence of Banach Norms). *Let a vector space X be a Banach space relative to norms $\|\cdot\|_1$ and $\|\cdot\|_2$.*
If

$$\exists c > 0 : c\|x\|_1 \leq \|x\|_2, \ x \in X,$$

then

$$\exists C > 0 : \|x\|_2 \leq C\|x\|_1, \ x \in X,$$

i. e., if one of two Banach norms on a vector space is stronger than the other, the norms are equivalent.

Exercise 6.48. Prove.

Hint. Apply the *Inverse Mapping Theorem* (Theorem 6.14) to the *bijective linear operator*

$$(X, \| \cdot \|_2) \ni x \mapsto Ix := x \in (X, \| \cdot \|_1)$$

(see Remarks 3.20).

Remarks 6.21.
–   The result is consistent with the *Norm Equivalence Theorem* for finite-dimensional Banach spaces (Theorem 3.10).

    Exercise 6.49. Explain.

–   The requirement of the *completeness* of X relative to both norms is essential. Indeed, the vector space $C[a, b]$ $(-\infty < a < b < \infty)$ is *incomplete* relative to the *integral norm*

$$C[a, b] \ni x \mapsto \|x\|_1 = \int_a^b |x(t)| \, dt$$

and is *complete* relative to the *maximum norm*

$$C[a, b] \ni x \mapsto \|x\|_\infty = \max_{a \leq t \leq b} |x(t)|,$$

the latter being *stronger* than the former, but the norms are *not equivalent* (see Exercise 3.42).

6.6.2.3 Application: Boundedness of the Schauder Coordinate Functionals
Recall that, in a Banach space $(X, \| \cdot \|)$ over $\mathbb{F}$ with a *Schauder basis* $E := \{e_n\}_{n \in \mathbb{N}}$, each element $x \in X$ allows a *unique Schauder expansion*

$$x = \sum_{k=1}^\infty c_k e_k,$$

the coefficients $c_n \in \mathbb{F}$, $n \in \mathbb{N}$, called the *coordinates* of x relative to E (see Definition 3.25 and Examples 3.17).

For each $n \in \mathbb{N}$, the mapping

$$X \ni x = \sum_{k=1}^\infty c_k e_k \mapsto c_n(x) := c_n \in \mathbb{F}$$

is a well-defined *linear functional* on X, called the nth *Schauder coordinate functional* relative to E.

Exercise 6.50. Verify.

We prove that, on a Banach space $(X, \| \cdot \|)$ with a Schauder basis, all Schauder coordinate functionals are *bounded*, unlike the linear *Hamel coordinate functionals* (cf. the *Unboundedness of Hamel Coordinate Functionals Proposition* (Proposition 5.11), Section 5.4, Problem 6).

Proposition 6.8 (Boundedness of Schauder Coordinate Functionals). *On a Banach space $(X, \| \cdot \|)$ over $\mathbb{F}$ with a Schauder basis $E := \{e_n\}_{n \in \mathbb{N}}$, all Schauder coordinate functionals*

$$X \ni x = \sum_{k=1}^{\infty} c_k e_k \mapsto c_n(x) := c_n \in \mathbb{F}, \ n \in \mathbb{N},$$

relative to E are bounded, i. e.,

$$\forall n \in \mathbb{N}: \ c_n(\cdot) \in X^*.$$

Proof. Consider the set of $\mathbb{F}$-termed sequences defined as follows:

$$Y := \left\{ y := \{c_n\}_{n=1}^{\infty} \ \middle| \ c_n \in \mathbb{F}, \ n \in \mathbb{N}, \ \sum_{k=1}^{\infty} c_k e_k \text{ converges in } X \right\}.$$

The set Y is a normed vector space relative to the termwise linear operations and the norm

$$Y \ni y := \{c_n\}_{n=1}^{\infty} \mapsto \|y\|_Y := \sup_{n \in \mathbb{N}} \left\| \sum_{k=1}^{n} c_k e_k \right\|.$$

Exercise 6.51. Verify.

Furthermore, the space $(Y, \| \cdot \|_Y)$ is *Banach*. Indeed, for any *fundamental sequence*

$$\{y_n := \{c_k^{(n)}\}_{k=1}^{\infty}\}_{n=1}^{\infty}$$

in $(Y, \| \cdot \|_Y)$,

$$\forall \varepsilon > 0 \ \exists N \in \mathbb{N} \ \forall m, n \geq N: \ \|y_n - y_m\|_Y = \sup_{p \in \mathbb{N}} \left\| \sum_{k=1}^{p} \left(c_k^{(n)} - c_k^{(m)} \right) e_k \right\| < \varepsilon,$$

and hence,

$$\forall p \in \mathbb{N}, \ \forall m, n \geq N: \ \left\| \sum_{k=1}^{p} \left(c_k^{(n)} - c_k^{(m)} \right) e_k \right\| < \varepsilon.$$

By *subadditivity* of norm, we have:

$$\forall j \in \mathbb{N}, \, \forall m, n \geq N : \left\| \left(c_j^{(n)} - c_j^{(m)} \right) e_j \right\|$$

$$= \left\| \sum_{k=1}^{j} \left(c_k^{(n)} - c_k^{(m)} \right) e_k - \sum_{k=1}^{j-1} \left(c_k^{(n)} - c_k^{(m)} \right) e_k \right\|$$

$$\leq \left\| \sum_{k=1}^{j} \left(c_k^{(n)} - c_k^{(m)} \right) e_k \right\| + \left\| \sum_{k=1}^{j-1} \left(c_k^{(n)} - c_k^{(m)} \right) e_k \right\| < 2\varepsilon,$$

which, by *absolute scalability* of norm, implies that

$$\forall j \in \mathbb{N}, \, \forall m, n \geq N : \left| c_j^{(n)} - c_j^{(m)} \right| = \frac{1}{\|e_j\|} \left\| \left(c_j^{(n)} - c_j^{(m)} \right) e_j \right\| < \frac{2\varepsilon}{\|e_j\|}.$$

Whence, we conclude that, for each $j \in \mathbb{N}$, the numeric sequence $\{c_j^{(n)}\}_{n=1}^{\infty}$ is *fundamental* in $(\mathbb{F}, |\cdot|)$, and thus, *converges*, i. e.,

$$\forall j \in \mathbb{N} \, \exists c_j \in \mathbb{F} : c_j^{(n)} \to c_j, \, j \to \infty.$$

Whereby, we obtain a numeric sequence $y := \{c_n\}_{n=1}^{\infty}$.
It can be shown that $y \in Y$ and

$$y_n \to y, \, n \to \infty, \, \text{in } (Y, \|\cdot\|_Y)$$

(see, e. g., [16, 31]), which proves the *completeness* of $(Y, \|\cdot\|_Y)$.
Since $E := \{e_n\}_{n \in \mathbb{N}}$ is a *Schauder basis* of $(X, \|\cdot\|)$, the mapping

$$Y \ni y := \{c_n\}_{n=1}^{\infty} \mapsto Ay := \sum_{k=1}^{\infty} c_k e_k \in X$$

is a *bijective linear operator* from $(Y, \|\cdot\|_Y)$ onto $(X, \|\cdot\|)$, which is *bounded*:

$$\forall y \in Y : \|Ay\| = \left\| \sum_{k=1}^{\infty} c_k e_k \right\| = \lim_{n \to \infty} \left\| \sum_{k=1}^{n} c_k e_k \right\| \leq \sup_{n \in \mathbb{N}} \left\| \sum_{k=1}^{n} c_k e_k \right\| = \|y\|_Y.$$

Hence, by the *Inverse Mapping Theorem* (Theorem 6.14), the inverse operator

$$A^{-1} : X \to Y$$

is *bounded* and, for each $x \in X$,

$$x = \sum_{k=1}^{\infty} c_k e_k = Ay,$$

with some $y := \{c_n\}_{n=1}^{\infty} \in Y$ and any fixed $j \in \mathbb{N}$, by the *norm axioms*,

$$|c_j(x)| = |c_j| = \frac{\|c_j e_j\|}{\|e_j\|} = \frac{1}{\|e_j\|} \left\| \sum_{k=1}^{j} c_k e_k - \sum_{k=1}^{j-1} c_k e_k \right\|$$

$$\leq \frac{1}{\|e_j\|}\left[\left\|\sum_{k=1}^{j}c_ke_k\right\| + \left\|\sum_{k=1}^{j-1}c_ke_k\right\|\right] \leq \frac{1}{\|e_j\|}2\sup_{n\in\mathbb{N}}\left\|\sum_{k=1}^{n}c_ke_k\right\|$$

$$= \frac{2}{\|e_j\|}\|y\|_Y = \frac{2}{\|e_j\|}\|A^{-1}x\|_Y \leq \frac{2\|A^{-1}\|}{\|e_j\|}\|x\|,$$

which proves that each Schauder coordinate functional $c_j(\cdot), j \in \mathbb{N}$, is *bounded*, i. e., $c_j(\cdot) \in X^*, j \in \mathbb{N}$. ☐

Remark 6.22. A Schauder basis $E := \{e_n\}_{n\in\mathbb{N}}$ and the set of the Schauder coordinate functionals $\{c_n\}_{n\in\mathbb{N}}$ relative to E are *biorthogonal*, i. e.,

$$c_i(e_j) = \delta_{ij}, \; i, j \in \mathbb{N},$$

where δ_{ij} is the *Kronecker delta*, which, in particular, implies that the Schauder coordinate functionals are *linearly independent* (cf. the *Biorthogonal Sets Proposition* (Proposition 6.12), Section 6.7, Problem 2).

Exercise 6.52. Verify.

6.6.3 Closed Graph Theorem and Application

6.6.3.1 Closed Graph Theorem
The *Inverse Mapping Theorem* (Theorem 6.14) underlies the proof of the following important statement.

Theorem 6.16 (Closed Graph Theorem). *Let $(X, \|\cdot\|_X)$ and $(Y, \|\cdot\|_Y)$ be Banach spaces over $\mathbb{F}$ and $A : X \to Y$ be a closed linear operator. Then A is bounded ($A \in L(X, Y)$).*

Proof. By the *completeness* of the spaces $(X, \|\cdot\|_X)$ and $(Y, \|\cdot\|_Y)$, the product space $X \times Y$ is a *Banach space* relative to the norm

$$X \times Y \ni (x, y) \mapsto \|(x, y)\| := \|x\|_X + \|y\|_Y$$

(see Section 3.6, Problem 5 and Remark 5.12) and so is the *graph* G_A of A, being a *closed subspace* in $(X \times Y, \|\cdot\|)$ (Proposition 2.4).

The mapping

$$G_A \ni (x, Ax) \mapsto P(x, Ax) := x \in X$$

is a *bijective linear operator* from $(G_A, \|\cdot\|)$ onto $(X, \|\cdot\|_X)$.

Exercise 6.53. Explain.

The operator P is *bounded* since

$$\forall\, x \in X : \; \|P(x, Ax)\|_X = \|x\|_X \leq \|x\|_X + \|Ax\|_Y = \|(x, Ax)\|.$$

By the *Inverse Mapping Theorem* (Theorem 6.14), the *inverse operator*

$$X \ni x \rightarrow P^{-1}x = (x, Ax) \in G_A$$

is *bounded*, and hence,

$$\forall x \in X : \|Ax\|_Y \leq \|x\|_X + \|Ax\|_Y = \|(x, Ax)\| = \|P^{-1}x\| \leq \|P^{-1}\|\|x\|_X,$$

which implies the *boundedness* for A, completing the proof. ☐

Remarks 6.23.
- The condition of the *completeness* of the target space $(Y, \|\cdot\|_Y)$ in the *Closed Graph Theorem* (*CGT*), provided other conditions hold, is *not essential* since a closed linear operator $A : X \rightarrow Y$ remains closed when considered as a linear operator $A : X \rightarrow \tilde{Y}$, where $\tilde{Y}$ is a *completion* of Y (see Theorem 3.9).

  Exercise 6.54. Explain.

- The condition of the *completeness* of the domain space $(X, \|\cdot\|_X)$ in the *CGT* is essential and cannot be dropped. Indeed, as is known (see Examples 5.5), the differentiation operator

$$C^1[a, b] \ni x \rightarrow [Ax](t) := \frac{d}{dt}x(t) \in C[a, b]$$

  $(-\infty < a < b < \infty)$ is a *closed unbounded linear operator* from the *incomplete normed vector space* $(C^1[a, b], \|\cdot\|_\infty)$ onto the Banach space $(C[a, b], \|\cdot\|_\infty)$.
- As is shown below (see Section 6.6.4), the *Inverse Mapping Theorem* (Theorem 6.14) is equivalent to the *Closed Graph Theorem* (Theorem 6.16).
- Thus far, we have the following chain of implications

$$OMT \Rightarrow IMT \Rightarrow CGT$$

  (see Section 6.6.2.1).

6.6.3.2 Application: Projection Operators
We now study the important class of linear operators called *projections*.

Recall that each subspace Y in a vector space X has a *complementary subspace Z*:

$$X = Y \oplus Z,$$

i.e., every $x \in X$ allows a *unique decomposition*

$$x = y + z$$

with $y \in Y$ and $z \in Z$ (see Theorem 3.6 and Proposition 3.5).

Remarks 6.24.
-   The complementary subspaces Y and Z are necessarily *disjoint*, i. e.,

$$Y \cap Z = \{0\}$$

(see Definition 3.14).
-   Except when $Y = \{0\}$, the complementary subspace Z need not be unique (see Remark 3.14).

Exercise 6.55. In the space $C[-a, a]$ $(0 < a < \infty)$,
(a) show that the subspace

$$Y := \{y \in C[-a, a] \,|\, y(-t) = -y(t)\}$$

of all *odd* continuous on $[-a, a]$ functions and the subspace

$$Z := \{y \in C[-a, a] \,|\, y(-t) = y(t)\}$$

of all *even* continuous on $[-a, a]$ functions are *complementary*;
(b) for each $x \in C[-a, a]$, find the *unique decomposition*

$$x = y + z$$

with $y \in Y$ and $z \in Z$.

With every *decomposition* of a vector space X into a direct sum of complementary subspaces Y and Z, there is an associated is a linear operator called the *projection onto Y along Z*.

Definition 6.9 (Projection Operator on a Vector Space). Let Y and Z be complementary subspaces in a vector space X over $\mathbb{F}$. The linear operator P on X, defined as follows:

$$X \ni x = y + z, \; y \in Y, z \in Z \mapsto Px := y \in Y,$$

is called the *projection operator* (or *projection*) onto Y along Z.

Exercise 6.56. Verify that P is *well defined* and *linear*.

Example 6.5. For the direct product $X := \prod_{i \in I} X_i$ of a nonempty collection $\{X_i\}_{i \in I}$ of vector spaces (see Definition 3.10) and each $j \in I$, the linear operator

$$X \ni x = (x_i)_{i \in I} \mapsto P_j x := (\delta_{ij} x_i)_{i \in I},$$

where δ_{ij} is the *Kronecker delta*, is a *projection* onto the subspace

$$Y := \{x = (x_i)_{i \in I} \in X \mid x_i = 0, \, i \in I, i \neq j\}$$

along the subspace

$$Z := \{x = (x_i)_{i \in I} \in X \mid x_j = 0\}.$$

Proposition 6.9 (Properties of Projection Operators). *Let Y and Z be complementary subspaces in a vector space X. The projection P onto Y along Z has the following properties:*

1. $Px = x \Leftrightarrow x \in Y$;
2. $Px = 0 \Leftrightarrow x \in Z$;
3. $P^2 = P$, *i. e., the operator P is idempotent.*

Exercise 6.57. Prove.

We immediately obtain the following corollary.

Corollary 6.12 (Properties of Projection Operators). *Let Y and Z be complementary subspaces in a vector space X. For the projection P onto Y along Z,*

(1) $R(P) = Y$;
(2) $\ker P = Z$;
(3) $I - P$ *(I is the identity operator on X) is the projection onto Z along Y, with*

$$R(I - P) = Z = \ker P \text{ and } \ker(I - P) = Y = R(P).$$

Exercise 6.58.

(a) Prove.
(b) Prove that, on a vector space X, P is a projection operator *iff* $I - P$ is a projection operator.

Remarks 6.25.

– Hence, projection operators on a vector space X occur in complementary pairs, P, $I - P$, adding up to the identity operator I.
– There are always at least two complementary projections on a vector space X: the *zero operator* 0 and the *identity operator* I.

Thus, a projection operator on a vector space X is an *idempotent linear operator* (see Proposition 6.9). The converse is true as well.

Proposition 6.10 (Characterization of Projections on a Vector Space). *A linear operator P on a vector space X is a projection iff P is idempotent, in which case*

$$X = Y \oplus Z$$

with $Y = R(P)$ and $Z = \ker P$.

Exercise 6.59. Prove.

Remark 6.26. Thus, one can define a projection operator P on a vector space X as an idempotent linear operator on X. To define a projection operator P on a normed vector space $(X, \| \cdot \|)$, we add the boundedness condition.

Definition 6.10 (Projection Operator on a Normed Vector Space). Let $(X, \| \cdot \|)$ be a normed vector space. A *projection operator* (or *projection*) P on $(X, \|\cdot\|)$ is an *idempotent bounded linear operator* on $(X, \| \cdot \|)$ (i. e., $P^2 = P$ and $P \in L(X)$).

Exercise 6.60. Prove that, on a normed vector space $(X, \|\cdot\|)$, P is a projection operator *iff* $I - P$ is a projection operator.

Remarks 6.27.
- Hence, projection operators on a normed vector space $(X, \| \cdot \|)$ occur in complementary pairs, $P, I - P$, adding up to the identity operator I.
- There are always at least two complementary projection on a normed vector space $(X, \| \cdot \|)$: the *zero operator* 0 and the *identity operator* I.

Examples 6.6.
1. For a closed subspace Y in a Hilbert space $(X, (\cdot, \cdot), \| \cdot \|)$, the operator

$$X \ni x \mapsto Px := y \in Y,$$

where y is the *orthogonal projection* of x on Y, i. e., the nearest point to x in Y (see the *Projection Theorem* 4.6), is a projection operator called the *orthogonal projection operator onto* Y (cf. Section 6.7, Problems 15 and 16), with $I - P$ being the orthogonal projection onto $Y^{\perp}$.
2. On l_p $(1 \le l \le \infty)$,

$$P(x_n)_{n \in \mathbb{N}} := (x_1, 0, x_3, 0, \dots), \ (x_n)_{n \in \mathbb{N}} \in l_p,$$

and

$$(I - P)(x_n)_{n \in \mathbb{N}} := (0, x_2, 0, x_4, 0, \dots), \ (x_n)_{n \in \mathbb{N}} \in l_p,$$

is a complementary pair of *projection operators* (*orthogonal* ones for $p = 2$).

Exercise 6.61. Verify.

Proposition 6.11 (Norm of a Projection Operator). *For a nontrivial projection* $P \ne 0$ *on a normed vector space* $(X, \| \cdot \|)$,

$$\|P\| \ge 1.$$

Exercise 6.62. Prove (cf. the *Characterization of Orthogonal Projections* (Proposition 6.19) Section 6.7, Problem 16).

Theorem 6.17 (Projections on a Normed Vector Space).
1. *If* $(X, \|\cdot\|)$ *is a normed vector space and* P *is a projection operator on* X, *then* $Y := R(P)$ *and* $Z := \ker P$ *are closed complementary subspaces:*

$$X = Y \oplus Z.$$

2. *Conversely, if $(X, \| \cdot \|)$ is a Banach space, and Y and Z are closed complementary subspaces:*

$$X = Y \oplus Z, \tag{6.30}$$

then the projection P onto Y along Z in the vector space sense is a projection in the normed vector space sense, i. e., $P \in L(X)$.

Proof.
1. The proof of this part immediately follows from the facts that

$$Z := \ker P \quad \text{and} \quad Y := R(P) = \ker(I - P)$$

and that the kernel an operator in $L(X)$ is a *closed subspace* in $(X, \| \cdot \|)$ (Proposition 5.2).
2. To prove this part, let us to show that the *projection operator* $P : X \to X$ onto Y along Z is a *closed* linear operator.
   Indeed, let $\{x_n\}_{n=1}^{\infty}$ be an arbitrary sequence in $(X, \| \cdot \|)$ such that

$$\lim_{n \to \infty} x_n = x \in X \text{ and } \lim_{n \to \infty} Px_n = y \in X.$$

Then, by the *closedness* of Y in $(X, \| \cdot \|)$, $y \in Y$ and, since

$$x_n = Px_n + (I - P)x_n, \; n \in \mathbb{N}, \tag{6.31}$$

in view of direct sum decomposition (6.30) and the *closedness* of Z in $(X, \| \cdot \|)$,

$$Z \ni (I - P)x_n = x_n - Px_n \to x - y =: z \in Z, \; n \to \infty.$$

Hence, passing to the limit in (6.31) as $n \to \infty$, we arrive at

$$x = y + z$$

with $y \in Y$ and $z \in Z$, which implies that $y = Px$.
By the *Sequential Characterization of Closed Linear Operators* (Proposition 5.3), we infer that the operator P is *closed*, which, considering that $(X, \| \cdot \|)$ is a Banach space, by the *Closed Graph Theorem* (Theorem 6.16), implies that $P \in L(X)$. □

Remarks 6.28.
– The condition of the *completeness* of the space $(X, \| \cdot \|)$ in part 2 of the prior theorem, provided other conditions hold, is *not essential* since, the space X allowing a direct sum decomposition

$$X = Y \oplus Z,$$

where Y and Z are closed complementary subspaces in $(X, \|\cdot\|)$ and P is a projection operator onto Y along Z, we have:

$$\tilde{X} = \tilde{Y} \oplus \tilde{Z},$$

where $\tilde{X}$, $\tilde{Y}$, and $\tilde{Z}$ are *completions* of X, Y, and Z (see Theorem 3.9), respectively. And hence, the operator P is *bounded* on X, being a restriction to X of the *bounded* on $\tilde{X}$ by the prior theorem projection operator $\tilde{P}$ onto $\tilde{Y}$ along $\tilde{Z}$.

Exercise 6.63. Explain.

- Thus, as is the case for projections on vector spaces, every *decomposition* of a normed vector space $(X, \|\cdot\|)$ into a direct sum of complementary closed subspaces Y and Z generates a *projection operator* on $(X, \|\cdot\|)$ and vice versa.

6.6.4 Equivalence of *OMT*, *IMT*, and *CGT*

Here, we prove the equivalence of the *Open Mapping Theorem* (*OMT*), the *Inverse Mapping Theorem* (*IMT*), and the *Closed Graph Theorem* (*CGT*).

Theorem 6.18 (Equivalence Theorem). *The Open Mapping Theorem (OMT), the Inverse Mapping Theorem (IMT), and the Closed Graph Theorem (CGT) are equivalent statements.*

Proof. Let us prove the following closed chain of implications:

$$OMT \Rightarrow IMT \Rightarrow CGT \Rightarrow OMT.$$

Observe that, we already have

$$OMT \Rightarrow IMT \Rightarrow CGT$$

(see Remarks 6.23), and hence, it remains to prove the last implication in the above chain:

$$CGT \Rightarrow OMT.$$

Suppose that *CGT* holds and let

$$A : (X, \|\cdot\|_X) \to (Y, \|\cdot\|_Y)$$

be a surjective bounded linear operator from X onto Y, where $(X, \|\cdot\|_X)$ and $(Y, \|\cdot\|_Y)$ are Banach spaces.

By the *boundedness* of the operator A, $\ker A$ is a *closed subspace* in $(X, \|\cdot\|_X)$ (Proposition 5.2). Hence, the quotient space $(X/\ker A, \|\cdot\|)$ with

$$X/\ker A \ni [x] := x + \ker A \mapsto \|[x]\| = \inf_{y \in \ker A} \|x + y\| = \rho(x, \ker A)$$

is a well-defined *Banach space* (the *Quotient Space Norm Theorem* (Theorem 3.19), Section 3.6, Problem 10) and the *bijective* linear operator

$$(X/\ker A, \|\cdot\|) \ni [x] := x + \ker A \mapsto \hat{A}[x] := Ax \in (Y, \|\cdot\|_Y)$$

is *bounded* (see Section 5.4, Problem 10).

By the *Characterization of Closedness for Bounded Linear Operators Proposition* (Proposition 5.4), the operator $\hat{A}$ is *closed*, and hence, by the *Closedness of Inverse Operator Proposition* (Proposition 5.14) (Section 5.4, Problem 9), so is the inverse operator

$$\hat{A}^{-1} : (Y, \| \cdot \|_Y) \to (X/\ker A, \| \cdot \|),$$

which, by the *Closed Graph Theorem* (Theorem 6.16), implies that the operator $\hat{A}^{-1}$ is *bounded*, and hence, *Lipschitz continuous* on $(Y, \| \cdot \|_Y)$ (Theorem 5.4).

For each *open set* G in $(X, \| \cdot \|_X)$, the image $A(G)$ is *open* in $(Y, \| \cdot \|_Y)$ since

$$A(G) = \left(\hat{A}^{-1}\right)^{-1}(T(G)),$$

where $T : X \to X/\ker A$ is the *canonical homomorphism* (see Definition 3.13), under which the image $T(G)$ is *open* in the quotient space $(X/\ker A, \| \cdot \|_1)$.

Exercise 6.64. Verify the *openness* of $T(G)$ in $(X/\ker A, \| \cdot \|_1)$.

Therefore, $A : X \to Y$ is an *open mapping*, which completes the proof. $\qquad\square$

6.7 Problems

1. For the bounded linear functional $f(x) := x_1 + x_2$ defined on the subspace

$$D(f) := \left\{ x = (x_1, x_2) \in \mathbb{R}^2 \,\middle|\, x_1 - x_2 = 0 \right\},$$

of the Euclidean space $l_2^{(2)}(\mathbb{R})$, describe all bounded linear extensions of f to the entire space $l_2^{(2)}(\mathbb{R})$.

2. Prove

    Proposition 6.12 (Biorthogonal Sets). *For every finite set* $\{x_1, \dots, x_n\}$ ($n \in \mathbb{N}$) *of linearly independent elements in a normed vector space* $(X, \| \cdot \|)$, *there exists a set* $\{f_1, \dots, f_n\} \subset X^*$ *of linearly independent bounded linear functionals such that*

    $$f_i(x_j) = \delta_{ij}, \ i, j = 1, \dots, n,$$

    where δ_{ij} *is the Kronecker delta.*
    The sets $\{x_1, \dots, x_n\}$ *and* $\{f_1, \dots, f_n\}$ *are called* biorthogonal *to each other.*

    Hint. Apply the *Separation of Element from Closed Subspace Corollary* (Corollary 6.1).

    Give an example showing that the statement cannot be stretched to biorthogonal countably infinite sets of linearly independent elements $\{x_k\}_{k \in \mathbb{N}} \subset X$ and functionals $\{f_k\}_{k \in \mathbb{N}} \subset X^*$.

3. Prove

Proposition 6.13 (Closed Subspace is Weakly Sequentially Closed). *Let Y be a closed subspace in a normed vector space $(X, \| \cdot \|)$; then Y is weakly sequentially closed, i. e., Y contains the limits of all its weakly convergent sequences.*

4. * Prove

Proposition 6.14 (Weakly* Convergent Subsequence). *Let $(X, \| \cdot \|)$ be a separable normed vector space. Then every bounded sequence $\{f_n\}_{n=1}^{\infty}$ in X^* contains a weakly* convergent subsequence.*

Hint. Develop a *"diagonal subsequence"* argument similar to that in the proof of the *"if"* part of the *Arzelà–Ascoli Theorem* (Theorem 2.47).

5. Prove

Proposition 6.15 (Continuity of a Bilinear Functional). *Let $(X_1, \| \cdot \|_1)$ and $(X_2, \| \cdot \|_2)$ be Banach spaces and $X = X_1 \times X_2$ be their Cartesian product, which is also a Banach space relative to the product norm*

$$X \ni (x,y) \mapsto \|(x,y)\|_{X_1 \times X_2} = \sqrt{\|x\|_1^2 + \|y\|_2^2}$$

(see Section 3.6, Problem 5), and $B(\cdot, \cdot) : X_1 \times X_2 \to \mathbb{F}$ be a bilinear functional continuous relative to each argument. Then $B(\cdot, \cdot)$ is jointly continuous on $X_1 \times X_2$, i. e.,

$$\forall (x,y) \in X \ \forall \{(x_n, y_n)\}_{n=1}^{\infty} \subseteq X, \ (x_n, y_n) \to (x,y), \ n \to \infty, \ in \ X :$$
$$B(x_n, y_n) \to B(x,y), \ n \to \infty,$$

Hint. Apply the *Uniform Boundedness Principle* (Theorem 6.7).

6. Determine the *Cesàro limit* of the *divergent* sequence $\{(-1)^n\}_{n=1}^{\infty}$.
7. Show that

$$\sum_{n=0}^{\infty} e^{int} = \frac{1}{1 - e^{it}} (C, 1) \quad (t \neq 2\pi k, \ k \in \mathbb{Z}).$$

In particular, for $t = \pi$, we have the *divergent* series $\sum_{n=0}^{\infty}(-1)^n$, for which

$$\sum_{n=0}^{\infty}(-1)^n = \frac{1}{2}(C,1).$$

8. Using the *regularity* of the *Cesàro method*, show that, for any sequence of positive numbers $\{x_n\}_{n=1}^{\infty}$ convergent to a positive number,

$$\lim_{n \to \infty} \sqrt[n]{\prod_{k=1}^{n} x_k} = \lim_{n \to \infty} x_n.$$

9. Let $(X, \| \cdot \|_X)$ be a Banach space, $(Y, \| \cdot \|_Y)$ be a normed vector space over $\mathbb{F}$ and $A : X \to Y$ be a surjective bounded linear operator from X onto Y (i. e., $A \in L(X, Y)$ and $R(A) = Y$). Prove that, if A is an open mapping, then $(Y, \| \cdot \|_Y)$ is a Banach space.

    Hint. Show that, if A is an open mapping, the *bijective* bounded linear operator

    $$(X / \ker A, \| \cdot \|) \ni [x] := x + \ker A \mapsto \hat{A}[x] := Ax \in (Y, \| \cdot \|_Y),$$

    where $\| \cdot \|$ is the *quotient-space norm* (see Section 5.4, Problem 10) has a bounded inverse $\hat{A}^{-1} \in L(Y, X / \ker A)$, which, by the *Completeness of the Range Proposition* (Proposition 5.13) (see Section 5.4, Problem 8), makes $(Y, \| \cdot \|_Y)$ to be a Banach space.

10. Let $(X, \| \cdot \|_X)$ be a Banach space, $(Y, \| \cdot \|_Y)$ be a normed vector space over $\mathbb{F}$ and $A \in L(X, Y)$. Show that either $R(A) = Y$, or $R(A)$ is of the *first category* in $(Y, \| \cdot \|_Y)$.

    Hint. Use some ideas of the proof of the *Open Mapping Theorem* (Theorem 6.13).

11. Prove

    Proposition 6.16 (Sufficient Condition of Boundedness). *Let $(X, \| \cdot \|_X)$ and $(Y, \| \cdot \|_Y)$ be Banach spaces over $\mathbb{F}$ and $A : X \to Y$ be a surjective linear operator from X onto Y such that*

    $$\exists c > 0 \, \forall x \in X : \, \|Ax\|_Y \geq c\|x\|_X.$$

    Then A is bounded, i. e., $A \in L(X, Y)$.

    Hint. Apply the *Closedness of Inverse Operator Proposition* (Proposition 5.14) (Section 5.4, Problem 9) and the *Closed Graph Theorem* (Theorem 6.16).

12. Let Y and Z be closed disjoint subspaces in a Banach space $(X, \| \cdot \|)$. Prove that the subspace $Y \oplus Z$ is closed in $(X, \| \cdot \|)$ *iff*

    $$\exists c > 0 \, \forall y \in Y, \, \forall z \in Z : \, c \left[\|y\| + \|z\|\right] \leq \|y + z\|.$$

    Hint. Apply the *Equivalence of Banach Norms Theorem* (Theorem 6.15).

13. Prove

    Proposition 6.17 (Another Sufficient Condition of Boundedness). *Let $(X, \| \cdot \|_X)$ and $(Y, \| \cdot \|_Y)$ be Banach spaces over $\mathbb{F}$ and $A : X \to Y$ be a linear operator. If, for any sequence $\{x_n\}_{n=1}^{\infty} \subset X$ such that*

    $$\lim_{n \to \infty} x_n = 0 \text{ in } (X, \| \cdot \|_X) \quad \text{and} \quad \lim_{n \to \infty} Ax_n = y \text{ in } (Y, \| \cdot \|_Y),$$

    the following is true:

    $$y = 0,$$

    then the operator A is bounded, i. e., $A \in L(X, Y)$.

Hint. Use the *Sequential Characterization of Closed Linear Operators* (Proposition 5.3) to show that A is closed and apply the *Closed Graph Theorem* (Theorem 6.16).

14. Prove

Theorem 6.19 (Hellinger–Toeplitz Theorem). *Let* $(X, (\cdot, \cdot), \| \cdot \|)$ *be a Hilbert space and* $A : X \to X$ *be a linear operator. If A is self-adjoint, i. e.,*[4]

$$\forall x, y \in X : (Ax, y) = (x, Ay),$$

then $A \in L(X)$.

Hint. Use the *Sequential Characterization of Closed Linear Operators* (Proposition 5.3) and the *Inner Product Separation Property* (Proposition 4.14) to show that A is closed and apply the *Closed Graph Theorem* (Theorem 6.16).

15. Prove

Proposition 6.18 (Characterization of Orthogonal Projections). *A projection operator P on a Hilbert space* $(X, (\cdot, \cdot), \| \cdot \|)$ *is an orthogonal projection iff P is self-adjoint, i. e.,*

$$\forall x, y \in X : (Px, y) = (x, Py).$$

16. Prove

Proposition 6.19 (Characterization of Orthogonal Projections). *A nontrivial projection operator P on a Hilbert space* $(X, (\cdot, \cdot), \| \cdot \|)$ *is an orthogonal projection iff*

$$\|P\| = 1.$$

Hint. To prove the *"if"* part, reason *by contrapositive* using the *Characterization of the Orthogonal Complement of a Subspace* (Proposition 4.6).

---

[4] Ernst David Hellinger (1883–1950).

7 Duality and Reflexivity

In this chapter, we discuss *duality* and *reflexivity* and consider representation theorems describing all bounded linear functionals on certain normed vector spaces, hence, unveiling the structure of their duals and enabling us to better understand the nature of weak convergence.

7.1 Self-Duality of Hilbert Spaces

We start our discourse with Hilbert spaces discussed in Chapter 4. The following celebrated representation theorem (due to Frigyes Riesz) describes all bounded linear functionals on such spaces and shows that they are, so to speak, *self-dual*.

7.1.1 Riesz Representation Theorem

Theorem 7.1 (Riesz Representation Theorem). *Let* $(X, (\cdot, \cdot), \|\cdot\|)$ *be a Hilbert space. Then*

$$\forall f \in X^* \; \exists! \, y_f \in X \, \forall x \in X : f(x) = (x, y_f).$$

The mapping

$$X^* \ni f \mapsto y_f \in X$$

is an isometric linear (if the space is real) or conjugate linear (if the space is complex) isomorphism between X^* *and* X, *in which sense we can write* $X^* = X$, *and thus, regard* X *to be self-dual.*

Proof. Consider an arbitrary $f \in X^*$.

If $f = 0$, $y_f = 0$. Indeed,

$$\forall x \in X : f(x) = 0 = (x, 0) \tag{7.1}$$

and, by the *Orthogonal Characterization of the Zero Vector* (Proposition 4.4), $y_f = 0$ is the only vector in X satisfying (7.1). Observe that, in this case, $\|f\| = \|y_f\| = 0$.

Exercise 7.1. Verify.

If $f \neq 0$,

$$Y := \ker f := \{x \in X \mid f(x) = 0\}$$

is a *closed proper subspace (hyperplane)* in X (see Propositions 5.1 and 5.10) and, by the *Projection Theorem* (Theorem 4.6),

$$X = Y \oplus Y^\perp$$

with $Y^\perp \neq \{0\}$.

https://doi.org/10.1515/9783110614039-007

Choosing a *nonzero* vector $z \in Y^{\perp}$, without loss of generality, we can regard that $f(z) = 1$.

Exercise 7.2. Explain.

Then, for any $x \in X$,

$$x = [x - f(x)z] + f(x)z$$

and, since

$$f(x - f(x)z) = f(x) - f(x)f(z) = f(x) - f(x) = 0,$$

i. e., $x - f(x)z \in Y$, we infer that

$$\forall x \in X : \; x - f(x)z \perp z, \; \text{i. e.,} \; (x - f(x)z, z) = 0,$$

which, by *linearity* of inner product in the first argument, implies that

$$\forall x \in X : \; (x, z) = (f(x)z, z) = f(x)\|z\|^2.$$

Let

$$y_f := \frac{1}{\|z\|^2} z \in X \setminus \{0\},$$

then, by *linearity/conjugate linearity* of inner product in the second argument,

$$\forall x \in X : \; (x, y_f) = \frac{1}{\|z\|^2}(x, z) = f(x),$$

which proves the *existence part*.

The *uniqueness* immediately follows by the *Inner Product Separation Property* (Proposition 4.14).

Exercise 7.3. Explain.

By the *Cauchy–Schwarz Inequality* (Theorem 4.2), for any $f \in X^* \setminus \{0\}$ with the corresponding $y_f \in X \setminus \{0\}$,

$$\forall x \in X : \; |f(x)| = |(x, y_f)| \le \|y_f\|\|x\|,$$

which shows that

$$\|f\| \le \|y_f\|.$$

Since further, for the *unit vector* $x := \frac{1}{\|y_f\|}y_f$ ($\|x\| = 1$),

$$\|f\| \ge |f(x)| = \frac{1}{\|y_f\|}|f(y_f)| = \frac{1}{\|y_f\|}(y_f, y_f) = \|y_f\|,$$

we conclude that

$$\forall f \in X^* \setminus \{0\} : \; \|f\| = \|y_f\|,$$

the equality being trivially true for the pair $f = 0$, $y_f = 0$ as well.

As follows from the *linearity/conjugate linearity* of inner product in the second argument, the mapping

$$X^* \ni f \mapsto y_f \in X \tag{7.2}$$

is *linear* when X is real and *conjugate linear* when X is complex.

Exercise 7.4. Verify.

It remains to be shown that the mapping given by (7.2) is *surjective* (i.e., *onto*). Indeed, for each $y \in X$, by *linearity* of inner product in the first argument and the *Cauchy–Schwarz Inequality* (Theorem 4.2),

$$f(x) := (x, y), \ x \in X,$$

is a *bounded linear functional* on X, and hence, $y = y_f$.

Thus, the mapping given by (7.2) is an isometric linear (if the space is real) or conjugate linear (if the space is complex) isomorphism between X^* and X, in which sense we can write $X^* = X$, and thus, regard X to be *self-dual*. □

Remarks 7.1.

- The *Riesz Representation Theorem* is also known as the *Fréchet[1]–Riesz Theorem*.
- By the *Riesz Representation Theorem*, the dual X^* of a Hilbert space $(X, (\cdot, \cdot), \| \cdot \|)$ is also a Hilbert space relative to the inner product

$$X^* \ni f, g \mapsto (f, g)_{X^*} := \overline{(y_f, y_g)} = (y_g, y_f), \ f, g \in X^*,$$

where

$$f(x) = (x, y_f) \text{ and } g(x) = (x, y_g), \ x \in X.$$

- By the *Riesz Representation Theorem*, each bounded linear (if the space is real) or conjugate linear (if the space is complex) functional g on a Hilbert space $(X, (\cdot, \cdot), \| \cdot \|)$ is of the form:

$$g(x) = (y_g, x)$$

with some *unique* $y_g \in X$, and the mapping

$$g \mapsto y_g$$

is an isometric linear isomorphism between the space of bounded linear/conjugate linear functionals on X and X.
- The condition of the *completeness* of the space in the *Riesz Representation Theorem* is essential and cannot be dropped.
Indeed, on $(c_{00}, (\cdot, \cdot), \| \cdot \|_2)$, which is an incomplete inner product space, when treated as a subspace of the Hilbert space l_2 (see Examples 4.3),

---

1 Maurice Fréchet (1878–1973).

$$c_{00} \ni x := (x_n)_{n \in \mathbb{N}} \mapsto f(x) := \sum_{n=1}^{\infty} \frac{x_n}{n}$$

is a bounded linear functional, being a restriction of such on l_2 (see the *Linear Bounded Functionals on Certain Hilbert Spaces Corollary* (Corollary 7.1)), but

$$\nexists y_f \in c_{00} : f(x) = (x, y_f)$$

(cf. Example 4.5).
– As is seen in Remarks 7.3, a normed vector space need not be Hilbert to be self-dual.

Exercise 7.5. Verify.

7.1.2 Linear Bounded Functionals on Certain Hilbert Spaces

The following description is an immediate corollary of the *Riesz Representation Theorem* (Theorem 7.1) (see Examples 4.3).

Corollary 7.1 (Linear Bounded Functionals on Certain Hilbert Spaces).
1. *For each $f \in l_2^{(n)*}$ ($n \in \mathbb{N}$), there is a unique element $y_f := (y_1, \dots, y_n) \in l_2^{(n)}$ such that*

$$f(x) = \sum_{k=1}^{n} x_k y_k, \ x := (x_1, \dots, x_n) \in l_2^{(n)},$$

    and $\|f\| = \|y_f\|_2 = [\sum_{k=1}^{n} |y_k|^2]^{1/2}$.
2. *For each $f \in l_2^*$, there is a unique element $y_f := (y_n)_{n \in \mathbb{N}} \in l_2$ such that*

$$f(x) = \sum_{k=1}^{\infty} x_k y_k, \ x := (x_n)_{n \in \mathbb{N}} \in l_2,$$

    and $\|f\| = \|y_f\|_2 = [\sum_{k=1}^{\infty} |y_k|^2]^{1/2}$.
3. *For each $f \in L_2^*(a, b)$ ($-\infty \le a < b \le \infty$) (see Examples 4.3), there is a unique element $y_f \in L_2(a, b)$ such that*

$$f(x) = \int_a^b x(t) y_f(t) \, dt, \ x \in L_2(a, b),$$

    and $\|f\| = \|y_f\|_2 = [\int_a^b |y_f(t)|^2 \, dt]^{1/2}$.

7.1.3 Weak Convergence in Hilbert Spaces

By the *Riesz Representation Theorem* (Theorem 7.1), we immediately obtain the following description of weak convergence in a Hilbert space.

Proposition 7.1 (Characterization of Weak Convergence in Hilbert Spaces).   *A se-quence of elements $\{x_n\}_{n=1}^{\infty}$ in a Hilbert space $(X, (\cdot, \cdot), \|\cdot\|)$ weakly converges to an element $x \in X$ iff*

$$\forall y \in X : (x_n, y) \to (x, y), \ n \to \infty.$$

Remark 7.2. The example given in Remark 6.8 shows that a weakly convergent sequence in a Hilbert space need not (strongly) converge. More generally, by the *Weak Convergence of an Orthonormal Sequence Proposition* (Proposition 7.6) (see Section 7.6, Problem 1), an *orthonormal sequence* in a Hilbert space weakly converges to 0, but does not (strongly) converge.

By the prior proposition along with the *Characterization of Convergence in Pre-Hilbert Spaces* (Proposition 4.15) (see Section 4.13, Problem 3), we obtain the following description of (strong) convergence in a Hilbert space.

Proposition 7.2 (Characterization of Convergence in Hilbert Spaces). *For a sequence $\{x_n\}_{n=1}^{\infty}$ in a Hilbert space $(X, (\cdot, \cdot), \|\cdot\|)$,*

$$x_n \to x \in X, \ n \to \infty, \ in \ (X, (\cdot, \cdot), \|\cdot\|)$$

iff
(1) $x_n \xrightarrow{w} x, \ n \to \infty$, *and*
(2) $\|x_n\| \to \|x\|, \ n \to \infty$.

7.2 Duality of Finite-Dimensional Spaces

We proceed with the important case of finite-dimensional normed vector spaces.

7.2.1 Representation Theorem

Theorem 7.2 (Representation Theorem for Duals of Finite-Dimensional Spaces). *Let $(X, \|\cdot\|)$ be normed vector space over $\mathbb{F}$ with $\dim X = n$ $(n \in \mathbb{N})$ and a basis $B :=\{x_1, \ldots, x_n\}$. Then*

$$\forall f \in X^* \ \exists! \ y_f := (y_1, \ldots, y_n) \in \mathbb{F}^n \ \forall x := \sum_{k=1}^{n} c_k x_k \in X$$

with some $c_k \in \mathbb{F}, \ k = 1, \ldots, n$:

$$f(x) = \sum_{k=1}^{n} c_k y_k. \tag{7.3}$$

The mapping

$$X^* \ni f \mapsto y_f \in \mathbb{F}^n \tag{7.4}$$

is an isomorphism between X^* and $\mathbb{F}^n$, which, if X is a equipped with the p-norm:

$$X \ni x = \sum_{k=1}^{n} c_k x_k = x \mapsto \|x\|_p := \|(c_1, \dots, c_n)\|_p \quad (1 \le p \le \infty), \tag{7.5}$$

and $\|x_k\|_p = 1, k = 1, \dots, n$, is an isometric isomorphism between X^* and $l_q^{(n)}(\mathbb{F})$, where q is the conjugate index to p ($1/p + 1/q = 1$), i. e.,

$$\forall f \in X^* : \|f\| = \|y_f\|_q. \tag{7.6}$$

In this sense, we can write $l_p^{(n)^*} = l_q^{(n)}$.

Proof. Let $1 \le p \le \infty$ be arbitrary and q be the *conjugate index* to p ($1/p + 1/q = 1$).

The fact that, relative to the basis $B := \{x_1, \dots, x_n\}$, (7.3) represents an arbitrary linear functional f on X, all linear functionals on $(X, \|\cdot\|)$ being automatically *bounded* (see Section 5.4, Problem 3), is easily verified. And so is the fact that the mapping given by (7.4) is an *isomorphism* between X^* and $\mathbb{F}^n$.

Exercise 7.6. Verify.

Suppose that X is equipped with the p-norm given by (7.5) and

$$\|x_k\|_p = 1, \; k = 1, \dots, n. \tag{7.7}$$

Observe that (7.6) is, obviously, true for when $f = 0$ and, respectively, $y_f = 0$. Henceforth, let us regard that $f \ne 0$ and $y_f \ne 0$.

As immediately follows from *Hölder's Inequality for n-Tuples* (Theorem 2.2),

$$\forall x = \sum_{k=1}^{n} c_k x_k \in X : |f(x)| \le \sum_{k=1}^{n} |c_k y_k| \le \|x\|_p \|y_f\|_q,$$

and hence,

$$\forall f \in X^* : \|f\| \le \|y_f\|_q. \tag{7.8}$$

It remains to be shown that

$$\forall f \in X^* : \|f\| \ge \|y_f\|_q, \tag{7.9}$$

Indeed, since

$$y_k = f(x_k), \; k = 1, \dots, n,$$

in view of (7.7), we have:

$$|y_k| = |f(x_k)| \le \|f\| \|x_k\|_p = \|f\|, \; k = 1, \dots, n.$$

Whence, we infer that

$$\|y_f\|_\infty := \max_{1 \le k \le n} |y_k| \le \|f\|,$$

which, along with (7.8) for $p = 1$ and $q = \infty$, proves (7.6) for the conjugate pair $p = 1$, $q = \infty$.

If $1 < p \le \infty$, for an arbitrary *nonzero* $f \in X^*$ with the corresponding $y_f :=$ $(y_1, \ldots, y_n) \ne 0$, we can choose the element

$$x := \sum_{k=1}^{n} c_k x_k$$

in $(X, \| \cdot \|_p)$ with

$$c_k := \begin{cases} |y_k|^{q-1} \frac{\overline{y_k}}{|y_k|} & \text{for } k = 1, \ldots, n \text{ with } y_k \ne 0, \\ 0 & \text{for } k = 1, \ldots, n \text{ with } y_k = 0. \end{cases}$$

Then, for the conjugate pair $p = \infty$, $q = 1$,

$$\forall x = \sum_{k=1}^{n} c_k x_k \in X : f(x) = f\left(\sum_{k=1}^{n} c_k x_k\right) = \sum_{k=1}^{n} c_k y_k = \sum_{k=1}^{n} |y_k| = \|y_f\|_1.$$

Hence, since $|c_k| = \frac{|y_k|}{|y_k|} = 1$ for $k = 1, \ldots, n$ with $y_k \ne 0$,

$$\|y_f\|_1 = \sum_{k=1}^{n} |y_k| = |f(x)| \le \|f\| \|x\|_\infty = \|f\| \max_{1 \le k \le n} |c_k| = \|f\|,$$

which, along with (7.8) for $p = \infty$ and $q = 1$, proves (7.6) for the conjugate pair $p = \infty$, $q = 1$.

It remains for us to consider the case of $1 < p, q < \infty$. In view of

$$\overline{y_k} y_k = |y_k|^2, \ k = 1, \ldots, n,$$

we have:

$$f(x) = f\left(\sum_{k=1}^{n} c_k x_k\right) = \sum_{k=1}^{n} c_k y_k = \sum_{k=1}^{n} |y_k|^{q-1} |y_k| = \sum_{k=1}^{n} |y_k|^q = \|y_f\|_q^q.$$

Hence, considering that $(q - 1)p = q$ (see Remark 2.5),

$$\sum_{k=1}^{n} |y_k|^q = |f(x)| \le \|f\| \|x\|_p = \|f\| \left[\sum_{k=1}^{n} |y_k|^{(q-1)p}\right]^{1/p} = \|f\| \left[\sum_{k=1}^{n} |y_k|^q\right]^{1/p}.$$

Whence, dividing through by

$$\left[\sum_{k=1}^{n} |y_k|^q\right]^{1/p} > 0,$$

in view of $1 - \frac{1}{p} = \frac{1}{q}$, we arrive at

$$\|y_f\|_q := \left[\sum_{k=1}^{n} |y_k|^q\right]^{1/q} \le \|f\|.$$

Therefore (7.9) holds for any $f \in X^*$, which along with (7.8) proves (7.6) for any conjugate pair $1 < p, q < \infty$, completing the proof of the fact that the *isomorphism* described by (7.4) is *isometric*.

In this sense, we can write $l_p^{(n)*} = l_q^{(n)}$. $\qquad\qquad\qquad\qquad\qquad\qquad\qquad\square$

Remarks 7.3.
- For $p = 2$, we have: $l_2^{(n)*} = l_2^{(n)}$, which is consistent with the *Linear Bounded Functionals on Certain Hilbert Spaces Corollary* (Corollary 7.1).
- For $\mathbb{F} = \mathbb{R}$, $n = 2$, and $p = 1$, by the prior theorem, $l_1^{(2)*}(\mathbb{R}) = l_\infty^{(2)}(\mathbb{R})$. The mapping

$$l_1^{(2)}(\mathbb{R}) \ni (x_1, x_2) \mapsto (x_1 + x_2, x_1 - x_2) \in l_\infty^{(2)}(\mathbb{R})$$

is an *isometric isomorphism*, which makes the space $l_1^{(2)}(\mathbb{R})$ to be *isometrically isomorphic* to $l_\infty^{(2)}(\mathbb{R})$, and hence, to its dual $l_1^{(2)*}(\mathbb{R})$. Thus, the real space $l_1^{(2)}$ is *self-dual* without being a Hilbert space (see Exercise 4.9, cf. Remarks 7.1).

Exercise 7.7. Verify.

7.2.2 Weak Convergence in Finite-Dimensional Spaces

The prior theorem underlies the following profound fact.

Proposition 7.3 (Characterization of Weak Convergence in Finite-Dimensional Spaces). *A sequence of elements $\{x_n\}_{n=1}^\infty$ in a finite-dimensional space $(X, \|\cdot\|)$ weakly converges to an element $x \in X$ iff it (strongly) converges to x:*

$$x_n \xrightarrow{w} x,\ n \to \infty \Leftrightarrow x_n \to x,\ n \to \infty.$$

Exercise 7.8. Prove.

7.3 Duality of Sequence Spaces

Let us now consider certain sequence spaces.

7.3.1 Representation Theorem for l_p^*

Theorem 7.3 (Representation Theorem for l_p^* $(1 \le p < \infty)$). *Let $1 \le p < \infty$ and q be its conjugate $(1/p + 1/q = 1)$. Then*

$$\forall f \in l_p^*\ \exists! y_f := \{y_k\}_{k=1}^\infty \in l_q\ \forall x := \{x_k\}_{k=1}^\infty \in l_p : f(x) = \sum_{k=1}^\infty x_k y_k.$$

The mapping

$$l_p^* \ni f \mapsto y_f \in l_q$$

is an isometric isomorphism between l_p^* and l_q, in which sense we can write $l_p^* = l_q$.

Proof. Let $1 \le p < \infty$ be arbitrary and q be the *conjugate index* to p ($1/p + 1/q = 1$). As is known in (see Examples 3.17), the set

$$E := \{e_n := \{\delta_{nk}\}_{k=1}^{\infty}\}_{n \in \mathbb{N}},$$

where δ_{nk} is the *Kronecker delta*, is a *Schauder basis* of l_p and

$$\forall x := \{x_k\}_{k=1}^{\infty} \in l_p : x = \sum_{k=1}^{\infty} x_k e_k$$

(see Exercise 3.45).
  By *linearity* and *continuity*,

$$\forall f \in l_p^* \, \forall x := \{x_k\}_{k=1}^{\infty} \in l_p : f(x) = \sum_{k=1}^{\infty} x_k f(e_k) \in \mathbb{F}.$$

Let us show that

$$\forall f \in l_p^* : y_f := \{y_k := f(e_k)\}_{k=1}^{\infty} \in l_q \text{ and } \|y_f\|_q \le \|f\|. \tag{7.10}$$

Indeed, since, in view of $\|e_k\|_p = 1$, $k \in \mathbb{N}$,

$$|y_k| = |f(e_k)| \le \|f\| \|e_k\|_p = \|f\|, \, k \in \mathbb{N},$$

we infer that

$$y_f \in l_\infty \text{ and } \|y_f\|_\infty := \sup_{k \in \mathbb{N}} |y_k| \le \|f\|,$$

which proves (7.10) for the conjugate pair $p = 1$, $q = \infty$.
  Suppose now that $1 < p, q < \infty$.
  Observe that (7.10) is, obviously, true for when $f = 0$ and, respectively, $y_f = 0$. Henceforth, let us regard that $f \ne 0$ and $y_f \ne 0$.
  For an arbitrary *nonzero* $f \in l_p^*$ with the corresponding $y_f := \{y_k\}_{k=1}^{\infty} \ne 0$, we can choose the sequence $\{x^{(n)} := \{x_k^{(n)}\}_{k=1}^{\infty}\}_{n=1}^{\infty}$ in l_p defined as follows:

$$\forall n \in \mathbb{N} : x_k^{(n)} := \begin{cases} |y_k|^{q-1} \frac{\overline{y_k}}{|y_k|} & \text{for } k = 1, \ldots, n \text{ with } y_k \ne 0, \\ 0 & \text{for } k = 1, \ldots, n \text{ with } y_k = 0, \\ 0 & \text{for } k > n, \end{cases}$$

in view of

$$\overline{y_k} y_k = |y_k|^2, \; k \in \mathbb{N},$$

we have:

$$\forall n \in \mathbb{N} : f(x^{(n)}) = f\left(\sum_{k=1}^{n} x_k^{(n)} e_k\right) = \sum_{k=1}^{n} x_k^{(n)} f(e_k) = \sum_{k=1}^{n} x_k^{(n)} y_k$$

$$= \sum_{k=1}^{n} |y_k|^{q-1} |y_k| = \sum_{k=1}^{n} |y_k|^q.$$

Hence, considering that $(q-1)p = q$ (see Remark 2.5),

$$\forall n \in \mathbb{N} : \sum_{k=1}^{n} |y_k|^q = |f(x^{(n)})| \le \|f\| \|x^{(n)}\|_p = \|f\| \left[\sum_{k=1}^{n} |y_k|^{(q-1)p}\right]^{1/p} = \|f\| \left[\sum_{k=1}^{n} |y_k|^q\right]^{1/p},$$

Since $y_f \ne 0$, for all sufficiently large $n \in \mathbb{N}$,

$$\sum_{k=1}^{n} |y_k|^q > 0,$$

and hence, dividing through by

$$\left[\sum_{k=1}^{n} |y_k|^q\right]^{1/p} > 0,$$

for all sufficiently large $n \in \mathbb{N}$, in view of $1 - \frac{1}{p} = \frac{1}{q}$, we have:

$$\left[\sum_{k=1}^{n} |y_k|^q\right]^{1/q} \le \|f\|.$$

Whence, passing to the limit as $n \to \infty$, we arrive at

$$\left[\sum_{k=1}^{\infty} |y_k|^q\right]^{1/q} \le \|f\|,$$

which proves (7.10) for $1 < p, q < \infty$.

As can be easily seen, the mapping

$$l_p^* \ni f \mapsto y_f \in l_q \tag{7.11}$$

is *linear*.

Exercise 7.9. Verify.

Let us show that it is also *surjective* (i. e., *onto*) and *isometric*.

Indeed, for any $y := \{y_k\}_{k=1}^{\infty} \in l_q$, by *Hölder's Inequality for Sequences* (Theorem 2.5), the mapping

$$l_p \ni x \mapsto f(x) = \sum_{k=1}^{\infty} x_k y_k \in F$$

is a well-defined *bounded linear functional* on l_p, i. e., $f \in l_p^*$, with

$$\|f\| \leq \|y\|_q.$$

Since

$$\forall n \in \mathbb{N} : f(e_n) = y_n,$$

we conclude that $y = y_f$ and, in view of (7.10),

$$\|f\| = \|y_f\|_q.$$

Thus, the mapping defined by (7.11) is indeed an *isometric isomorphism* between l_p^* and l_q, in which sense we can write $l_p^* = l_q$. □

Remarks 7.4.

– If $1 \leq p < \infty$ and, for a numeric sequence $\{y_k\}_{k=1}^{\infty}$ such that

$$\forall \{x_k\}_{k=1}^{\infty} \in l_p : \sum_{k=1}^{\infty} |x_k y_k| < \infty,$$

the linear multiplication operator

$$l_p \ni x := \{x_k\}_{k=1}^{\infty} \mapsto Ax := \{y_k x_k\}_{k=1}^{\infty} \in l_1$$

is *closed*.

Exercise 7.10. Verify (cf. Examples 5.5)

By the *Closed Graph Theorem* (Theorem 6.16), A is *bounded* (i. e., $A \in L(l_p, l_1)$), and hence, so is the linear functional

$$l_p \ni x := \{x_k\}_{k=1}^{\infty} \mapsto f(x) := s(Ax) = \sum_{k=1}^{\infty} x_k y_k \in \mathbb{F},$$

where $s \in l_1^*$ is the *sum functional* (see Examples 5.2), with

$$\|f\| \leq \|A\|.$$

Exercise 7.11. Verify.

Whence, as in the proof of the prior theorem, one can show that

$$y \in l_q \quad \text{and} \quad \|y\|_q = \|f\|,$$

where q is the *conjugate index* to p ($1/p + 1/q = 1$).

– For $p = 2$, we have: $l_2^* = l_2$, which is consistent with the *Linear Bounded Functionals on Certain Hilbert Spaces Corollary* (Corollary 7.1).

– The structure of the space l^*_∞ ($p = \infty$) is more complicated and involves the machinery of measure theory (see, e. g., [12]). However, it can be shown (see Section 7.6, Problem 2) that l_1 is *isometrically embedded* in l^*_∞ as follows:

$$l_1 \ni y := \{y_k\}_{k=1}^\infty \mapsto f_y \in l^*_\infty : f_y(x) := \sum_{k=1}^\infty x_k y_k, \; x = \{x_k\}_{k=1}^\infty \in l_\infty.$$

– As is observed in Remark 6.5, the example of l_1 with $l_1^* = l_\infty$ shows that the dual of a separable normed vector need not be separable, i. e., the converse to the *Sufficiency for Separability in Terms of Dual Space* (Theorem 6.3) is not true.

7.3.2 Weak Convergence in l_p ($1 \le p < \infty$)

7.3.2.1 Weak Convergence in l_p ($1 < p < \infty$)

By the *Representation Theorem for l_p^** ($1 \le p < \infty$) (Theorem 7.3), we obtain the following description of weak convergence in l_p ($1 < p < \infty$) as a particular case of the *Characterization of Weak Convergence in Reflexive Spaces* (Theorem 7.18) (the "*if*" part is to be proved in Section 7.6, Problem 7).

Theorem 7.4 (Characterization of Weak Convergence in l_p ($1 < p < \infty$)). *Let* $\{x_n := \{x_k^{(n)}\}_{k=1}^\infty\}_{n=1}^\infty$ *be a sequence in* l_p ($1 < p < \infty$) *over* $\mathbb{F}$.
For each $f \in l_p^*$, *there exists* $\lim_{n\to\infty} f(x_n) \in \mathbb{F}$ *iff*
(1) $\forall k \in \mathbb{N} : \exists \lim_{n\to\infty} x_k^{(n)} \in \mathbb{F}$ *and*
(2) $\sup_{n\in\mathbb{N}} \|x_n\|_p < \infty$,

in which case

$$\exists x := \{x_k\}_{k=1}^\infty \in l_p : x_n \overset{w}{\longrightarrow} x, \; n \to \infty,$$

with

$$\|x\|_p \le \varliminf_{n\to\infty} \|x_n\|_p := \lim_{n\to\infty} \inf_{k\ge n} \|x_k\|_p \le \sup_{n\in\mathbb{N}} \|x_n\|_p.$$

Exercise 7.12. Prove the "*only if*" part.

Hint. Consider the convergence on the *Schauder coordinate functionals*

$$c_n(x) := x_n \in \mathbb{F}, \; n \in \mathbb{N}, x := \{x_k\}_{k=1}^\infty \in l_p,$$

relative to the standard Schauder basis $E := \{e_n := \{\delta_{nk}\}_{k=1}^\infty\}_{n\in\mathbb{N}}$, where δ_{nk} is the *Kronecker delta*, in l_p ($1 < p < \infty$) (see Section 6.6.2.3 and Remarks 7.5).

Remarks 7.5.

-   Thus, *componentwise convergence* is a *necessary condition*, but *not a sufficient one*, for weak convergence in l_p $(1 < p < \infty)$, which is also true for the cases of $p = 1$ and $p = \infty$.

Exercise 7.13. Verify directly for $1 \le p \le \infty$ and provide a *counterexample* that works for all $1 \le p \le \infty$.

Hint. Consider the convergence on the functionals $c_n \in l_p^*$, $n \in \mathbb{N}$, defined as follows:

$$\forall n \in \mathbb{N}: \ c_n(x) := x_n \in \mathbb{F}, \quad x := \{x_k\}_{k=1}^{\infty} \in l_p,$$

which, for $1 \le p < \infty$, are the *Schauder coordinate functionals* relative to the standard Schauder basis E (cf. Exercise 7.12).
To produce a universal counterexample, modify the sequence $\{e_n := \{\delta_{nk}\}_{k=1}^{\infty}\}_{n=1}^{\infty}$, where δ_{nk} is the *Kronecker delta*.

Examples 7.1.
1. In l_p $(1 < p < \infty)$, the sequence $\{e_n\}_{n=1}^{\infty}$ weakly converges to 0, but does not (strongly) converge (cf. Remark 6.8).

    Exercise 7.14. Verify.

2. In l_1 the same sequence $\{e_n\}_{n=1}^{\infty}$ is not even weakly convergent.

    Exercise 7.15. Verify.

    The following statement explains why this fact is not coincidental.

7.3.2.2 Weak Convergence in l_1
The following amazing infinite-dimensional analogue of the *Characterization of Weak Convergence in Finite-Dimensional Spaces* (Theorem 7.3) shows that a normed vector space need not be finite-dimensional to ensure that weak convergence in it coincides with strong one.

Theorem 7.5 (Characterization of Weak Convergence in l_1). *A sequence of elements $\{x_n\}_{n=1}^{\infty}$ in l_1 converges weakly to an element $x \in l_1$ iff it (strongly) converges to x:*[2]

$$x_n \xrightarrow{w} x, \ n \to \infty \Leftrightarrow x_n \to x, \ n \to \infty.$$

Proof. *"If"* part. As is known (Proposition 6.4), (strong) convergence implies weak in any normed vector space.

---
2 Due to Issai Schur (1875–1941).

"Only if" part. Suppose that a sequence $\{x_n := \{x_k^{(n)}\}_{k=1}^\infty\}_{n=1}^\infty$ *weakly converges* in l_1. Without the loss of generality, we can regard that

$$x_n \xrightarrow{w} 0, \ n \to \infty, \tag{7.12}$$

and are to show that

$$x_n \to 0, \ n \to \infty. \tag{7.13}$$

Exercise 7.16. Explain.

Assume that with premise (7.12) in place,

$$x_n \nrightarrow 0, \ n \to \infty, \text{ in } l_1.$$

Then, for some $\varepsilon > 0$, there exists a *subsequence* $\{x_{n(i)} := \{x_k^{(n(i))}\}_{k=1}^\infty\}_{i=1}^\infty$ such that

$$\forall i \in \mathbb{N}: \ \|x_{n(i)}\|_1 := \sum_{k=1}^\infty \left|x_k^{(n(i))}\right| \geq 3\varepsilon. \tag{7.14}$$

Since $x_{n(1)} \in l_1$,

$$\exists M_1 \in \mathbb{N}: \ \sum_{k=M_1+1}^\infty \left|x_k^{(n(1))}\right| < \varepsilon,$$

which, in view of (7.14) with $i = 1$, implies that

$$\sum_{k=1}^{M_1} \left|x_k^{(n(1))}\right| > 2\varepsilon$$

and we set $i(1) := 1$.

Since weak convergence in l_1 necessarily implies *componentwise convergence* (see Remarks 7.5),

$$\exists i(2) > i(1) \ \forall i \geq i(2): \ \sum_{k=1}^{M_1} \left|x_k^{(n(i))}\right| < \varepsilon/2,$$

and, since $x^{(n(i(2)))} \in l_1$,

$$\exists M_2 > M_1: \ \sum_{k=M_2+1}^\infty \left|x_k^{(n(i(2)))}\right| < \varepsilon/2,$$

which, in view of (7.14) with $i = i(2)$, implies that

$$\sum_{k=M_1+1}^{M_2} \left|x_k^{(n(i(2)))}\right| = \sum_{k=1}^\infty \left|x_k^{(n(i(2)))}\right| - \sum_{k=1}^{M_1} \left|x_k^{(n(i(2)))}\right| - \sum_{k=M_2+1}^\infty \left|x_k^{(n(i(2)))}\right|$$

$$> 3\varepsilon - \varepsilon/2 - \varepsilon/2 = 2\varepsilon.$$

Continuing inductively, we obtain a *subsequence* $\{x_{n(i(j))}\}_{j=1}^{\infty}$ of $\{x_n\}_{n=1}^{\infty}$ and a *strictly increasing* sequence of nonnegative integers $\{M_j\}_{j=1}^{\infty}$:

$$0 =: M_0 < M_1 < M_2 < \cdots$$

such that, for any $j \in \mathbb{N}$,

$$\sum_{k=1}^{M_{j-1}} \left| x_k^{(n(i(j)))} \right| < \varepsilon/2 \quad \text{(vacuous for } j = 1\text{)},$$

$$\sum_{k=M_j+1}^{\infty} \left| x_k^{(n(i(j)))} \right| < \varepsilon/2, \quad \text{and}$$

$$\sum_{k=M_{j-1}+1}^{M_j} \left| x_k^{(n(i(j)))} \right| > 2\varepsilon, \tag{7.15}$$

which implies, in particular, that

$$\forall j \in \mathbb{N} \, \exists k = M_{j-1} + 1, \dots, M_j : x_k^{(n(i(j)))} \neq 0.$$

Let

$$y_k := \begin{cases} \dfrac{\overline{x_k^{(n(i(j)))}}}{\left| x_k^{(n(i(j)))} \right|} & \text{for } k = M_{j-1} + 1, \dots, M_j \text{ with } x_k^{(n(i(j)))} \neq 0, \\ 0 & k = M_{j-1} + 1, \dots, M_j \text{ with } x_k^{(n(i(j)))} = 0, \end{cases} \quad j \in \mathbb{N}. \tag{7.16}$$

the numeric sequence $y := \{y_k\}_{k=1}^{\infty}$ being well-defined since the collection of *pairwise disjoint* sets

$$N_j := \{\{M_{j-1} + 1, \dots, M_j\}\}_{j \in \mathbb{N}}$$

is a *partition* of $\mathbb{N}$, i. e.,

$$\mathbb{N} := \bigcup_{j=1}^{\infty} N_j.$$

As is easily seen $y := \{y_k\}_{k=1}^{\infty} \in l_{\infty}$ with $\|y\|_{\infty} := \sup_{k \in \mathbb{N}} |y_k| = 1$, and hence by the *Representation Theorem for l_p^* ($1 \le p < \infty$)* (Theorem 7.3) with $p = 1$,

$$f_y(z) := \sum_{k=1}^{\infty} z_k y_k, \ z := \{z_k\}_{k=1}^{\infty} \in l_1$$

is a bounded linear functional on l_1, i. e., $f \in l_1^*$, with $\|f_y\| = \|y\|_{\infty} = 1$.

By the definition of $y := \{y_k\}_{k=1}^{\infty}$ and in view of (7.15), for any $j \in \mathbb{N}$,

$$\left| f_y(x_{n(i(j))}) \right| = \left| \sum_{k=1}^{\infty} x_k^{(n(i(j)))} y_k \right|$$

$$\geq \left| \sum_{k=M_{j-1}+1}^{M_j} x_k^{(n(i(j)))} y_k \right| - \left| \sum_{k=1}^{M_{j-1}} x_k^{(n(i(j)))} y_k \right| - \left| \sum_{k=M_j}^{\infty} x_k^{(n(i(j)))} y_k \right|$$

$$\geq \sum_{k=M_{j-1}+1}^{M_j} \left| x_k^{(n(i(j)))} \right| - \|f_y\| \sum_{k \neq M_{j-1}+1,\dots,M_j} \left| x_k^{(n(i(j)))} \right| > 2\varepsilon - \varepsilon = \varepsilon.$$

Hence, the subsequence $\{x_{n(i(j))}\}_{j=1}^{\infty}$ does not weakly converge to 0, which contradicts the premise given by (7.12) and the obtained contradiction implying the desired conclusion given by (7.13). □

7.3.3 Duality and Weak Convergence for $(c_0, \|\cdot\|_\infty)$

7.3.3.1 Representation Theorem for c_0^*
Theorem 7.6 (Representation Theorem for c_0^*).

$$\forall f \in c_0^* \; \exists! \, y_f := \{y_k\}_{k=1}^{\infty} \in l_1 \; \forall x := \{x_k\}_{k=1}^{\infty} \in c_0 : f(x) = \sum_{k=1}^{\infty} x_k y_k.$$

The mapping

$$c_0^* \ni f \mapsto y_f \in l_1$$

is an isometric isomorphism between c_0^* and l_1, in which sense we can write $c_0^* = l_1$.

Exercise 7.17. Prove.

Hint. Modify the proof of the *Representation Theorem for l_p^* $(1 \leq p < \infty)$* (Theorem 7.3).

7.3.3.2 Weak Convergence in $(c_0, \|\cdot\|_\infty)$
By the *Representation Theorem for c_0^** (Theorem 7.6) and *Lebesgue's Dominated Convergence Theorem* (see, e. g., [11, 19, 33]), one can prove the *"if"* part of the following characterization of weak convergence in $(c_0, \|\cdot\|_\infty)$.

Theorem 7.7 (Characterization of Weak Convergence in $(c_0, \|\cdot\|_\infty)$). *A sequence $\{x_n = \{x_k^{(n)}\}_{k=1}^{\infty}\}_{n=1}^{\infty}$ is weakly convergent to an $x := \{x_k\}_{k=1}^{\infty}$ in $(c_0, \|\cdot\|_\infty)$ iff*
(1) $\forall k \in \mathbb{N} : \lim_{n \to \infty} x_k^{(n)} = x_k$ *and*
(2) $\sup_{n \in \mathbb{N}} \|x_n\|_\infty < \infty$.

Exercise 7.18. Prove the *"only if"* part.

Examples 7.2. In $(c_0, \|\cdot\|_\infty)$,
1. the sequence $\{e_n := \{\delta_{nk}\}_{k=1}^{\infty}\}_{n=1}^{\infty}$, where δ_{nk} is the *Kronecker delta*, weakly converges to the *zero sequence* $(0, 0, 0, \dots)$, but does not (strongly) converge;

2. the sequence

$$x_n := (\underbrace{1, \ldots, 1}_{n \text{ terms}}, 0, 0, \ldots), \; n \in \mathbb{N},$$

does not weakly converge, although it is, obviously, bounded and termwise convergent.

Exercise 7.19. Verify.

7.4 Duality and Weak Convergence for $(C[a, b], \| \cdot \|_\infty)$

The case of the space $(C[a, b], \| \cdot \|_\infty)$ is considered here, largely without proof.

7.4.1 Riesz Representation Theorem for $C^*[a, b]$

Theorem 7.8 (Riesz Representation Theorem for $C^*[a, b]$). *For any $F \in C^*[a, b]$ ($-\infty < a < b < \infty$), there is a unique function $g \in BV[a, b]$ vanishing at a and right-continuous on $(a, b]$ such that*

$$F(f) = \int_a^b f(t) \, dg(t),$$

where the latter is the Riemann–Stieltjes integral of f relative to g, and

$$\|F\| = \|g\| = V_a^b(g),$$

i. e., the dual $(C^*[a, b], \| \cdot \|)$ is isometrically embedded in $BV[a, b]$ with the norm

$$BV[a, b] \ni g \mapsto \|g\| := |g(a)| + V_a^b(g).$$

For a proof, see, e. g., [12, 16].

Remark 7.6. In particular, for each $t \in [a, b]$, the bounded linear *fixed-value functional*

$$C[a, b] \ni f \mapsto F_t(f) := f(t) \in \mathbb{F}$$

on $(C[a, b], \| \cdot \|_\infty)$ (see Examples 5.2) is generated by the function

$$g_a(s) := \begin{cases} 0, & s = a, \\ 1, & a < s \le b, \end{cases} \quad \text{for } t = a, \quad g_t(s) := \begin{cases} 0, & a \le s < t, \\ 1, & t \le s \le b, \end{cases} \quad \text{for } a < t \le b.$$

Exercise 7.20. Verify.

7.4.2 Weak Convergence in $(C[a, b], \| \cdot \|_\infty)$

The following statement characterizes weak convergence in $(C[a, b], \| \cdot \|_\infty)$.

Theorem 7.9 (Characterization of Weak Convergence in $(C[a, b], \| \cdot \|_\infty)$). *A sequence* $\{f_n\}_{n=1}^\infty$ *is weakly convergent to an f in* $(C[a, b], \| \cdot \|_\infty)$ $(-\infty < a < b < \infty)$ *iff*
(1) $\forall t \in [a, b] : \lim_{n \to \infty} f(t) = f(t)$ *and*
(2) $\sup_{n \in \mathbb{N}} \|f_n\|_\infty < \infty$.

For a proof, see, e. g., [12, 31].

Exercise 7.21. Prove the *"only if"* part.

Examples 7.3. In $(C[0, 1], \| \cdot \|_\infty)$,
1. the sequence

$$f_n(t) := nte^{-nt}, \ n \in \mathbb{N}, t \in [0, 1],$$

weakly converges to the zero function $f(t) := 0$, $t \in [0, 1]$, but does not (strongly) converge (i. e., *does not uniformly converge* on $[0, 1]$).
2. the sequence $\{t^n\}_{n=1}^\infty$ *does not weakly converge*, although it is, obviously, bounded and pointwise convergent.

Exercise 7.22. Verify.

7.5 Reflexivity

7.5.1 Definition and Examples

As defined in Section 6.2.5, for a normed vector space $(X, \| \cdot \|)$, the dual $(X^*)^*$ of the dual X^* is called the *second dual space* (or *bidual space*) of X and denoted X^{**}.

By the *Canonical Isomorphism Theorem* (Theorem 6.5), for each fixed element x in a normed vector space $(X, \| \cdot \|)$, the mapping

$$X \ni x \mapsto Jx := F_x \in X^{**} : F_x(f) := f(x), \ f \in X^*,$$

is an isometric isomorphism from $(X, \| \cdot \|)$ to the second dual $(X^{**}, \| \cdot \|)$, called the *canonical isomorphism* (or the *natural embedding*), i. e., $(X, \| \cdot \|)$ is *isometrically embedded* in $(X^{**}, \| \cdot \|)$.

Remarks 7.7.
- We use the same symbol $\|\cdot\|$ to designate the norm of in the dual and bidual spaces, such an economy of symbols being a rather common practice (cf. Remarks 6.2).

– Henceforth, we adopt the juxtaposition notation

$$x^*x$$

for the action of a functional $x^* \in X^*$ on an element $x \in X$, and thus, for the *canonical isomorphism*

$$X \ni x \mapsto Jx =: \hat{x} \in X^{**},$$

we have:

$$\hat{x}x^* = x^*x, \; x \in X, \; x^* \in X^*,$$

and

$$\|\hat{x}\| = \sup_{x^* \in X^*, \, \|x^*\|=1} |\hat{x}x^*| = \sup_{x^* \in X^*, \, \|x^*\|=1} |x^*x| = \|x\|$$

(see the proof of the *Canonical Isomorphism Theorem* (Theorem 6.5)).

Definition 7.1 (Reflexive Space). A normed vector space $(X, \|\cdot\|)$ is called *reflexive* if the canonical isomorphism J is a *surjective* mapping, i.e., maps X onto X^{**}:

$$\hat{X} := J(X) = X^{**},$$

i.e.,

$$\forall x^{**} \in X^{**} \; \exists! x \in X \; \forall x^* \in X^* : x^{**}x^* = x^*x,$$

and hence, the spaces X and X^{**} are *isometrically isomorphic* under the canonical isomorphism, in which sense we can write $X = X^{**}$.

The following statements produce examples of various reflexive and nonreflexive spaces.

Theorem 7.10 (Reflexivity of Hilbert Spaces). *A Hilbert space is reflexive.*

Proof. Let $(X, (\cdot, \cdot), \|\cdot\|)$ be a (real or complex) Hilbert space.

By the *Riesz Representation Theorem* (Theorem 7.1), there is an *isometric* linear (if X is real) or conjugate linear (if X is complex) *isomorphism*

$$Q : X^* \to X$$

between X^* and X defined as follows:

$$\forall x^* \in X^* \; \exists! Qx^* \in X \; \forall x \in X : x^*x = (x, Qx^*)$$

and the dual X^* is a Hilbert space relative to the inner product

$$(x^*, y^*)_{X^*} := \overline{(Qx^*, Qy^*)} = (Qy^*, Qx^*), \; x^*, y^* \in X^* \tag{7.17}$$

(see Remarks 7.1).

Whence, by the *Riesz Representation Theorem* (Theorem 7.1), we infer that there is an *isometric* linear (if X, and hence X^* and X^{**}, are real) or conjugate linear (if X, and hence X^* and X^{**}, are complex) *isomorphism*

$$R : X^{**} \to X^*,$$

defined as follows:

$$\forall x^{**} \in X^{**} \; \exists! \, Rx^{**} \in X^* \; \forall x^* \in X^* : \; x^{**}x^* = (x^*, Rx^{**})_{X^*}.$$

Thus, in view of (7.17),

$$\forall x^{**} \in X^{**}, \; \forall x^* \in X^* : \; x^{**}x^* = (x^*, Rx^{**})_{X^*} = (QRx^{**}, Qx^*) = x^*(QRx^{**}) = x^* x,$$

where

$$x := QRx^{**} \in X,$$

which shows that the space $(X, (\cdot, \cdot), \| \cdot \|)$ is *reflexive*. □

Remark 7.8. In particular, the (real or complex) Hilbert spaces $l_2^{(n)}$ ($n \in \mathbb{N}$), l_2, and $L_2(a, b)$ ($-\infty \le a < b \le \infty$) (see Examples 4.3) are reflexive,

Theorem 7.11 (Reflexivity of Finite-Dimensional Spaces). *A finite-dimensional normed vector space is reflexive.*

Proof. Let $(X, \| \cdot \|)$ be normed vector space over $\mathbb{F}$ with $\dim X = n$ ($n \in \mathbb{N}$) and a basis $B := \{x_1, \dots, x_n\}$. Then, by the *Representation Theorem for Duals of Finite-Dimensional Spaces* (Theorem 7.2), there is an *isomorphism*

$$T : X^* \to \mathbb{F}^n$$

between X^* and $\mathbb{F}^n$ defined as follows:

$$\forall x^* \in X^* \; \exists! \, Tx^* := (y_1, \dots, y_n) \in \mathbb{F}^n \; \forall x := \sum_{k=1}^{n} c_k x_k \in X$$

with some $c_k \in \mathbb{F}$, $k = 1, \dots, n$:

$$x^* x = \sum_{k=1}^{n} c_k y_k,$$

and hence, by the *Isomorphism Theorem* (Theorem 3.5),

$$\dim X^* = n.$$

Similarly, for $X^{**} := (X^*)^*$,

$$\dim X^{**} = n.$$

Considering that, by the *Isomorphism Theorem* (Theorem 3.5), for the *canonical isomorphism* $J : X \to X^{**}$,

$$\dim J(X) = \dim X = n,$$

we conclude that

$$J(X) = X^{**},$$

and hence, the space $(X, \| \cdot \|)$ is *reflexive*. □

Remarks 7.9.
-   In particular, for any $n \in \mathbb{N}$ and $1 \le p \le \infty$, we have:

    $$l_p^{(n)\,**} = l_p^{(n)},$$

    the equality understood in the sense of the spaces' being isometrically isomorphic under the canonical isomorphism.
-   For $p = 2$, we have the reflexive Hilbert space $l_2^{(n)}$ ($n \in \mathbb{N}$), which is consistent with the *Reflexivity of Hilbert Spaces Theorem* (Theorem 7.10) (see Remark 7.8).

Theorem 7.12 (Reflexivity of l_p ($1 < p < \infty$)). *The space l_p ($1 < p < \infty$) is reflexive.*

Proof. Let $1 < p < \infty$ be a arbitrary and $1 < q < \infty$ be its *conjugate index* ($1/p + 1/q = 1$).

By the *Representation Theorem for l_p^* ($1 \le p < \infty$)* (Theorem 7.3), there is an *isometric isomorphism*

$$T : l_p^* \to l_q$$

between l_p^* and l_q defined as follows:

$$\forall x^* \in l_p^* \; \exists! \, Tx^* := \{y_k\}_{k=1}^{\infty} \in l_q \; \forall x := \{x_k\}_{k=1}^{\infty} \in l_p : x^*x = \sum_{k=1}^{\infty} x_k y_k =: \langle x, Tx^* \rangle. \quad (7.18)$$

Further (see Section 7.6, Problem 6), there is an *isometric isomorphism*

$$T^* : l_q^* \to l_p^{**}$$

between l_q^* and l_p^{**} defined as follows:

$$l_q^* \ni y^* \mapsto T^*y^* := y^*T \in l_p^{**},$$

where the juxtaposition y^*T designates the composition, i. e.,

$$(y^*T)x^* := y^*(Tx^*), \quad x^* \in l_p^*.$$

By the *Representation Theorem for* l_p^* $(1 \le p < \infty)$ (Theorem 7.3), there is an *isometric isomorphism*

$$R : l_q^* \to l_p$$

between l_q^* and l_p defined as follows:

$$\forall y^* \in l_q^* \ \exists! \, Ry^* := \{x_k\}_{k=1}^\infty \in l_p \ \forall y := \{y_k\}_{k=1}^\infty \in l_q : y^*y = \sum_{k=1}^\infty y_k x_k =: \langle y, Ry^* \rangle. \quad (7.19)$$

Thus,

$$\forall x^{**} \in l_p^{**} \ \exists! \, y^* \in l_q^* : x^{**} = T^* y^*$$

and, for any $x^* \in l_p^*$, in view of (7.19) and (7.18), we have:

$$x^{**}x^* = (T^*y^*)x^* = (y^*T)x^* = y^*(Tx^*) = \langle Tx^*, Ry^* \rangle = \langle Ry^*, Tx^* \rangle = x^*(Ry^*) = x^*x,$$

where

$$x := Ry^* \in X,$$

which shows that the space l_p is *reflexive*. □

Remark 7.10. In particular, for $p = 2$, we have the reflexive Hilbert space l_2, which is consistent with the *Reflexivity of Hilbert Spaces Theorem* (Theorem 7.10) (cf. Remark 7.8).

Proposition 7.4 (Nonreflexivity of l_1). *The space l_1 is not reflexive.*

Exercise 7.23. Prove *by contradiction*.

Hint. Use the *Representation Theorem for* l_p^* $(1 \le p < \infty)$ (Theorem 7.3) for $p = 1$ and the *Sufficiency for Separability in Terms of Dual Space* (Theorem 6.3).

Proposition 7.5 (Nonreflexivity of $(c_0, \|\cdot\|_\infty)$). *The space $(c_0, \|\cdot\|_\infty)$ is not reflexive.*

Exercise 7.24. Prove.

Hint. Use the *Representation Theorems* for c_0^* and l_1^* and (Theorems 7.6 and 7.3 $(p = 1)$) and the result formulated as Problem 6 of Section 7.6.

Theorem 7.13 (Nonreflexivity of $(C[a,b], \|\cdot\|_\infty)$). *The space $(C[a,b], \|\cdot\|_\infty)$ $(-\infty < a < b < \infty)$ is not reflexive.*

For a proof, see, e. g., [12].

Henceforth, we consider a number of important properties inherent to reflexivity.

7.5.2 Completeness of a Reflexive Space

As is easily seen all the reflexive normed vector spaces discussed above happen to be *complete*. The following statement explains why this is not a mere coincidence.

Theorem 7.14 (Completeness of a Reflexive Space). *A reflexive normed vector space is a Banach space.*

Exercise 7.25. Prove.

Hint. Use the *Completeness of the Dual Space Corollary* (Corollary 5.1).

Remarks 7.11.
- Thus, the *completeness* of a normed vector space is a *necessary condition* for its reflexivity, and hence, any incomplete normed vector space, in particular $(c_{00}, \|\cdot\|_\infty)$, is not reflexive.
- However, as the examples of the Banach spaces l_1, $(c_0, \|\cdot\|_\infty)$, and $(C[a,b], \|\cdot\|_\infty)$ demonstrate, the completeness of a normed vector space is *not sufficient* for its reflexivity.

7.5.3 Reflexivity of a Closed Subspace

Theorem 7.15 (Reflexivity of a Closed Subspace). *A closed subspace of a reflexive normed vector space is reflexive.*

Proof. Let $(Y, \|\cdot\|)$ be a *closed subspace* of a reflexive normed vector space $(X, \|\cdot\|)$.
Consider the mapping

$$X^* \ni x^* \mapsto Tx^* \in Y^* : (Tx^*)y := x^*y, \; y \in Y,$$

i. e., Tx^* is the *restriction* of the functional $x^* \in X^*$ to Y.
As is easily seen, $T : X^* \to Y^*$ is a *bounded linear operator* with

$$\|Tx^*\| \le \|x^*\|, \; x^* \in X^*. \tag{7.20}$$

Exercise 7.26. Verify.

Thus,

$$T \in L(X^*, Y^*) \text{ with } \|T\| \le 1.$$

Also, consider the mapping

$$Y^{**} \ni y^{**} \mapsto Sy^{**} := y^{**}T \in X^{**},$$

where the juxtaposition $y^{**}T$ designates the composition, i. e.,

$$(y^{**}T)x^* := y^{**}(Tx^*), \; x^* \in X^*.$$

Clearly, $S : Y^{**} \to X^{**}$ is a *linear operator* and, in view of (7.20), for arbitrary $y^{**} \in Y^{**}$ and $x^* \in X^*$,

$$\|(Sy^{**})x^*\| = \|(y^{**}T)x^*\| = \|y^{**}(Tx^*)\| \le \|y^{**}\|\|Tx^*\| \le \|y^{**}\|\|x^*\|,$$

which implies that

$$S \in L(Y^{**}, X^{**}) \text{ with } \|S\| \le 1.$$

Since $(X, \|\cdot\|)$ is *reflexive*,

$$\forall x^{**} \in X^{**} \; \exists! x \in X : \; x^{**} = Jx,$$

where $J : X \to X^{**}$ is the *canonical isomorphism*.
Let us show that

$$J^{-1}S(Y^{**}) \subseteq Y. \tag{7.21}$$

Indeed, by the *Separation of Element from Closed Subspace Corollary* (Corollary 6.1), the assumption that

$$\exists y^{**} \in Y^{**} : \; x := J^{-1}Sy^{**} \notin Y$$

implies that

$$\exists x^* \in X^* : \; x^*x \ne 0 \text{ and } x^*y = 0, \; y \in Y. \tag{7.22}$$

Hence,

$$Tx^* = 0$$

and, in view of

$$Jx = Sy^{**},$$

we have:

$$0 = y^{**}(Tx^*) = (y^{**}T)x^* = (Sy^{**})x^* = (Jx)x^* = x^*x,$$

which *contradicts* (7.22), proving inclusion (7.21).

For an arbitrary $y^* \in Y^*$, let $x^* \in X^*$ be an extension of y^*, which exists (need not be unique) according to the *Hahn–Banach Theorem for Normed Vector Spaces* (Theorem 6.2). Then

$$y^* = Tx^*$$

and, in view of the fact that

$$\forall y^{**} \in Y^{**} : y := J^{-1}Sy^{**} \in Y,$$

which implies that $Jy = Sy^{**} \in X^{**}$.

For arbitrary $y^{**} \in Y^{**}$ and $y^* \in Y^*$, we have:

$$y^{**}y^* = y^{**}(Tx^*) = (y^{**}T)x^* = (Sy^{**})x^* = (Jy)x^* = x^*y = Tx^*y = y^*y,$$

which implies that the space $(Y, \| \cdot \|)$ is *reflexive*. □

Remark 7.12. As the example of the subspace c_{00} of the reflexive Hilbert space l_2 shows, the condition of the *closedness* of the subspace in the prior theorem is essential and cannot be dropped.

Exercise 7.27. Explain.

7.5.4 Isometric Isomorphism and Reflexivity

We are show now that an isometric isomorphism preserves reflexivity.

Theorem 7.16 (Isometric Isomorphism and Reflexivity). *A normed vector space* $(Y, \| \cdot \|_Y)$ *isometrically isomorphic to a reflexive normed vector space* $(X, \| \cdot \|_X)$ *is reflexive.*

Proof. Let $T : X \to Y$ be an *isometric isomorphism* between X and Y, then the mapping

$$Y^* \ni y^* \mapsto T^*y^* := y^*T \in X^*,$$

where the juxtaposition y^*T designates the composition, i. e.,

$$(y^*T)x := y^*(Tx), \ x \in X,$$

is an *isometric isomorphism* between $(Y^*, \| \cdot \|)$ and $(X^*, \| \cdot \|)$ and the mapping

$$X^{**} \ni x^{**} \mapsto T^{**}x^{**} := x^{**}T^* \in Y^{**},$$

where the juxtaposition $x^{**}T^*$ designates the composition, i. e.,

$$(x^{**}T^*)y^* := x^{**}(T^*y^*), \ y^* \in Y^*,$$

is an *isometric isomorphism* between $(X^{**}, \| \cdot \|)$ and $(Y^{**}, \| \cdot \|)$ (see Section 7.6, Problem 6).

By the *reflexivity* of $(X, \| \cdot \|)$,

$$\forall y^{**} \in Y^{**} \ \exists x \in X : y^{**} = T^{**}\hat{x},$$

where $\hat{x} := Jx$, $J : X \to X^{**}$ being the *canonical isomorphism*.

Then, for arbitrary $y^{**} \in Y^{**}$ and $y^* \in Y^*$, we have:

$$y^{**}y^* = (T^{**}\hat{x})y^* = (\hat{x}T^{**})y^* = \hat{x}(T^*y^*) = (T^*y^*)x = (y^*T)x = y^*(Tx) = y^*y,$$

where

$$y := Tx \in Y,$$

which implies that the space $(Y, \|\cdot\|_Y)$ is *reflexive*. □

7.5.5 Characterization of Reflexivity

The following statement characterizes the reflexivity of a Banach space.

Theorem 7.17 (Characterization of Reflexivity). *A Banach space is reflexive iff its dual space is reflexive.*

Proof. "Only if" part. Let a Banach space $(X, \|\cdot\|)$ be *reflexive*, with $J : X \to X^{**}$ being the *canonical isomorphism*. Then

$$\forall x^{***} \in (X^*)^{**} := (X^{**})^* : x^* := x^{***}J \in X^*,$$

where the juxtaposition $x^{***}J$ designates the composition, i. e.,

$$(x^{***}J)x := x^{***}(Jx), \ x \in X.$$

Exercise 7.28. Explain.

By the *reflexivity* of $(X, \|\cdot\|)$,

$$\forall x^{**} \in X^{**} \ \exists! x \in X : x^{**} = Jx$$

and we have:

$$\forall x^{***} \in X^{***}, \ \forall x^{**} \in X^{**} : x^{***}x^{**} = x^{***}(Jx) = (x^{***}J)x = x^*x = (Jx)x^* = x^{**}x^*,$$

which implies that the dual $(X^*, \|\cdot\|)$ is *reflexive*.

"If" part. Suppose that, for a Banach space $(X, \|\cdot\|)$, the dual space $(X^*, \|\cdot\|)$ is *reflexive*. Then, by the "only if" part, the bidual space $(X^{**}, \|\cdot\|)$ is *reflexive*, and hence, by the *Reflexivity of a Closed Subspace Theorem* (Theorem 7.15), so is its *closed subspace* $(\hat{X}, \|\cdot\|)$, where $\hat{X} := J(X)$.

Exercise 7.29. Explain why the subspace $\hat{X}$ is *closed* in $(X^{**}, \|\cdot\|)$ (the *completeness* of the space $(X, \|\cdot\|)$ is essential here).

Since an isometric isomorphism preserves reflexivity (Theorem 7.16), the space $(X, \|\cdot\|)$, being *isometrically isomorphic* to the reflexive space $(\hat{X}, \|\cdot\|)$, is *reflexive*. □

By the *Isometric Isomorphism and Reflexivity Theorem* (Theorem 7.16) along with the fact that, by the *Representation Theorem for* l_p^* $(1 \leq p < \infty)$ (Theorem 7.3), l_∞ is *isometrically isomorphic* to l_1^*, which is *not reflexive* by the *Characterization of Reflexivity* (Theorem 7.17) since the space l_1 is *not reflexive* (Proposition 7.4), we obtain the following statement.

Corollary 7.2 (Nonreflexivity of l_∞). *The space l_∞ is not reflexive.*

7.5.6 Weak Convergence and Weak Completeness

Based on the *Banach–Steinhaus Theorem for Functionals* (Theorem 6.9) and the *Canonical Isomorphism Theorem* (Theorem 6.5), we obtain the following version of the *Characterization of Weak Convergence* (Theorem 6.10) for reflexive spaces.

Theorem 7.18 (Characterization of Weak Convergence in Reflexive Spaces). *Let* $\{x_n\}_{n=1}^\infty$ *be a sequence in a reflexive normed vector space* $(X, \|\cdot\|)$ *over* $\mathbb{F}$.
For each $x^* \in X^*$, *there exists* $\lim_{n\to\infty} x^* x_n \in \mathbb{F}$ *iff*
(1) *there exist a fundamental set S in the dual space* $(X^*, \|\cdot\|)$ *such that*

$$\forall x^* \in S : \exists \lim_{n\to\infty} x^* x_n \in \mathbb{F}$$

and
(2) $\sup_{n\in\mathbb{N}} \|x_n\| < \infty$,

in which case

$$\exists x \in X : x_n \xrightarrow{w} x, \ n \to \infty,$$

with

$$\|x\| \leq \varliminf_{n\to\infty} \|x_n\| := \lim_{n\to\infty} \inf_{k\geq n} \|x_k\| \leq \sup_{n\in\mathbb{N}} \|x_n\|.$$

Exercise 7.30.
(a) Prove.

> **Hint.** As in the proof of the *Characterization of Weak Convergence* (Theorem 6.10), replace the sequence $\{x_n\}_{n=1}^\infty$ with the sequence $\{Jx_n\}_{n=1}^\infty$ in the bidual space $(X^{**}, \|\cdot\|)$, where $J : X \to X^{**}$ is the *canonical isomorphism*, then apply the *Banach–Steinhaus Theorem for Functionals* (Theorem 6.9) using the *reflexivity* of $(X, \|\cdot\|)$.

(b) Give an example showing that the *reflexivity* requirement for the space $(X, \|\cdot\|)$ is essential and cannot be dropped (cf. Examples 7.2).

Remarks 7.13.
– What makes the prior characterization substantially different from the seemingly more general *Characterization of Weak Convergence* (Theorem 6.10) is that, al-

though formulated in a more restrictive setting, with the requirement of *reflexivity* on the space, it does not presuppose *a priori* the existence of a weak limit in the space (cf. Remark 6.13), the latter established based on the *reflexivity*.

– As a particular case of the *Characterization of Weak Convergence in Reflexive Spaces* (Theorem 7.18), via the *Representation Theorem for l_p^* (1 ≤ p < ∞)* (Theorem 7.3), we obtain the *Characterization of Weak Convergence in l_p (1 < p < ∞)* (Theorem 7.4) (cf. Section 7.6, Problem 7).

Definition 7.2 (Weak Fundamentality/Weak Completeness). A sequence $\{x_n\}_{n=1}^{\infty}$ in a normed vector space $(X, \|\cdot\|)$, such that, for each $x^* \in X^*$, $\{x^*x_n\}_{n=1}^{\infty}$ is a Cauchy (fundamental) numeric sequence, is called a *weak Cauchy sequence* or a *weakly fundamental sequence*.

A normed vector space $(X, \|\cdot\|)$, in which every weakly fundamental sequence weakly converges, is called weakly complete.

Remark 7.14. Every fundamental sequence in a normed vector space is weakly fundamental, but not vice versa.

Exercise 7.31. Verify and provide a corresponding *counterexample*.

The following statement strengthens the *Completeness of a Reflexive Space Theorem* (Theorem 7.14).

Theorem 7.19 (Weak Completeness of Reflexive Spaces). *A reflexive normed vector space is weakly complete.*

Exercise 7.32. Prove.

Hint. Use the *Characterization of Weak Convergence in Reflexive Spaces* (Theorem 7.18) and the fact that a weakly fundamental sequence is necessarily *weakly bounded*, and hence, by the *Equivalence of Weak Boundedness and Boundedness Theorem* (Theorem 6.11), is *bounded*.

7.5.7 Bounded Sequence Property

The following is a very important property of reflexive spaces.

Theorem 7.20 (Bounded Sequence Property). *Every bounded sequence $\{x_n\}_{n=1}^{\infty}$ in a reflexive normed vector space $(X, \|\cdot\|)$ has a subsequence $\{x_{n(k)}\}_{k=1}^{\infty}$ weakly convergent in $(X, \|\cdot\|)$.*

Proof. Let $\{x_n\}_{n=1}^{\infty}$ be an arbitrary *bounded* sequence in a reflexive normed vector space $(X, \|\cdot\|)$.

Consider the *closed subspace*

$$Y := \overline{\operatorname{span}\left(\{x_n\}_{n\in\mathbb{N}}\right)}$$

in $(X, \|\cdot\|)$.

As is easily seen, the subspace $(Y, \|\cdot\|)$ is *separable*.

Exercise 7.33. Explain.

Also, by the *Reflexivity of a Closed Subspace Theorem* (Theorem 7.15), the subspace $(Y, \|\cdot\|)$ is *reflexive*.

Hence, the bidual space $(Y^{**}, \|\cdot\|)$, being *isometrically isomorphic* to $(Y, \|\cdot\|)$, is *separable* and, by the *Sufficiency for Separability in Terms of Dual Space* (Theorem 6.3), the dual space $(Y^*, \|\cdot\|)$ is also *separable*.

Let $\{y_n^*\}_{n\in\mathbb{N}}$ be a *countable dense*, and hence, *fundamental* set in $(Y^*, \|\cdot\|)$, i. e.,

$$\overline{\operatorname{span}\left(\{y_n^*\}_{n\in\mathbb{N}}\right)} = Y^*$$

(see Definition 6.1).

The numeric sequence $\{y_1^* x_n\}_{n=1}^\infty$ is *bounded*.

Exercise 7.34. Explain.

Hence, by the *Bolzano–Weierstrass Theorem* (Theorem 2.40), there exists a subsequence $\{x_{1,k}\}_{k=1}^\infty$ of $\{x_n\}_{n=1}^\infty$ such that the subsequence $\{y_1^* x_{1,k}\}_{k=1}^\infty$ converges.

Similarly, $\{x_{1,k}\}_{k=1}^\infty$ has a subsequence $\{x_{2,k}\}_{k=1}^\infty$ such that the subsequence $\{y_2^* x_{2,k}\}_{k=1}^\infty$ converges.

Continuing inductively, we obtain a collection of nested subsequences of $\{x_n\}_{n=1}^\infty$,

$$\left\{\{x_{n,k}\}_{k=1}^\infty\right\}_{n\in\mathbb{N}},$$

each one being a subsequence of the preceding one.

By the construction, for the *"diagonal subsequence"* $\{x_{n(k)}\}_{k=1}^\infty$ of $\{x_n\}_{n=1}^\infty$ defined as follows:

$$x_{n(k)} := x_{k,k}, \; k \in \mathbb{N},$$

we have:

$$\forall m \in \mathbb{N}: \; \exists \lim_{k\to\infty} y_m^* x_{n(k)} \in \mathbb{F}.$$

This, in view of the facts that the set $\{y_n^*\}_{n=1}^\infty$ is *fundamental* in $(Y^*, \|\cdot\|)$ and that the subsequence $\{x_{n(k)}\}_{k=1}^\infty$ is *bounded* in $(Y, \|\cdot\|)$:

$$\sup_{n\in\mathbb{N}} \|x_{n(k)}\| < \infty,$$

by the *Characterization of Weak Convergence in Reflexive Spaces* (Theorem 7.18), imply that the subsequence $\{x_{n(k)}\}_{k=1}^{\infty}$ *weakly converges* in $(Y, \|\cdot\|)$, i.e.,

$$\exists x \in Y \; \forall y^* \in Y^* : \lim_{k \to \infty} y^* x_{n(k)} = y^* x.$$

For every $x^* \in X^*$, let $y^* \in Y^*$ be the restriction of x^* to Y, i.e.,

$$y^* y := x^* y, \; y \in Y.$$

Hence,

$$\forall x^* \in X^* : \lim_{k \to \infty} x^* x_{n(k)} = \lim_{k \to \infty} y^* x_{n(k)} = y^* x = x^* x,$$

i.e., the subsequence $\{x_{n(k)}\}_{k=1}^{\infty}$ weakly converges to $x \in Y$ in $(X, \|\cdot\|)$, which completes the proof. $\qquad \square$

Remarks 7.15.
- The *Bounded Sequence Property*, which, in fact, *characterizes* reflexive spaces in the class of Banach spaces, implies that, in a reflexive normed vector space, each bounded set is precompact in the weak sense (see, e.g., [12]).
- The condition of reflexivity in the *Bounded Sequence Property* is essential and cannot be dropped.

Exercise 7.35.
(a) Provide a corresponding *counterexample*.
(b) Apply the *Bounded Sequence Property* to prove that the space l_1 is not reflexive.

7.6 Problems

1. Prove

    Proposition 7.6 (Weak Convergence of an Orthonormal Sequence). *For a countably infinite orthonormal set* $\{e_n\}_{n \in \mathbb{N}}$ *in a Hilbert space* $(X, (\cdot, \cdot), \|\cdot\|)$,

    $$e_n \xrightarrow{w} 0, \; n \to \infty,$$

    but the sequence $(e_n)_{n \in \mathbb{N}}$ *does not (strongly) converge.*

2. Prove that l_1 is *isometrically embedded* in l_{∞}^* as follows:

    $$l_1 \ni y := \{y_k\}_{k=1}^{\infty} \mapsto f_y \in l_{\infty}^* : f_y(x) = \sum_{k=1}^{\infty} x_k y_k, \; x := \{x_k\}_{k=1}^{\infty} \in l_{\infty}.$$

3. If $(Y, \|\cdot\|)$ is *dense subspace* of a normed vector space $(Y, \|\cdot\|)$, show that duals $(X^*, \|\cdot\|)$ and $(Y^*, \|\cdot\|)$ are *isometrically isomorphic*.

Hint. Show that the mapping

$$X^* \ni x^* \mapsto Tx^* \in Y^* : (Tx^*)y := x^*y, \, y \in Y,$$

i. e., Tx^* is the *restriction* of the functional $x^* \in X^*$ to Y, is an *isometric isomorphism* between $(X^*, \|\cdot\|)$ and $(Y^*, \|\cdot\|)$.

4. Use the prior hint and the *Representation Theorem for c_0^** (Theorem 7.6) to prove

   Theorem 7.21 (Representation Theorem for c_{00}^*).

   $$\forall f \in c_{00}^* \, \exists! \, y_f := \{y_k\}_{k=1}^\infty \in l_1 \, \forall x := \{x_k\}_{k=1}^\infty \in c_{00} : f(x) = \sum_{k=1}^\infty x_k y_k.$$

   The mapping

   $$c_{00}^* \ni f \mapsto y_f \in l_1$$

   is an isometric isomorphism between c_{00}^* and l_1, in which sense we can write $c_{00}^* = l_1$.

5. Describe all bounded linear functionals on $(c, \|\cdot\|_\infty)$.

   Hint. Apply the *Representation Theorem for $(c_0, \|\cdot\|_\infty)$* (Theorem 7.6) and obtain all functionals of c^* as extensions of all functionals of c_0^*, i. e., of all bounded linear functionals on the *hyperplane* $c_0 = \ker l$ of c, where $l \in c^*$ is the *limit functional* on c:

   $$c \ni x = (x_n)_{n\in\mathbb{N}} \mapsto l(x) := \lim_{n\to\infty} x_n \in \mathbb{F}.$$

6. Let $T : X \to Y$ be an *isometric isomorphism* between normed vector spaces $(X, \|\cdot\|)$ and $(Y, \|\cdot\|)$. Prove that the mapping

   $$Y^* \ni y^* \mapsto T^*y^* := y^*T \in X^*,$$

   where the juxtaposition y^*T designates the composition, i. e.,

   $$(y^*T)x := y^*(Tx), \, x \in X,$$

   is an *isometric isomorphism* between $(Y^*, \|\cdot\|)$ and $(X^*, \|\cdot\|)$ and the mapping

   $$X^{**} \ni x^{**} \mapsto T^{**}x^{**} := x^{**}T^* \in Y^{**},$$

   where the juxtaposition $x^{**}T^*$ designates the composition, i. e.,

   $$(x^{**}T^*)y^* := x^{**}(T^*y^*), \, y^* \in Y^*,$$

   is an *isometric isomorphism* between $(X^{**}, \|\cdot\|)$ and $(Y^{**}, \|\cdot\|)$.

7. Prove the *"if"* part of *Characterization of Weak Convergence in l_p (1 < p < \infty)* Theorem (Theorem 7.4) as a particular case of the *Characterization of Weak Convergence in Reflexive Spaces Theorem* (Theorem 7.18) via the *Representation Theorem for l_p^* (1 \le p < \infty)* (Theorem 7.3).

A The Axiom of Choice and Equivalents

A.1 The Axiom of Choice

> To choose one sock from each of infinitely many pairs of socks requires the Axiom of Choice, but for shoes the Axiom is not needed.
> Bertrand Russell

Here, we give a concise discourse on the celebrated *Axiom of Choice*, its equivalents, and *ordered sets*.

A.1.1 The Axiom of Choice

Expository Reference to a Set by Cantor. *By a set X, we understand "a collection into a whole of definite, well-distinguished objects, called the elements of X, of our perception or of our thought."*

Axiom of Choice (1904). *For each nonempty collection $\mathscr{F}$ of nonempty sets, there is a function $f : \mathscr{F} \to \bigcup_{X \in \mathscr{F}} X$ such that[1]*

$$\mathscr{F} \ni X \mapsto f(X) \in X.$$

Or equivalently.

For each nonempty collection $\{X_i\}_{i \in I}$ of nonempty sets, there is a function $f : I \to \bigcup_{i \in I} X_i$ such that

$$I \ni i \mapsto f(i) \in X_i.$$

The function f is called a choice function on $\mathscr{F}$, respectively, on I.

See, e. g., [18, 21, 22, 36].

A.1.2 Controversy

The *Axiom of Choice* enables one to prove the following counterintuitive statements.

Theorem A.1 (Vitali Theorem (1905)). *There exists a set in $\mathbb{R}$ that is not Lebesgue measurable.[2]*

See, e. g., [11, 19, 33].

---

1 Due Ernst Zermelo (1871–1953).
2 Giuseppe Vitali (1875–1932).

https://doi.org/10.1515/9783110614039-008

Theorem A.2 (Banach–Tarski Paradox (1924)). *Given a solid ball in 3-dimensional space, there exists a decomposition of the ball into a finite number of disjoint pieces, which can be reassembled, using only rotations and translations, into two identical copies of the original ball. The pieces involved are nonmeasurable, i. e., one cannot meaningfully assign volumes to them.*[3]

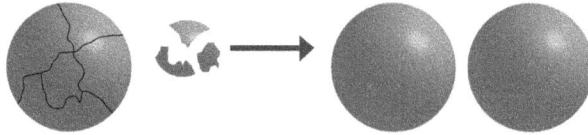

A.1.3 Timeline

- 1904: Ernst Zermelo formulates the *Axiom of Choice* in order to prove the *Well-Ordering Principle* (see Theorem A.7).
- 1939: Kurt Gödel[4] proves that, if the other standard set-theoretic *Zermelo-Fraenkel*[5] *Axioms* (see, e. g., [18, 21, 22, 36]) are consistent, they do not disprove the *Axiom of Choice*.
- 1963: Paul Cohen[6] completes the picture by showing that, if the other standard *Zermelo-Fraenkel Axioms* are consistent, they do not yield a proof of the *Axiom of Choice*, i. e., the *Axiom of Choice* is *independent*.

A.2 Ordered Sets

Here, we introduce and study various types of *order* on a set.

Definition A.1 (Partially Ordered Set). A *partially ordered set* is a nonempty set X with a *binary relation* $\leq$ of *partial order*, which satisfies the following *partial order axioms*:

1. For any $x \in X$, $x \leq x$. *Reflexivity*
2. For any $x, y \in X$, if $x \leq y$ and $y \leq x$, then $x = y$. *Antisymmetry*
3. For any $x, y, z \in X$, if $x \leq y$ and $y \leq z$, then $x \leq z$. *Transitivity*

If $x \leq y$, we say that x is a *predecessor* of y and that y is a *successor* of x.

Notation. $(X, \leq)$.

---

3 Alfred Tarski (1901–1983).
4 Kurt Gödel (1906–1978).
5 Abraham Fraenkel (1891–1965).
6 Paul Cohen (1934–2007).

Examples A.1.
1. An arbitrary nonempty set X is partially ordered by the *equality* (*coincidence*) relation $=$.
2. The set $\mathbb{R}$ is partially ordered by the usual order $\leq$.
3. The *power set* $\mathscr{P}(X)$ of a nonempty set X (see Section 1.1.1) is partially ordered by the set-theoretic inclusion $\subseteq$.

Remark A.1. Elements x, y of a partially ordered set $(X, \leq)$ are called *comparable* if $x \leq y$ or $y \leq x$. In a partially ordered set, incomparable elements may occur.

Exercise A.1.
(a) Verify the prior examples and remark.
(b) Give two more examples.
(c) Give an example of a partially ordered set $(X, \leq)$, in which no two distinct elements are comparable.

Remarks A.2.
–   If $\leq$ is partial order on X, then the relation $\geq$ defined as follows

$$x \geq y \iff y \leq x$$

is also a partial order on X.
–   If $x \leq y$ and $x \neq y$, we write $x < y$ or $y > x$.

Definition A.2 (Upper and Lower Bounds). Let Y be a nonempty subset of a *partially ordered set* $(X, \leq)$.
–   An element $x \in X$ is called an *upper bound* of Y if

$$\forall y \in Y : y \leq x.$$

–   An element $x \in X$ is called an *lower bound* of Y if

$$\forall y \in Y : x \leq y.$$

Remark A.3. Upper/lower bounds of a set Y need not exist.

Exercise A.2. Give corresponding examples.

Definition A.3 (Maximal and Minimal Elements). Let Y be a nonempty subset of a *partially ordered set* $(X, \leq)$.
–   An element $x \in Y$ is called a *maximal element* of Y if

$$\nexists y \in Y : x < y,$$

i. e., x has no successors in Y.

– An element $x \in Y$ is called a *minimal element* of Y if

$$\nexists y \in Y : y < x,$$

i. e., x has no predecessors in Y.

Remarks A.4.
– If an element $x \in Y$ is not comparable with all other elements of Y, it is automatically both a maximal and minimal element of Y.
– A *maximal element* of Y need not be greater than all other elements in Y.
– A *minimal element* of Y need not be less than all other elements in Y.
– Maximal and minimal elements of Y need not exist nor be unique.

Exercise A.3. Give corresponding examples.

Definition A.4 (Greatest and Least Elements). Let Y be a nonempty subset of a *partially ordered set* $(X, \leq)$.
– An element $x \in Y$ is called the *greatest element* (also the *last element*) of Y if it is an *upper bound* of Y, i. e.,

$$\forall y \in Y : y \leq x.$$

– An element $x \in Y$ is called the *least element* (also the *first element*) of Y if it is a *lower bound* of Y, i. e.,

$$\forall y \in Y : x \leq y.$$

Remark A.5. The greatest/least element of a set need not exist.

Exercise A.4. Give corresponding examples.

Exercise A.5. Let $\mathscr{P}(X)$ be the power set of a set X consisting of more than one element and partially ordered by the set-theoretic inclusion $\subseteq$ and

$$\mathscr{Y} := \{A \in \mathscr{P}(X) \,|\, A \neq \emptyset, X\}.$$

(a) What are the *lower* and *upper bounds* of $\mathscr{Y}$?
(b) What are the *minimal* and *maximal elements* of $\mathscr{Y}$?
(c) What are the *least* and *greatest elements* of $\mathscr{Y}$?

Proposition A.1 (Properties of the Greatest and Least Elements). *Let Y be a nonempty subset of a partially ordered set $(X, \leq)$.*
(1) *If the greatest/least element of Y exists, it is unique.*
(2) *If the greatest/least element of Y exists, it is also the unique maximal/minimal element of Y.*

Exercise A.6.

(a) Prove.

(b) Give an example showing that a maximal/minimal element of Y need not be its greatest/least element.

Definition A.5 (Least Upper and Greatest Lower Bounds). Let Y be a nonempty subset of a *partially ordered set* $(X, \leq)$.

–   If the set U of all upper bounds of Y is nonempty and has the least element u, u is called the *least upper bound* (the *supremum*) of Y, and we write $u = \sup Y$.

–   If the set L of all lower bounds of Y is nonempty and has the greatest element l, l is called the *greatest lower bound* (the *infimum*) of Y and we write $l = \inf Y$.

Remark A.6. For a subset Y in a partially ordered set $(X, \leq)$, $\sup Y$ and $\inf Y$ need not exist and, when they do, need not belong to Y.

Exercise A.7. Give corresponding examples.

Definition A.6 (Totally Ordered Set). A *totally ordered set* (also a *linearly ordered set* or a *chain*) is a partially ordered set $(X, \leq)$, in which any two elements are comparable, i. e.,

$$\forall x, y \in X : x \leq y \text{ or } y \leq x.$$

Exercise A.8.

(a) When is the *power set* $\mathscr{P}(X)$ of a nonempty set X partially ordered by the set-theoretic inclusion $\subseteq$ a chain?

(b) Show that, for a nonempty subset Y of a *totally ordered set* $(X, \leq)$, a *maximal/minimal element* of Y, when it exists, is the *greatest/least element* of Y.

Definition A.7 (Well-Ordered Set). A *well-ordered set* is a totally ordered set $(X, \leq)$, in which every nonempty subset has the first element.

Examples A.2.

1. $(\mathbb{N}, \leq)$, as well as its arbitrary nonempty subset, is well ordered.

2. $(\mathbb{Z}, \leq)$ is totally ordered, but *not* well ordered.

Remarks A.7.

–   As follows from *Zermello's Well-Ordering Principle* (Theorem A.7), every nonempty set can be well-ordered.

–   Each well-ordered set $(X, \leq)$ is similar to $(\mathbb{N}, \leq)$ in the sense that each nonempty subset $Y \subseteq X$ has the *first element* and each element $x \in X$, except for the last one, if any, has the *unique immediate successor* (the *unique next element*) $s(x) \in X$, i. e., there is a *unique* element $s(x) \in X$ such that

$$x < s(x) \text{ and } \nexists y \in X : x < y < s(x).$$

This fact affords the possibility of inductive proofs (*transfinite induction*) and constructions over such sets similar to those over $\mathbb{N}$.

Exercise A.9.
(a) Verify the latter.
(b) Prove that, if a nonempty subset Y of a well-ordered set $(X, \leq)$ has an upper bound, it has the *least upper bound*.
(c) Prove that the usual order $\leq$ of the real line $\mathbb{R}$ restricted to any *uncountable* subset $Y \subseteq \mathbb{R}$ is a total order, but not a well order on Y.

A.3 Equivalents

The Axiom of Choice is obviously true, the well-ordering principle obviously false, and who can tell about Zorn's lemma?
Jerry Bona

Here, we are to prove the equivalence of the *Axiom of Choice* to following three fundamental set-theoretic principles.

Theorem A.3 (Hausdorff Maximal Principle). *In a partially ordered set, there exists a maximal chain.*

Theorem A.4 (Hausdorff Maximal Principle (Precise Version)). *In a partially ordered set, every chain is contained in a maximal chain.*

Theorem A.5 (Zorn's Lemma). *In a partially ordered set, whose every chain has an upper bound, there is a maximal element.*[7]

Theorem A.6 (Zorn's Lemma (Precise Version)). *For each element x in a partially ordered set $(X, \leq)$, whose every chain has an upper bound, there is a maximal element u in $(X, \leq)$ such that $x \leq u$.*

Theorem A.7 (Zermello's Well-Ordering Principle). *Every nonempty set can be well ordered.*

Proof. We are to prove the following closed chain of implications:

$$AC \Rightarrow HMP \Rightarrow ZL \Rightarrow ZWOP \Rightarrow AC,$$

where the abbreviations *AC*, *HMP*, *ZL*, and *ZWOP* stand for the *Axiom of Choice*, the *Hausdorff Maximal Principle*, *Zorn's Lemma*, and *Zermello's Well-Ordering Principle*, respectively.

---

7 Max Zorn (1906–1993).

AC ⇒ HMP

Assume the *Axiom of Choice* and let C be an arbitrary chain in a partially ordered set $(X, \le)$ and let $\mathscr{C}$ be the collection of all chains in X containing C partially ordered by the set-theoretic inclusion $\subseteq$.

Observe that $\mathscr{C} \neq \emptyset$ since, obviously, $C \in \mathscr{C}$.

Our goal is to prove that $(\mathscr{C}, \subseteq)$ has a *maximal element U*.

Let f be a *choice function* assigning to every nonempty subset A of X one of its elements $f(A)$.

For each $A \in \mathscr{C}$, let $\hat{A}$ be the set of all those elements in X whose *adjunction* to A produces a chain in $\mathscr{C}$:

$$\hat{A} := \{x \in X \mid A \cup \{x\} \in \mathscr{C}\}.$$

Clearly, $A \subseteq \hat{A}$, the equality holding *iff* A is a maximal element in $(\mathscr{C}, \subseteq)$.

Consider a function $g : \mathscr{C} \mapsto \mathscr{C}$ defined as follows:

$$\mathscr{C} \ni A \mapsto g(A) := \begin{cases} A \cup \{f(\hat{A} \setminus A)\} & \text{if } A \subset \hat{A}, \\ A & \text{if } A = \hat{A}. \end{cases}$$

Observe that, for each $A \in \mathscr{C}$, the set $g(A)$ differs from A by *at most one element*.

Thus, to prove that $U \in \mathscr{C}$ is a maximal element in $(\mathscr{C}, \subseteq)$, one needs to show that $U = \hat{U}$, i. e., that $g(U) = U$.

Let us introduce the following temporary definition.

Definition A.8 (Tower). We call a subcollection $\mathscr{I}$ of $\mathscr{C}$ a *tower* if it satisfies the conditions

(1) $C \in \mathscr{I}$.
(2) If $A \in \mathscr{I}$, then $g(A) \in \mathscr{I}$.
(3) If $\mathscr{D}$ is a chain in $\mathscr{I}$, then $\bigcup_{A \in \mathscr{D}} A \in \mathscr{I}$.

Observe that *towers* exist since, as is easily seen, $\mathscr{C}$ is a *tower* itself.

Furthermore, the intersection of all towers $\mathscr{I}_0$ is also a tower and is, in fact, the *smallest tower*, the *nonemptiness* of $\mathscr{I}_0$ being ensured by condition (1) of the above definition.

Let us show that $\mathscr{I}_0$ is a *chain* in $(\mathscr{C}, \subseteq)$.

We call a set $B \in \mathscr{I}_0$ *comparable* if, for each $A \in \mathscr{I}_0$ either $A \subseteq B$ or $B \subseteq A$. Thus, proving that $\mathscr{I}_0$ is a chain amounts to showing that all its elements are comparable.

Observe that there is at least one comparable set in $\mathscr{I}_0$, which is C, since it is contained in any other set in $\mathscr{I}_0$.

Let B be an arbitrary comparable set in $\mathscr{I}_0$. Suppose that $A \in \mathscr{I}_0$ and A is a *proper subset* of B. Then $g(A) \subseteq B$. Indeed, since B is comparable, either $g(A) \subseteq B$, or $B \subset g(A)$. In the latter case, A is a proper subset of a proper subset of $g(A)$, which contradicts the fact that $g(A) \setminus A$ is *at most a singleton*.

Further, consider the collection $\mathcal{U}(B)$ of all those $A \in \mathcal{I}_0$ for which either $A \subseteq g(B)$, or $g(B) \subseteq A$.

It is not difficult to make sure that $\mathcal{U}(B)$ is a *tower* (the verification of the least trivial condition (2) uses the argument of the preceding paragraph).

Since $\mathcal{U}(B)$ is a tower and $\mathcal{U}(B) \subseteq \mathcal{I}_0$, $\mathcal{U}(B) = \mathcal{I}_0$ necessarily.

All these considerations imply that, for each comparable set $B \in \mathcal{I}_0$, the set $g(B)$ is also comparable. The latter, jointly with the facts that the set C is comparable and that, by condition (3), the union of the sets of a chain of comparable sets is also comparable, imply that all comparable sets of $\mathcal{I}_0$ constitute a tower, and hence, they exhaust the entire $\mathcal{I}_0$.

Since $\mathcal{I}_0$ is a chain in $\mathcal{I}$, by condition (3), the set

$$U := \bigcup_{A \in \mathcal{I}_0} A \in \mathcal{I} \subseteq \mathcal{C}$$

and, obviously, U contains every set of $\mathcal{I}_0$. In particular, by condition (2) applied to the tower $\mathcal{I}_0$, $g(U) \subseteq U$, which implies that $g(U) = U$ proving that U is a maximal chain containing C.

Hence, the *Axiom of Choice* does imply the *Hausdorff Maximal Principle*.

$HMP \Rightarrow ZL$

Assume the *Hausdorff Maximal Principle* and let x be an arbitrary element in a partially ordered set $(X, \leq)$, whose every chain has an upper bound.

By the *HMP*, there is a *maximal chain* U in $(X, \leq)$ containing the trivial chain $\{x\}$, i. e., $x \in U$. By the premise of *Zorn's Lemma*, U has an *upper bound* u in $(X, \leq)$. In particular, this implies that $x \leq u$.

From the maximality of the chain U, it follows that u is a *maximal element* in $(X, \leq)$. Otherwise, there would exist such an element $v \in X$ that $u < v$, which would imply that the chain U could be extended to a larger chain $U \cup \{v\}$ contradicting the maximality of U.

Thus the *Hausdorff Maximal Principle* implies *Zorn's Lemma*.

$ZL \Rightarrow ZWOP$

Assume *Zorn's Lemma* and let X be an arbitrary nonempty set.

Let

$$\mathcal{W} := \{A \subseteq X \mid \exists \leq_A \text{ a well order on } A\}.$$

Observe that $\mathcal{W} \neq \emptyset$ since it contains all *singletons* (more generally, all *finite subsets*) of X.

Exercise A.10. Explain.

The following defines a *partial order* on $\mathcal{W}$:

$$(A, \leq_A) \leq (B, \leq_B)$$

iff

$$A \subseteq B \text{ and } \forall x, y \in A : x \leq_A y \iff x \leq_B y \text{ and } \forall x \in A \; \forall y \in B \setminus A : x <_B y. \qquad (A.1)$$

Exercise A.11. Verify that $\preceq$ is a *partial order* on $\mathscr{W}$.

For an arbitrary *chain* $\mathscr{C}$ in $(\mathscr{W}, \preceq)$, let

$$D := \bigcup_{C \in \mathscr{C}} C \subseteq X.$$

We define a *total order* $\leq_D$ on D as follows:

For any $x, y \in D$, exists $C_x, C_y \in \mathscr{C}$ such that $x \in C_x, y \in C_y$. Considering that $(\mathscr{C}, \preceq)$ is a chain, $C_x \subseteq C_y$ or $C_y \subseteq C_x$, and hence, $x, y \in C$, where $C := \max(C_x, C_y)$, and we define

$$x \leq_D y \iff x \leq_C y.$$

Exercise A.12. Verify

(a) that $\leq_D$ is *well defined*, i.e., does not depend on the choice of $(C, \leq_C) \in \mathscr{C}$, which contains both x and y and

(b) that $(D, \leq_D)$ is a *chain*.

Thus, $(D, \leq_D)$ is a *chain*.

Let E be an arbitrary nonempty subset of D. For an $x \in E$, there is a $(C_x, \leq_{C_x}) \in \mathscr{C}$ is such that $x \in C_x$. Then $E \cap C_x$ is a nonempty subset of C_x and, since $(C_x, \leq_{C_x})$ is *well ordered*, without loss of generality, we can regard x to be the least element of $E \cap C_x$. Suppose $y \in E$ and $y <_D x$. Then, considering that $(\mathscr{C}, \preceq)$ is a *chain*, there is a $(C, \leq_C) \in \mathscr{C}$ containing both x and y with two possibilities:

1. $(C, \leq_C) \preceq (C_x, \leq_{C_x})$, in which case $y \in E \cap C_x$ implying that $y <_{C_x} x$, and hence, contradicting the choice of x.

2. $(C_x, \leq_{C_x}) \preceq (C, \leq_C)$, in which case, if $y \in C_x$, then $y \in E \cap C_x$ implying that $y <_{C_x} x$, and hence, contradicting the choice of x. If $y \in C \setminus C_x$, then, by (A.1), $x <_C y$ contradicting the fact that $y <_D x$.

The obtained contradictions show that x is the least element of E in $(D, \leq_D)$ proving that $(D, \leq_D)$ is *well ordered*.

The set $(D, \leq_D)$ is an *upper bound* of $\mathscr{C}$ in $(\mathscr{W}, \preceq)$.

Exercise A.13. Verify.

The latter, by *Zorn's Lemma*, implies that $(\mathscr{W}, \preceq)$ has a *maximal element* $(M, \leq_M)$. Then $M := X$ since, otherwise, choosing any $u \in X \setminus M$, we could extend the well order on M to a well order $\leq_N$ on $N := M \cup \{u\}$ as follows:

$$\forall x, y \in M : x \leq_N y \iff x \leq_M y \quad \text{and} \quad \forall x \in M : x <_N u,$$

which would imply that $(M, \leq_M) \prec (N, \leq_N)$ while contradicting the maximality of $(M, \leq_M)$ in $(\mathscr{W}, \preceq)$. Thus X can be well ordered, i. e., *Zorn's Lemma* implies *Zermello's Well-Ordering Principle*.

$ZWOP \Rightarrow AC$

Exercise A.14. Prove.

Hint. For a nonempty collection of nonempty sets $\{X_i\}_{i \in I}$, consider a *well order* $\preceq$ on $X := \bigcup_{i \in I} X_i$. ☐

As an additional demonstration of how a typical proof by *Zorn's Lemma* works, let us prove the implication

$$ZL \Rightarrow AC$$

Assume *Zorn's Lemma* and let $\{X_i\}_{i \in I}$ be a nonempty collection of nonempty sets. Consider the collection $\mathscr{F}$ of all possible functions $f : I \supseteq D(f) \rightarrow \bigcup_{i \in I} X_i$, such that, for any i in the *domain* $D(f)$ of f, $f(i) \in X_i$.

Such functions, obviously, exist on *singletons* (more generally, on *finite subsets*) of I.

Let us introduce a *partial order* $\preceq$ on $\mathscr{F}$ as follows:

$$f \preceq g \;\Leftrightarrow\; D(f) \subseteq D(g) \text{ and } f(x) = g(x), \; x \in D(f),$$

i. e., g is an *extension* of f.

Exercise A.15. Verify that $\preceq$ is a partial order on $\mathscr{F}$.

The premise of *Zorn's Lemma* holds in $(\mathscr{F}, \preceq)$.

Exercise A.16. Verify.

Hint. For an arbitrary chain $\mathscr{C}$ in $(\mathscr{F}, \preceq)$, show that $u \in \mathscr{F}$ with the domain

$$u(i) := g(i), \quad i \in D(u) := \bigcup_{g \in \mathscr{C}} D(g), \; g \in \mathscr{C},$$

is an *upper bound* of $\mathscr{C}$ in $(\mathscr{F}, \preceq)$.

Then, by *Zorn's Lemma*, there is a *maximal choice function* f in $(\mathscr{F}, \preceq)$. This, implies that the domain $D(f)$ of f is the entire indexing set I and completes the proof.

Exercise A.17. Explain.

Bibliography

[1]   N. I. Akhiezer, *Lectures in the Theory of Approximation*, 2nd ed., Nauka, Moscow, 1965
       (Russian).
[2]   N. I. Akhiezer and I. M. Glazman, *Theory of Linear Operators in Hilbert Space*, 2nd ed., Dover
       Publications, Inc., New York, 1993.
[3]   G. Bachman and L. Narici, *Functional Analysis*, 2nd ed., Dover Publications, Inc., Mineola, New
       York, 2000.
[4]   A. V. Balakrishnan, *Applied Functional Analysis*, 2nd ed., Springer-Verlag, New York, 1981.
[5]   Yu. M. Berezansky, G. F Us, and Z. G. Sheftel, *Functional Analysis. Lecture Course*, Vishcha
       Shkola, Kiev, 1990 (Russian).
[6]   H. Bresis, *Functional Analysis, Sobolev Spaces and Partial Differential Equations*, Universitext,
       Springer, New York, 2011.
[7]   G. Chacón, H. Rafeiro, and J. C. Vallejo, *Functional Analysis: A Terse Introduction*, Walter de
       Gruyter GmbH, Berlin/Boston, 2017.
[8]   J. B. Conway, *A Course in Functional Analysis, 2nd ed.*, Graduate Texts in Mathematics, vol. 96,
       Springer-Verlag, New York, 1990.
[9]   L. Debnath and P. Mikusiński, *Introduction to Hilbert Spaces with Applications*, 2nd ed.,
       Academic Press, San Diego, 1999.
[10]   A. Ya. Dorogovtsev, *Mathematical Analysis. A Reference Handbook*, Vishcha Shkola, Kiev, 1985
       (Russian).
[11]   A. Ya. Dorogovtsev, *Elements of the General Theory of Measure and Integral*, Vishcha Shkola,
       Kiev, 1989 (Russian).
[12]   N. Dunford and J. T. Schwartz with the assistance of W. G. Bade and R. G. Bartle, *Linear
       Operators. Part I: General Theory*, Wiley Interscience Publishers, New York, 1958.
[13]   A. Friedman, *Foundations of Modern Analysis*, Dover Publications, Inc., New York, 1982.
[14]   P. Enflo, *A counterexample to the approximation problem in Banach spaces*, Acta Math. **130**
       (1973), no. 2, 309–317.
[15]   B. R. Gelbaum and J. M. H. Olmsted, *Counterexamples in Analysis*, Dover Publications, Inc.,
       Mineola, New York, 2003.
[16]   C. Goffman and G. Pedrick, *First Course in Functional Analysis*, 2nd ed., Chelsea Publishing Co.,
       New York, 1983.
[17]   D. H. Griffel, *Applied Functional Analysis*, Dover Publications, Inc., New York, 2002.
[18]   P. R. Halmos, *Naive Set Theory*, Undergraduate Texts in Mathematics, Springer-Verlag, New
       York-Heidelberg-Berlin, 1974.
[19]   P. R. Halmos, *Measure Theory*, Graduate Texts in Mathematics, vol. 18, Springer-Verlag, New
       York-Heidelberg-Berlin, 1974.
[20]   P. R. Halmos, *A Hilbert Space Problem Book*, 2nd ed., Graduate Texts in Mathematics, vol. 19,
       Springer-Verlag, New York-Heidelberg-Berlin, 1982.
[21]   F. Hausdorff, *Set Theory*, 2nd ed., Chelsea Publishing Co., New York, 1962.
[22]   T. J. Jech, *The Axiom of Choice*, Dover Publications, Inc., New York, 2008.
[23]   I. Kaplansky, *Set Theory and Metric Spaces*, Allyn and Bacon, Inc., Boston, 1972.
[24]   A. N. Kolmogorov and S. V. Fomin, *Elements of the Theory of Functions and Functional Analysis*,
       6th ed., Nauka, Moscow, 1989 (Russian).
[25]   E. Kreyszig, *Introductory Functional Analysis with Applications*, John Wiley & Sons, Inc., New
       York-London-Sydney, 1978.
[26]   A. G. Kurosh, *Lectures on General Algebra*, 2nd ed., Nauka, Moscow, 1973 (Russian).
[27]   S. Lang, *Real and Functional Analysis*, 3rd ed., Springer, New York, 1993.

https://doi.org/10.1515/9783110614039-009

[28] P. D. Lax, *Functional Analysis*, Wiley-Interscience, New York, 2002.

[29] B. V. Limaye, *Linear Functional Analysis for Scientists and Engineers*, Springer, Singapore, 2016.

[30] J. Lindenstrauss and L. Tzafriri, *On the complemented subspaces problem*, Israel J. Math. **9** (1971), no. 2, 263–269.

[31] L. A. Lyusternik and V. I. Sobolev, *A Short Course in Functional Analysis*, Vysshaya Shkola, Moscow, 1982 (Russian).

[32] B. D. MacCluer, *Elementary Functional Analysis*, Springer, New York, 2009.

[33] M. V. Markin, *Real Analysis: Measure and Integration*, Walter de Gruyter GmbH, Berlin/Boston, 2019, ISBN 978-3-11-060097-1.

[34] M. V. Markin, *Elementary Operator Theory*, Walter de Gruyter GmbH, Berlin/Boston, 2019, ISBN 978-3-11-060096-4.

[35] A. N. Michel and C. J. Herget, *Applied Algebra and Functional Analysis*, Dover Publications, Inc., New York, 2011.

[36] G. H. Moore, *Zermelo's Axiom of Choice: Its Origins, Development, and Influence*, Dover Publications, Inc., Mineola, New York, 2013.

[37] T. J. Morrison, *Functional Analysis. An Introduction to Banach Space Theory*, Pure and Applied Mathematics, Wiley-Interscience, New York, 2001.

[38] J. R. Munkres, *Topology*, 2nd ed., Prentice Hall, Upper Saddle River, New Jersey, 2000.

[39] J. Muscat, *Functional Analysis. An Introduction to Metric Spaces, Hilbert spaces, and Banach Algebras*, Springer International Publishing, Switzerland, 2014.

[40] M. O'Nan, *Linear Algebra*, 2nd ed., Harcourt Brace Jovanovich, Inc., New York, 1976.

[41] C. W. Patty, *Foundations of Topology*, Waveland Press, Inc., Prospect Heights, Illinois, 1997.

[42] M. Reed and B. Simon, *Methods of Modern Mathematical Physics. I. Functional Analysis*, 2nd ed., Academic Press, Inc., New York-London, 1980.

[43] W. Rudin, *Principles of Mathematical Analysis*, 3rd ed., International Series in Pure and Applied Mathematics, McGraw-Hill Book Co., New York, 1976.

[44] W. Rudin, *Functional Analysis*, 2nd ed., McGraw-Hill, Inc., New York, 1991.

[45] M. Schechter, *Principles of Functional Analysis*, 2nd ed., Graduate Studies in Mathematics, vol. 36, American Mathematical Society, Providence, RI, 2002.

[46] G. E. Shilov, *Elementary Functional Analysis*, Dover Publications, Inc., New York, 1996.

[47] B. Simon, *Real Analysis. A Comprehensive Course in Analysis, Part 1*, American Mathematical Society, Providence, RI, 2015.

[48] W. A. Sutherland, *Introduction to Metric and Topological Spaces*, 2nd ed., Oxford University Press Inc., New York, 2009.

[49] A. E. Taylor and D. C. Lay, *Introduction to Functional Analysis*, Reprint of the 2nd ed., Robert E. Krieger Publishing Co., Inc., Melbourne, FL, 1986.

[50] V. A. Trenogin, *Functional Analysis*, Nauka, Moscow, 1980 (Russian).

[51] B. Z. Vulikh, *Introduction to Functional Analysis*, 2nd ed., Nauka, Moscow, 1967 (Russian).

[52] S. Willard, *General Topology*, Dover Publications, Inc., New York, 2004.

[53] K. Yosida, *Functional Analysis*, 6th ed., Springer-Verlag, Berlin-Heidelberg-New York, 1993.

[54] N. Young, *An Introduction to Hilbert Space*, 2nd ed., Cambridge University Press, Cambridge, UK, 1997.

[55] S. Zaidman, *Functional Analysis and Differential Equations in Abstract Spaces*, Chapman & Hall/CRC, Boca Raton, FL, 1999.

[56] R. J. Zimmer, *Essential Results of Functional Analysis*, Chicago Lectures in Mathematics Series, The University of Chicago Press, Chicago and London, 1980.

Index

www.ingramcontent.com/pod-product-compliance
Lightning Source LLC
Chambersburg PA
CBHW080923220326

41598CB00034B/5662